MT. DIABLO
UNIFIED SCHOOL DISTRICT

BSCS Biology
A Human Approach

BIOLOGY

Mt. Diablo Unified School District

2018-2019 Fall Semester

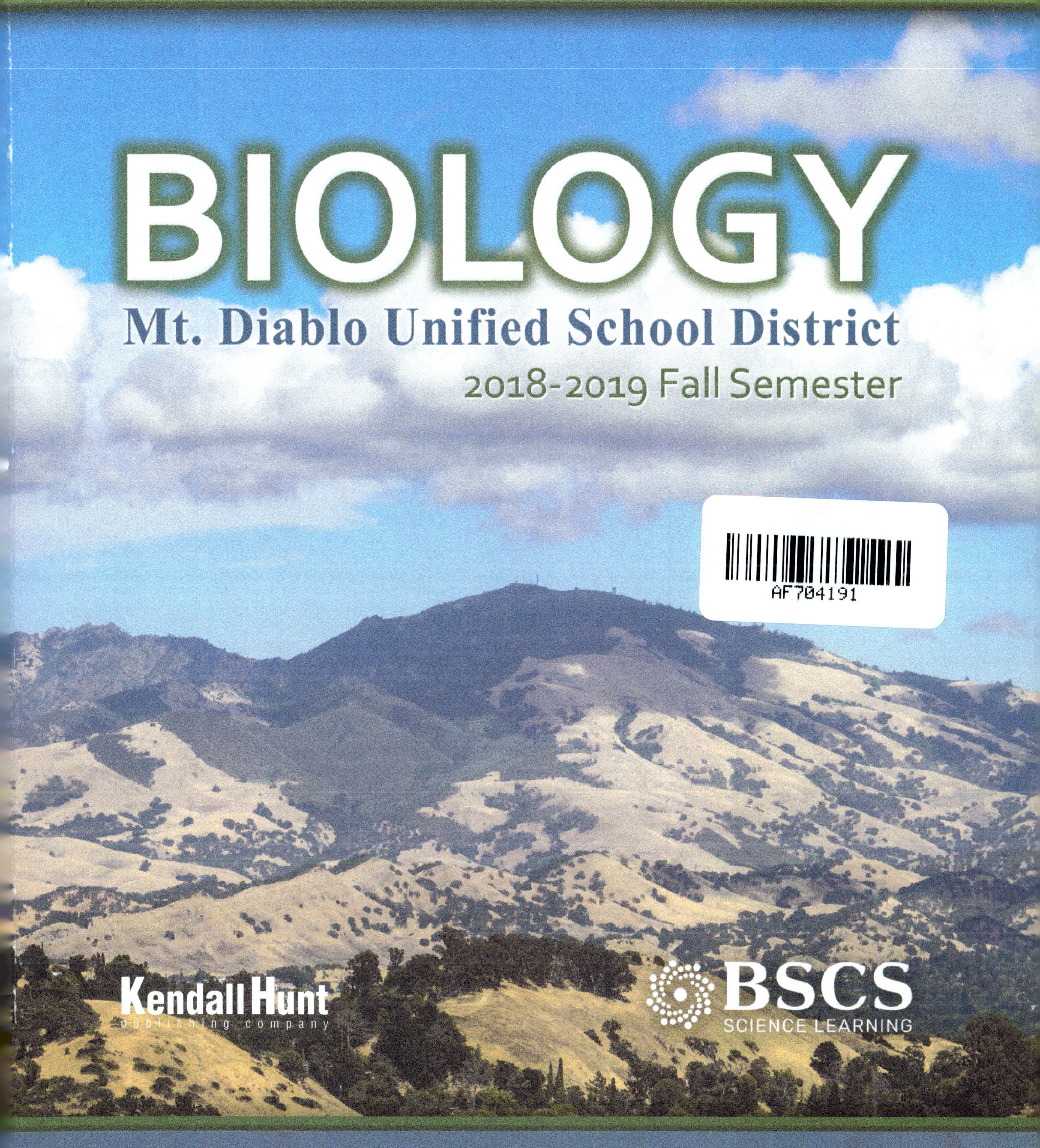

Kendall Hunt
publishing company

BSCS
SCIENCE LEARNING

ISBN 978-1-5249-6315-6

This material is based on work supported by the National Science Foundation under Grant No. ESI 9252974. Any opinions, findings, conclusions, or recommendations expressed in this publication are those of the authors and do not necessarily reflect the views of the granting agency.

Printed in the United States of America

Acknowledgments

The development of the ***SFUSD Science Core Curriculum for Biology*** has been a collaborative effort on the part of teachers across our district, SFUSD Office of Curriculum & Instruction departments, and science and education partners.

SFUSD Science Department

Jim Ryan, STEM Executive Director

Sarah Delaney, Science Supervisor

Dawn Rege, STEM HS Science Content Specialist

Katrina Rotter, STEM HS Science Content Specialist

Jeffrey Dowling, STEM HS Science Content Specialist

Katie Tobin, STEM HS Science Content Specialist

Shahram Mostarshed, STEM HS Science Content Specialist

Judith Gerber, STEM Teacher on Special Assignment

Vanessa Carter, Environmental Literacy Content Specialist

Theresa Hall, STEM Science Clerk

SFUSD Educational Technology Department

SFUSD Humanities Department

SFUSD Math Department

SFUSD Multilingual Pathways Department

BSCS—Biological Sciences Curriculum Study

California Academy of Sciences

Center to Support Excellence in Teaching—Stanford University

Exploratorium

HHMI BioInteractive

Kendall Hunt Publishing Company

Tuva Labs Inc.

UC Museum of Paleontology

CSF Science and Educational Partnership

Acknowledgments

We would like to thank the members of the **SFUSD HS Science Biology Response and Development Team** for their dedication to this work.

Amber Lancaster, *June Jordan School for Equity*
Andra Kimball, *Mission High School*
Bonnie E. Daley, *Lowell High School*
Brian Finley, *Thurgood Marshall High School*
Bryan Olney, *Thurgood Marshall High School*
Catherine S. Christensen, *Lowell High School*
Dawn Rege, *C&I STEM Department*
Deb Apple, *C&I Extended Learning & Support*
Eric Lewis, *C&I STEM Department*
Erin Bird, *Lowell High School*
Eugene Pearson, *Phillip & Sala Burton High School*
Grace Gould, *Thurgood Marshall High School*
Jan Bautista, *C&I Multilingual Pathways Department*
Jodi Andres, *George Washington High School*
John Schlauraff, *Phillip & Sala Burton High School*
Katherine Farrar, *UC San Francisco*
Katie Tobin, *C&I STEM Department*
Katrina Rotter, *C&I STEM Department*
Mark Wenning, *Lowell High School*
Marloes Sijstermans, *Wallenberg High School*
Rebecca M Fulop, *Mission High School*
Robert Coverdell, *Downtown High School*
Sabina King, *Galileo High School*
Suniqua Thomas, *Phillip & Sala Burton High School*

We would also like to thank SFUSD science teachers and students who participated in **Field Testing** and provided feedback on the SFUSD Science Core Curriculum.

Aljona Andrejeff, *George Washington High School*
Amy Trusso, *Ruth Asawa School of the Arts*
Becky Fulop, *Mission High School*
Bonnie Daley, *Lowell High School*
Denise Kwan, *Galileo High School*
Eugene Pearson, *Phillip & Sala Burton High School*
Freja Robinson, *Lowell High School*
Jodi Andres, *George Washington High School*
Kevin Hartzog, *Balboa High School*
Kyle Liu, *George Washington High School*
Leah Thomas, *Mission High School*
Lindsay Penrose, *Mission High School*
Resa Garcia, *Balboa High School*
Richard Gin, *Abraham Lincoln High School*
Rosanna Mariotti, *Abraham Lincoln High School*

SFUSD Biology Core Curriculum Course Description

In this NGSS Biology course, students explore biological concepts that build comprehension around two driving questions: *What connections exist between Earth's changing environment and the coevolution of life?* and *How and in what ways do organisms, including humans, depend on and impact the environment?* Using the guidance of the CA Science Framework's High School 3-Course Model, Biology integrates both Life Science and Earth and Space Science performance expectations. Earth and Space Science concepts are integrated in a strategic way, designed to deepen student understanding of life on Earth and the interactions between the biotic and abiotic systems. Throughout the course students build an understanding of life and how life changes over time in response to a changing environment. Central to this understanding is the study of interactions of living organisms and their environments on both macroscopic and microscopic scales.

The course utilizes a 5E Instructional Model in which teaching and learning proceeds through five different stages: Engage, Explore, Explain, Elaborate, and Evaluate. Using the 5E instructional model allows students the opportunity to ask questions and define problems about the natural and designed world, design investigations in which they collect and analyze trends and patterns in their data, develop models of physical phenomena and to communicate their findings from their investigations. This course emphasizes the use of evidence based reasoning for scientific explanations and engineering solutions in order to communicate recommendations to address real world problems. Students discuss and evaluate a variety of scientific texts and data from different sources, which include the analysis and interpretation of data sets used by the scientific community.

SFUSD Biology Scope & Sequence

Fall Semester			
Unit Title	**Unit 1: Ecology: Interdependent Relationships**	**Unit 2: Cycles of Energy & Matter**	**Unit 3: Evolution**
Essential Question(s)	• *How do Earth's dynamic systems impact interactions between living and nonliving things?* • *What happens to an ecosystem when the environment changes?*	• *How do organisms use energy and matter from the environment to survive?* • *How does the use of energy and matter by living organisms influence Earth's systems?*	• *How do differences among individuals affect the characteristics of a population over time? How do we know populations of organisms change over time?* • *How have the Earth's changing environments influenced where and how organisms survive and proliferate over time?*

Spring Semester			
Unit Title	**Unit 4: Biological Systems**	**Unit 5: Genetics & Development**	**Unit 6: Ecology: Human Impact, Climate Change and Biodiversity**
Essential Question(s)	• *How can the basic structure of the cell sustain life and provide the basis for all of life's diversity?* • *How do organisms live, grow and respond to their environment and reproduce?*	• *Why are all individuals within a population similar, yet unique?* • *How are the characteristics of one generation passed to the next?*	• *How can we apply our understanding of humans' influence on Earth's systems to develop and maintain sustainable and resilient societies?* • *How can we apply our understanding of Earth's systems to mitigate adverse impacts of human activity on natural resources, human populations, and biodiversity?*

Contents

Appendices

Acknowledgments

Fifth Edition Staff

BSCS Administrative Staff

Danny Edelson, *Executive Director*
Pamela Van Scotter, *Senior Associate Director*
Robert Foulk, *Director of Operations*
Susan Rust, *Director of Communications*
Richard Cardullo, PhD, *University of California-Riverside; Chairman, Board of Directors*

BSCS Project Staff

Pamela Van Scotter, *Senior Associate Director*
Brooke Bourdélat-Parks, PhD, *Senior Science Educator*
April Gardner, *Science Educator*
Stacey Luce, *Production Coordinator*
Chris Moraine, *Production Specialist*

Fourth Edition Contributors

Content Reviewers

Dr. Amy Angert, Colorado State University, Ft. Collins, CO; Dr. Andrew Ray, Oregon Institute of Technology, Klamath Falls, OR; Dr. Doug Dransfield, Cape Elizabeth, ME; Francoise Benay, University of Colorado, Boulder, CO; Dr. Scott Gilbert, Swarthmore College, Swarthmore, PA; Dr. Katharine Semsar, University of Colorado, Boulder, CO; Dr. Kevin Middleton, California State University, San Bernardino, CA; Kim Nichols, Biological Sciences Initiative, Boulder, CO; Ronald LaCoss, University of Colorado, Boulder, CO; Dr. Ann Mackenzie, Miami University, Oxford, OH; Dr. Maureen Munn, University of Washington, Seattle, WA; Dr. Chris Ray, University of Colorado, Boulder, CO; Dr. Sam Donovan, University of Pittsburgh, Pittsburgh, PA; Tammy A. Maldonado, University of Colorado, Boulder, CO

Advisors

Joshua Adams, Raymond S. Kellis HS, Glendale, AZ; Margaret Aguilar, Deer Valley USD, Phoenix, AZ; Joni Lee Bartholomew, Raymond S. Kellis HS, Glendale, AZ; Jessica LaNae Batty, Sunrise Mountain, Peoria, AZ; Alan Chintis, Mountain View HS, Tucson, AZ; Geraldine Fisher, Desert View HS, Tucson, AZ; Sylvia Janie Garcia-Lohr, Flagstaff HS, Flagstaff, AZ; Joni Giacomino, Buena HS, Bisbee, AZ; Mary Frances Giannola, Flagstaff HS, Flagstaff, AZ; Amanda Lee Grunden, Boulder Creek HS, Anthem, AZ; Erik Hanchett, Safford HS, Safford, AZ; Carolyn Sue Harris, Bisbee HS, Bisbee, AZ; Seth Hoopingarner, Peoria Transition Center, Peoria, AZ; Barry Horst, Douglas HS, Douglas, AZ; Judith Hughes, Peoria HS, Peoria, AZ; Kristen Lee Kaus, Cactus HS, Glendale, AZ; Jeffrey Landers, Page HS, Page, AZ; David Richard Lanning, Page HS, Page, AZ; Stephen Joseph Murray, Catalina Magnet HS, Tucson, AZ; Dwight Rawlings, Raymond S. Kellis HS, Glendale, AZ; Tom Reeder, Ironwood HS, Glendale, AZ; Eric Matthew Regh, Deer Valley USD, Phoenix, AZ; Donna Savill, Douglas HS, Douglas, AZ; Michele Marie Schiff, Ironwood HS, Glendale, AZ; David Serafin, Cactus HS, Glendale, AZ; David Edwin Simmons, Page HS, Page, AZ; William M. Sorensen, Cactus HS, Glendale, AZ; Erin Stuart, Boulder Creek HS, Anthem, AZ; Kelly Neil Thomas, St. Johns HS, St. Johns, AZ; Wendy West, Desert Ridge HS, Gilbert, AZ; Brooke Nicole White, Cactus HS, Glendale, AZ; Howard Yauney, Desert Ridge HS, Gilbert, AZ; Patricia Pauline Zint, Ironwood HS, Glendale, AZ

Second Edition Contributors

Edward Drexler, Pius XI High School, Milwaukee, WI; The Jane Goodall Institute, Silver Springs, MD; Tim Myles, University of Toronto, Toronto, Ontario; Doug Niles, Delavan, WI;

Pedagogical Advisors

Scott Charleton, Lebanon High School, Lebanon, OH; Frank Girolami, Mason High School, Mason, OH; Barbara Grosz, Pine Crest Preparatory School, Fort Lauderdale, FL; Melanie Hardel, Hartford Union High School, Hartford, WI; Terry Houchens, Joliet Central High School, Joliet, IL; Mary McClellan, Seattle School District, Seattle, WA; Kimberly Noethen, Cornell University, Ithaca, NY; Eugene O'Brien, Hartford Union High School, Hartford, WI

Content Reviewers

Dr. John G. Bailey, College of Veterinary Medicine, Mississippi State, MS (Chapter 7); Dr. Marvin Druger, Syracuse University, Syracuse, NY (Chapter 1); Dr. Diane Ebert-May, Michigan State University, East Lansing, MI (Chapter 9); Dr. James J. Gallagher, Michigan State University, East Lansing, MI (Chapter 3); Dr. Burnette W. Hamil, Mississippi State University, Mississippi State, MS (Chapter 16); Dr. Craig Hanke, University of Wisconsin, Green Bay,

WI (Chapter 5); Dr. Ross Jilk, Rochester University, Kansas City, MO (Chapter 12); Carolyn W. Keys, The University of Georgia, Athens, GA (Engage, Explain, Evaluate); Dr. Norman G. Lederman, Illinois Institute of Technology, Chicago, IL (Chapter 2); Dr. Sarah H. Martin, University of Southern Mississippi, Hattiesburg, MS (Chapters 13 and 14); Dr. Rhonda A. Patterson, University of Southern Mississippi, Hattiesburg, MS (Chapter 6); Dr. Montgomery Slatkin, University of California, Berkeley, CA (Chapter 11); Dr. Marshall Sundberg, Emporia State University, Emporia, KS (Chapter 15); Dr. James H. Wandersee, Louisiana State University, Baton Rouge, LA (Chapter 8); Dr. Richard E. Wilson, Rockhurst University, Kansas City, MO (Chapter 4)

Consultants

Dr. Robert D. Carlson, Marshfield Clinic, Marshfield, WI; Ms. Diana Gordon, Vernier Software & Technology, Beaverton, OR; Dr. Eileen M. Lento, PASCO Scientific, Roseville, CA; Dr. Gerald Saunders, University of Northern Colorado, Greeley, CO; Mr. Geof Smith, Ward's Natural Science Establishment, Inc., Rochester, NY; Dr. Edward B. Whitney, Colorado Department of Labor, Denver, CO

First Edition Contributors

Contributors

Robert A. Bouchard, College of Wooster, Wooster, OH; Edward Drexler, Pius XI High School, Milwaukee, WI; Kim Finer, Kent State University, Canton, OH; Ann Haley-Oliphant, Miami University, Oxford, OH; Laura J. Laughran, New Directions, Tucson, AZ

Consultants

Susan Speece, External Evaluator, Fresno City College, Fresno, CA; Constance Bouchard, College of Wooster, Wooster, OH; Ted Dunning, New Mexico State University, Las Cruces, NM; Irene Pepperberg, University of Arizona, Tucson, AZ; Bert Kempers, Media Design Associates, Inc., Boulder, CO; Will Allgood, Media Design Associates, Inc., Boulder, CO; Mark Viner, Media Design Associates, Inc., Boulder, CO; Larry N. Norton, Quantum Technology, Inc., Evergreen, CO; Chester Penk, Quantum Technology, Inc., Evergreen, CO; Ward's Natural Science Establishment, Inc., Rochester, NY

Advisory Board

Judy Capra, Jefferson County Public Schools, Golden, CO; Mack Clark, Academy School District 20, Colorado Springs, CO; Diane Ebert-May, Northern Arizona University, Flagstaff, AZ; Philip R. Elliott, The Colorado College, Colorado Springs, CO; April Gardner (*Executive Committee*), University of Northern Colorado, Greeley, CO; Michele Girard, Peyton High School, Peyton, CO; Eville Gorham, University of Minnesota, Minneapolis, MN; Joseph Graves, Arizona State University—West, Phoenix, AZ; Ann Haley-Oliphant (*Executive Committee*), Miami University, Oxford, OH; Paul DeHart Hurd, Prof. Emeritus, Stanford University, Stanford, CA; Mary Kiely, Stanford University, Stanford, CA; Douglas Kissler, Douglas County High School, Castle Rock, CO; Carole Kubota, University of Washington, Seattle, WA; Douglas Lundberg, Air Academy High School, United States Air Force Academy, CO; Michael E. Martinez (*Executive Committee*), University of California, Irvine, CA; Donald E. Mason, Mitchell High School, Colorado Springs, CO; Laurence McCullough, Baylor College of Medicine, Houston, TX; Martin K. Nickels, Illinois State University, Normal, IL; Floyd Nordland, Prof. Emeritus, Purdue University, West Lafayette, IN; S. Scott Obenshain, University of New Mexico, Albuquerque, NM; William O'Rourke, Harrison School District, Colorado Springs, CO; Ann Pollet, Pueblo County High School, Pueblo, CO; Jerry Resnick, Clara Barton High School, Brooklyn, NY; Parker A. Small, Jr., University of Florida, Gainesville, FL; Gordon E. Uno, University of Oklahoma, Norman, OK; Betty M. Vetter, Commission on Professionals in Science and Technology, Washington, DC; Bruce Wallace (*Executive Committee*), Virginia Polytechnic Institute and State University, Blacksburg, VA; Harry Zimbrick, School District 11, Colorado Springs, CO

Reviewers

Douglas Allchin, University of Texas, El Paso, TX; Tom Anderson, University of Illinois, Champaign, IL; James Botsford, New Mexico State University, Las Cruces, NM; Robert A. Bouchard, College of Wooster, Wooster, OH; Jack Carter, Prof. Emeritus, The Colorado College, Colorado Springs, CO; Frank Cassel, Prof. Emeritus, North Dakota St. University, Fargo, ND; Angelo Collins, Vanderbilt University, Nashville, TN; Robert Cook-Degan, Institute of Medicine, Washington, DC; David Corbin, Monsanto Company, Chesterfield, MO; Jorge Crisci, Museo de La Plata, Argentina; Mary Ann Cutter, University of Colorado, Colorado Springs, CO; Hans Dethlefs, The Neighborhood Health Center—South, Omaha, NE; Edward Drexler, Pius XI High School, Milwaukee, WI; James Ebersole, The Colorado College, Colorado Springs, CO; Diane Ebert-May, Northern Arizona University, Flagstaff, AZ; Philip R. Elliott, The Colorado College, Colorado Springs, CO; Michael Fatone, United States Air Force Academy, CO; Kim Finer, Kent State University, Canton, OH; Steven Fleck, United States Olympic Center, Colorado Springs, CO; Geoff Gamble, Washington State University, Pullman, WA; Barbara Grosz, Pine Crest School, Fort Lauderdale, FL; Topper Hagerman, Steadman-Hawkins Sports Medicine Foundation, Vail, CO; Jerry Harder, NOAA Aeronomy Laboratory, Boulder, CO;

Jeff Hays, United States Air Force Academy, CO; Werner Heim, Prof. Emeritus, The Colorado College, Colorado Springs, CO; Barry Hewlett, Washington State University, Pullman, WA; Michael Hoffman, The Colorado College, Colorado Springs, CO; Michael Keelan, Medical College of Wisconsin, Milwaukee, WI; Rich Kulmacz, University of Texas Health Science Center, Houston, TX; Linda Lundgren, Bear Creek High School, Lakewood, CO; Thomas Manney, Kansas State University, Manhattan, KS; Cheryl Mason, San Diego State University, San Diego, CA; Jeffry Mitton, University of Colorado, Boulder, CO; Adrian Morrison, University of Pennsylvania, Philadelphia, PA; Jamie Nekoba, Waiákea High School, Hilo, HI; Gene O'Brien, Hartford Union High School, Hartford, WI; John Opitz, Montana State University, Helena, MT; Carl Pierce, Harrington Cancer Center, Amarillo, TX; Tracy Posnanski, University of Wisconsin, Milwaukee, WI; Ken Rainis, Ward's Natural Science Establishment, Inc., Rochester, NY; Barbara Saigo, Saiwood Biology Resources, Montgomery, AL; Orwyn Sampson, Brigadier General Retired, United States Air Force Academy, CO; James Short, Packer Collegiate Institute, Brooklyn, NY; James Siedow, Duke University, Durham, NC; Fran Slowiczek, San Diego City Schools, San Diego, CA; Susan Speece, Fresno City College, Fresno, CA; Sam Stoler, National Institutes of Health, Washington, DC; Richard Storey, The Colorado College, Colorado Springs, CO; Gordon E. Uno, University of Oklahoma, Norman, OK; Jeff Velten, New Mexico State University, Las Cruces, NM; Mariana Wolfner, Cornell University, Ithaca, NY

Field Test Site Centers and Coordinators

Arizona
Northern Arizona University, Flagstaff, AZ: Diane Ebert-May, Julie McCormick, Brownie Sternberg

Colorado
BSCS, Colorado Springs, CO: Laura J. Laughran, Randall K. Backe

University of Northern Colorado, Greeley, CO: April Gardner, Alan Lennon, Brenda Zink

Florida
University of South Florida, Tampa, FL: Barbara Spector, Leslie Brackin, Craig Holm

Kansas
Kansas State University, Manhattan, KS: Gail Shroyer, Carol Arjona

Ohio
Miami University, Oxford, OH: Jane Butler Kahle, Rick Fairman

Washington
University of Washington, Seattle, WA: Carole Kubota, Claire McDaniel Orner

Field Test Schools and Teachers

Colombia, South America
Haydée Bejardno de Cadena, Marcela Melendez, Monica Sarmiento, Colegio Los Nogales, Bogotá

Arizona
Marcia Fisher, Arcadia High School, Scottsdale; Doug Davis, Dub Manis, Dee Schwartz, Chinle High School, Chinle; Geri Fisher, Jo Quintenz, Desert View High School, Tucson; Kathy Thayer, Ray High School, Ray; Clyde Christensen, Scott Greenhalgh, Ray Pool, Mary Southall, Elizabeth Stone, Tempe High School, Tempe; Willie Long Reed, Tuba City High School, Tuba City; Jack Johnson, Williams High School, Williams; Carlos Estrada, Karen Steele, Window Rock High School, Fort Defiance

Colorado
Don Born, Peggy Wickliff, Air Academy High School, United States Air Force Academy; Linda Lynch, Douglas County High School, Castle Rock; Doug Hewins, Liberty High School, Colorado Springs; Barbara Andrews, Mitchell High School, Colorado Springs; Rata Clarke, Ray Coddington, Jean Orton, Jim Snare, Palmer High School, Colorado Springs; Rod Baker, Michele Girard, Peyton High School, Peyton; Kathy Dorman, Malcom Hovde, Ponderosa High School, Parker; Ann Pollet, Deborah Walters, Pueblo County High School, Pueblo; Glen Smith, Sabin Junior High School, Colorado Springs; Jeff Cogburn, B.J. Stone, Valley High School, Gilcrest; Bill Bragg, Wasson High School, Colorado Springs; Larry Jakel, Doug Steward, Weld Central High School, Keenesburg; Jay Matheson, West Center for Intergenerational Learning, Colorado Springs; Christy Beauprez, Glenn Peterson, Windsor High School, Windsor

Florida
James Happel, Constance Hopkins, Manatee High School, Bradenton; Scott MacGregor, Joe Martin, Palmetto High School, Palmetto; Barbara Grosz, Pine Crest School, Fort Lauderdale

Hawaii
Jamie Nekoba, Waiákea High School, Hilo; Jennifer Busto, Maryknoll Schools, Honolulu

Illinois
Shelly Peretz, Thornridge High School, Dolton

Kansas
J.D. Hand, Chuck Mowry, Gina Whaley, Junction City Senior High School, Junction City

Minnesota
Clyde Cummins, St. Paul Academy Summit School, St. Paul

Missouri
David Jungmeyer, California R-1 High School, California

New Jersey
Judith Jones, Saint John Vianney High School, Holmdel; Margaret Sheldon, West Morris Central High School, Chester; Karen Martin, West Morris Mendham High School, Mendham

Ohio
Barbara Blackwell, Susan Keiffer-Barone, Aiken High School, Cincinnati; Scott Popoff, Sycamore High School, Cincinnati

Texas
Peter Mariner, Francis Mikan, Dean Mohlman,Tom Stege, St. Stephen's Episcopal School, Austin

Washington
Kathleen Heidenrich, Vicky Lamoreaux, River Ridge High School, Lacey; Larry Bencivengo, Mary Margaret Welch, Mercer Island High School, Mercer Island; Mary Ketchum, Jeannie Wenndorf, Lindberg High School, Renton; Gro Buer, Carol Nussbaum, B.E.S.T. Alternative School, Kirkland; Connie Kelly, Diane Lashinsky, Patrick Taylor, Shorecrest High School, Seattle

Wisconsin
Gene O'Brien, Hartford Union High School, Hartford

Program Overview

I. Unifying Themes

The fall semester is organized into four units. After beginning the school year by launching into Biology, three main themes exist in the fall semester SFUSD Biology Core Curriculum. These concepts are recurring themes that unify all of biology. Although you will see these themes in every unit, we focus on one theme in each unit.

Ecology: Interdependent Relationships

Unit 1 centers on ecological concepts and environmental issues. You will learn about interactions among populations, resources, and environments, and dilemmas that often arise from these interactions. You will examine the concepts involved in population dynamics. This sets the stage for studying the interactions between humans and their environment. You will spiral back to this at the end of the spring semester when you focus on how human actions can modify the environment. The final emphasis is on how humans can work together to generate policies that guide the use of common resources.

Cycles of Energy and Matter

Unit 2 begins by letting you explore the requirements of physical performance. You will explore the interplay between energy and matter by studying the metabolic processes photosynthesis and cellular respiration as well as through interactions in a community. Finally, you will consider the role of producers, consumers, and decomposers in the flow of energy and cycling of matter in a community.

Evolution

How do organisms, including humans, change through time? This is the central question of Unit 3. You will assess the unique qualities of humans and the diversity of life while trying to place humans in the scheme of living systems. Then you will consider how evolution provides an explanation for the adaptations organisms display and the existence of great biological diversity on Earth. Unity, diversity, common ancestry, and evolution, including cultural evolution, are the major conceptual themes in Unit 3.

II. Subthemes

Two subthemes, or background ideas, are woven through the entire program. They help to establish connections between biology and your life. They improve your reasoning ability. The Process of Science subtheme refers to the discovery process by which information about natural phenomena is obtained and evaluated. It also refers to the changing body of knowledge that characterizes scientific understanding. This theme systematically exposes you to the processes of science. This includes asking questions and defining problems; developing and using models; planning and carrying out investigations; analyzing and interpreting data; using mathematics and computational thinking; constructing explanations and defining solutions; engaging in argument in evidence; and obtaining, evaluating, and communicating information.

The Science and Humanity subtheme makes your study of biology more relevant and approachable. It does this by incorporating the critical elements of human culture; the history of science; the place of ethics, ethical analysis, and decision making in a world that is increasingly shaped by science and technology; and the importance of human technology as a way of adapting. We define technology as the use of knowledge to achieve a practical solution to a perceived problem. We also recognize that the ultimate effects of the technological process or product on society and the biosphere may extend beyond the intended effects.

III. Instructional Model

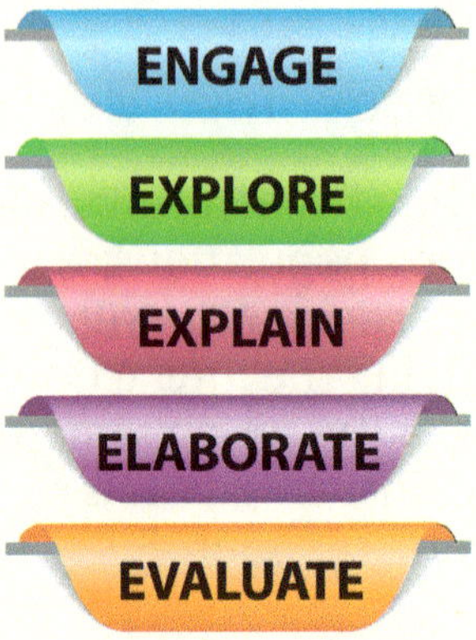

We organized the instruction of major concepts in this book around a model of learning that recognizes how individuals build or construct new ideas. This model is known as the BSCS 5E® Instructional Model. The program is organized around five phases of learning that we best can describe using words that begin with E: Engage, Explore, Explain, Elaborate, and Evaluate. The entire book is shaped by this model.

Each chapter begins with an activity which helps you *engage* the big idea of the chapter and articulate any initial ideas you have around the chapter's topics of study. Next, you will take part in activities that help you *explore* concepts and information related to the chapter. In the middle of the chapter, you will encounter an activity that will give you the opportunity to construct an *explanation* of what you have been learning. Later activities in the chapter are designed to help you *elaborate* your learning by either broadening your knowledge to include new examples, or deepening your understanding of the explanation you have developed. In each chapter, the final activity helps you bring together all of the learning you have done and *evaluate* your understanding.

IV. Cooperative Learning

Cooperative learning is an educational strategy that helps you increase your responsibility for your own learning. Cooperative learning also models the processes that scientists use when collaborating. It helps you develop the working relationship skills necessary for today's workforce.

V. Assessment

Assessment opportunities allow you to evaluate your progress. These activities are embedded throughout the program, and the assessments themselves are learning experiences. The following assessment strategies are included in the program:

- Assessments of your performance, such as experiments
- Written tests that have a variety of short-answer and essay questions

- Assessments of cooperative learning skills
- Presentations, both by teams and by individuals
- Written assignments, both by teams and by individuals
- Journal assignments that include short-term and long-term work
- Projects, both ongoing and one-time
- An ongoing activity about an organism you "adopt"
- Opportunities for self-assessment and peer assessment
- Discussions, both by teams and by the whole class

Symbols Used in This Program*

Science Notebook	SPANISH Audio and text available online	Poison
Online Resource Flourish Learning Network	Caution	Lab Apron
SEP Science & Engineering Practices	Irritant	Lab Gloves
ENGLISH Audio available online	Flammable	Safety Goggles

* Refer to the need to know, *Safety Symbols Used in This Program,* in appendix A, *SFUSD Science Laboratory Safety.*

Have you ever wondered about something you have seen in the world? Maybe you have thought about how a Great Dane and a Chihuahua can be so different but still both be dogs. Maybe you have thought about why you get tired when you do not eat for a while. You might have considered why your favorite pond in Golden Gate Park turned green.

There are so many questions you could ask about the world around you. Many of these questions could be about you as a human. What makes us human? How are we alike and different from the other living things around us? How similar are we to humans that lived hundreds of years ago?

When you wonder about a puzzle or have a problem, how do you figure out a solution? You might answer, "I think about them." But are you aware of *how* you think? San Francisco Unified School District's NGSS Biology Course focuses on science as one of the important ways that humans understand and explain their world. The activities in this program will encourage you to think in the ways a scientist does. You will use some of the same practices that scientists and engineers use to carry out their work. You will ask questions, analyze and interpret data, and construct explanations the same way that scientists do.

To help guide you through your work, there are several supports that you will use. As you start a new chapter, you will see a chapter organizer. This tool will help you remember what you have learned and where you are headed. You will see the key ideas that are important. Use the chapter organizer every day to help you keep track of your learning.

You will also use tables, reading strategies, and graphic organizers to help you organize and record important information. The program has you complete these in specific places to align with what you are learning. But, you might find that these organizers and strategies help you elsewhere, too!

You will also be asked to revise your work and think about how you learned new ideas. This is important because it helps you monitor your learning. You do not have to start every chapter knowing all the answers. But, as you learn more and more, you can think about what you have learned and how you have learned it. You will be able to look back in your science notebook and see what you have accomplished through the whole year.

UNIT

0

Launch into Biology

In this introductory unit, you will explore humans from a scientific viewpoint. The following goals will help you learn the big ideas in this unit. By the end of the unit, you should be able to

- ✔ explain how humans are both similar to and distinct from other organisms, and
- ✔ use evidence to develop explanations about the living world.

UNIT CONTENTS

CHAPTER 1

The Human Animal

And this is the weaving of human living: of whose fabric each individual is a part.

James Agee and Walker Evans

Consider for a moment the differences among humans. We differ in height, the shape of our eyes, the texture of our hair, and the color of our skin. We differ in our ways of living, the languages that we speak, the foods that we grow and eat, and our concepts of the universe. In other words, humans are diverse in physical appearance and cultural behavior. Yet all humans have common bonds, and we see some of ourselves reflected in each of a thousand other faces.

As you begin the first chapter of this unit, you will look at humans in a way that may be new to you. You will look at humans the way scientists do. To do this, you will build a range of knowledge and skills to help you make scientific observations and explanations. You will examine physical characteristics that humans share with primates. Primates are the group of animals that includes humans, apes, and monkeys. You will also examine the ways in which humans are different from the rest of the primates. Using evidence from ancient artifacts can help you learn about humans the same way scientists do. This chapter also explores humans as social and cultural animals that use language and have a great capacity for learning. Your experiences in these activities will help you begin to develop a scientific understanding of the human animal.

GOALS FOR THE CHAPTER

Throughout chapter 1, *The Human Animal*, you will work to gain an understanding of the following concepts:

- ✔ Humans share many characteristics, in varying degrees, with other organisms.
- ✔ A combination of characteristics distinguishes humans from other animals. Scientists' views of human uniqueness have changed over time.
- ✔ Scientific studies pose questions that can be tested with specific measurements and available resources.
- ✔ Scientific studies use evidence and logic to develop explanations.

To help you develop these understandings, you will complete the following activities.

- **ENGAGE** How Different Are We?
- **EXPLORE** Primates Exploring Primates
- **EXPLAIN** More Evidence
- **ELABORATE** How Similar Are We?
- **EVALUATE** First Encounter with the Critter

ENGAGE

How Different Are We?

Key Idea: Humans have distinctive characteristics.

Linking Question:
What can I learn from observations about the similarities and differences between humans and other primates?

CHAPTER 1

The Human Animal

EXPLORE

Primates Exploring Primates

Key Idea: Humans share many physical features with other primates, but their hands, feet, ways of moving, and brains are distinctive.

Linking Question:
How do scientists use evidence and inference to learn more about ancient humans and the development of humans?

EVALUATE

First Encounter with the Critter

Key Idea: Humans and other organisms share many features, but we have distinct physical, behavioral, and cultural characteristics.

Linking Question:
Given humans' similarities to and differences from other life-forms, what does it mean to be human?

MAJOR CONCEPTS

- ✔ Humans share many characteristics, in varying degrees, with other organisms.
- ✔ A combination of characteristics distinguishes humans from other animals. Scientists' views of human uniqueness have changed over time.
- ✔ Scientific studies pose questions that can be tested with specific measurements and available resources.
- ✔ Scientific studies use evidence and logic to develop explanations.

ELABORATE

How Similar Are We?

Key Idea: Human similarities with other organisms can be extended to include diverse forms of life.

Linking Question:
What features of humans are shared with organisms that look very different from them?

EXPLAIN

More Evidence

Part A: A Look at Ancient Humans
Part B: A Long Childhood
Key Idea: Scientists use evidence and inference to learn more about humans based on ancient artifacts and on their development.

ENGAGE

How Different Are We?

In the introduction to this unit, you were asked if you had ever wondered about something you have seen in the world. if so, you were doing two scientific processes. These processes—observing and thinking—always precede experimental work. Careful observations help you collect information about the world around you. Thinking about those observations helps you make sense of the information that you have gathered. Observing and thinking lead you to ask questions based on your careful observation of the phenomenon.

To begin your study of humans, you will make preliminary observations and think about what those observations mean. As you complete this Engage activity, *How Different Are We?,* reflect on the focus question, "What evidence do you have that humans are unique?"

Materials (per team)

flip chart paper
felt-tipped markers

PROCESS AND PROCEDURES

1. Conduct a brainstorming session with the team that your teacher assigns. In your session, consider the question, "What is it about humans that makes us unique?"

   Imagine that you are describing a human to an alien in outer space. How would you describe humans biologically, so that the alien could tell them apart from other animals?

2. Write your team's responses on a large piece of paper and post it in the front of the room so others can see them.
3. Write the following questions in your science notebook, leaving several lines of space between them. You will fill in ideas during the class discussion in step 4.
   a. What common ideas do all teams have?
   b. What ideas are unique to each team?
   c. Examine the characteristics that the teams listed as being unique to humans. What other animals show the same characteristics?
4. Participate in a class discussion by sharing the responses from your team and listening carefully to other teams. As you listen, record in your science notebook your answers to the discussion questions from step 3.

5. Choose at least 6 characteristics that all teams agree make humans different from other organisms, based on the class discussion. Record this list of 6 characteristics in your science notebook.
6. Consider the 6 human characteristics you listed in step 5 as you read the scenario, *Chimps at Gombe*. For each characteristic, record in your science notebook whether or not you think that the chimpanzees also possess the characteristic. Record this by writing "chimp" next to those characteristics that are shared.

   Chimps at Gombe describes Dr. Jane Goodall's observations of a family of chimpanzees at the Gombe Stream Chimpanzee Reserve in Tanzania, Africa. Jane Goodall is a British scientist who began studying chimpanzees in the wild in the 1960s (figure 1.1). If you would like to learn more about Jane Goodall, read the sidebar, *The Chimp Scientist*.
7. Discuss with your team the thoughts that crossed your mind as you read *Chimps at Gombe*. Record your ideas in your science notebook. Focus your discussion on the following questions.
   a. Were you surprised by anything that Dr. Jane Goodall observed? If so, what?
   b. Did you find yourself reflecting more on the similarities or on the differences between humans and chimps? Explain your response.
   c. If you had been Goodall observing this group of chimpanzees, what questions would you now have about these animals? Record in your science notebook at least 2 questions that you would like to research if you were Goodall.

Figure 1.1 Jane Goodall. As Goodall studied chimpanzees in the wild, she recorded as many movements, interactions, gestures, and grunts as she could observe.

SCENARIO

ENGLISH SPANISH

Chimps at Gombe

Gombe Stream Chimpanzee Reserve, Tanzania, Africa

1980: Nope and Pom, two wild chimpanzees, are feeding quietly. Suddenly, a chimp screams in the distance. Pom stares toward the sound, turns to Nope, and grimaces. Then the screams break out again. Now at least two chimps are screaming, and one is an infant. Above the screams, the observer can hear male barks and whoops of attack. Instantly, Pom leaps to her feet and charges toward the sounds of battle. By the time the observer arrives, it is quiet again, and Pom and her mother, Passion, are grooming each other intently. Pom's little brother, Pax, is close beside them. Both Passion and Pax have fresh, bleeding wounds, but the family is safe and together.

1981: On a blustery morning, Pom's new baby clings tightly to her hair. As the wind starts to die down, he begins to play, venturing out on the limb that he and his mother are sharing. Just then, a violent gust sweeps through the tree and his little body falls spread-eagle through the air. The observer hears a thud and then nothing but silence. Pom looks down at her infant son. Slowly she climbs out of the tree and gathers the tiny form into her arms. For the next two hours, she grooms and nurses him. He leans against her body with his eyes closed. Finally, she carries his battered body away. Three days later, he is dead.

1982: Pom is 17 and her brother Pax is still a youngster when their mother, Passion, dies. Passion has been ill for weeks, and now she trembles with every movement. One morning, she is dead. She must have fallen in the night; her body hangs in a tangle of vines. Pom and Pax sit staring at their mother's body. Little Pax repeatedly approaches her and tries to nurse from her cold breasts. Then he starts to scream and pull at her dangling hand. So frantic are his efforts that finally he succeeds in pulling her loose. As Passion sprawls lifeless on the wet ground, her children inspect her body many times. Pax cries softly. At last, just before darkness, Pom and Pax move off together.

1983: Having lost both her baby and her mother, a weak and listless Pom finally leaves the community and is not seen at Gombe again. Pax, on the other hand, attaches himself to an older brother, Prof, who provides the care that a mother chimpanzee would. One day, Pax sneezes loudly. Before the astonished observer can get out the camera, Prof hurries over to Pax and stares at his runny nose. Then, picking up a handful of leaves, Prof carefully wipes the mess away.

Analysis

Use the following tasks to analyze your list from step 5 of the 6 characteristics that make humans different from other animals. Record your responses in your science notebook.

1. For each characteristic you listed in step 5, identify and describe a nonhuman animal that you think may also display that characteristic to some degree.
2. Do you think there is any single characteristic that sets humans apart from all other animals? Explain your answer.
3. Chimps and humans share many characteristics. Write down your best ideas about why you think these 2 species share so many characteristics.

   It is OK not to know the right answer at this point. Just as it is important for a scientist to reflect on his or her observations, it is important for you to reflect back frequently on your own understanding. When you connect new information to what you already know, learning becomes easier. Frequent reflection helps you spend less time to learn more effectively. You will have a chance to revisit your answer at a later point.
4. Like a scientist, you will keep a science notebook throughout the course. The accuracy with which you describe events in your science notebook is important. This will allow you to compare your observations with those of others. It is also a vital way to keep track of your ideas and how they change. A page from Jane Goodall's notebook is shown in figure 1.2.
   a. How do her field notes show her work as a scientist? Observing and thinking are two key ways scientists work.
   b. How do you think careful notes and records of her work helped Goodall?
   c. Think about the ways you used a notebook in this lesson. What other ways can you think of that might be useful in the future?

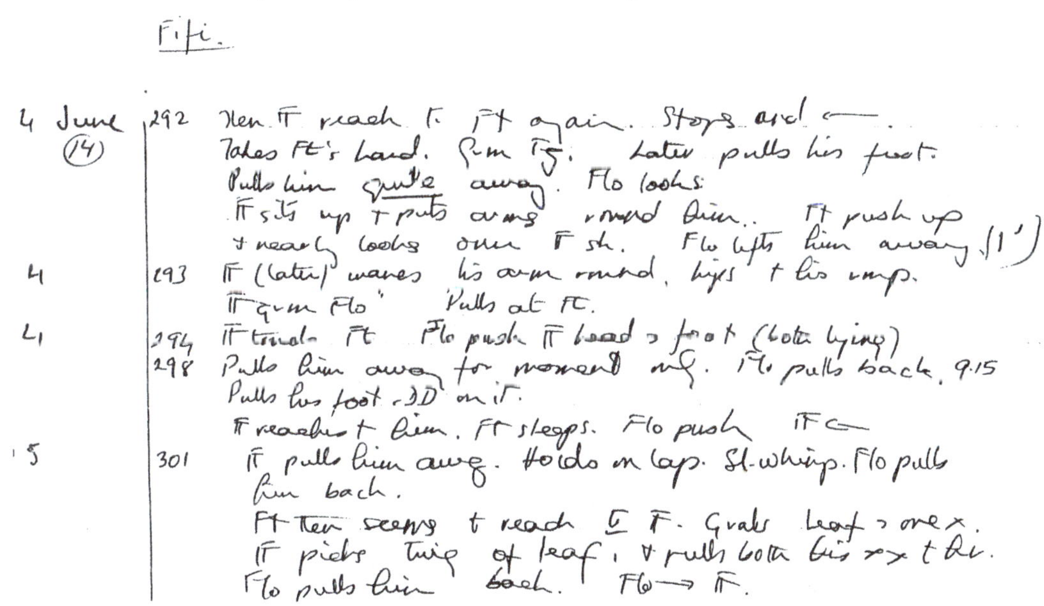

Figure 1.2 Jane Goodall's field notes. How does her notebook show that she is working like a scientist?

Further Challenges

To learn more about Jane Goodall, check your library for the many books that she has written and look for information at the Jane Goodall Institute Web site. You also may be interested in other scientists who have done or are doing similar work. The American Society of Primatologists maintains a Web site with links to many interesting research projects, scientists, and general information about primates.

EXPLORE

Primates Exploring Primates

In the Engage activity, *How Different Are We?*, you read and thought about observations that someone else made. You also considered your own ideas about how humans are similar to and different from other animals. Now you will have a chance to explore these ideas in more depth and to make observations of your own. You will observe humans and other related animals as they move about and use tools. Remember, your goal is to look at humans in the way scientists do. In this Explore activity, *Primates Exploring Primates*, your goal is to gather evidence about the characteristics that set humans apart from other animals and the characteristics that they share with other animals.

Materials (per team of 2)

online resource
assortment of objects to grip
1 padlock and key
1 stopwatch (optional)
masking tape (optional)

PROCESS AND PROCEDURES

1. In the last lesson, you learned about some of the work that scientist Jane Goodall has done with chimpanzees. Look at figure 1.3, How Science Works.
   a. What do you notice about the diagram?
   b. What parts of the diagram do you think would be most interesting?
2. Observe humans moving from place to place for 10 minutes. Make a record of these observations in your science notebook under the title "observations of humans moving."

   Record your observations with descriptive words and drawings. For example, after watching someone jog by, you might write:

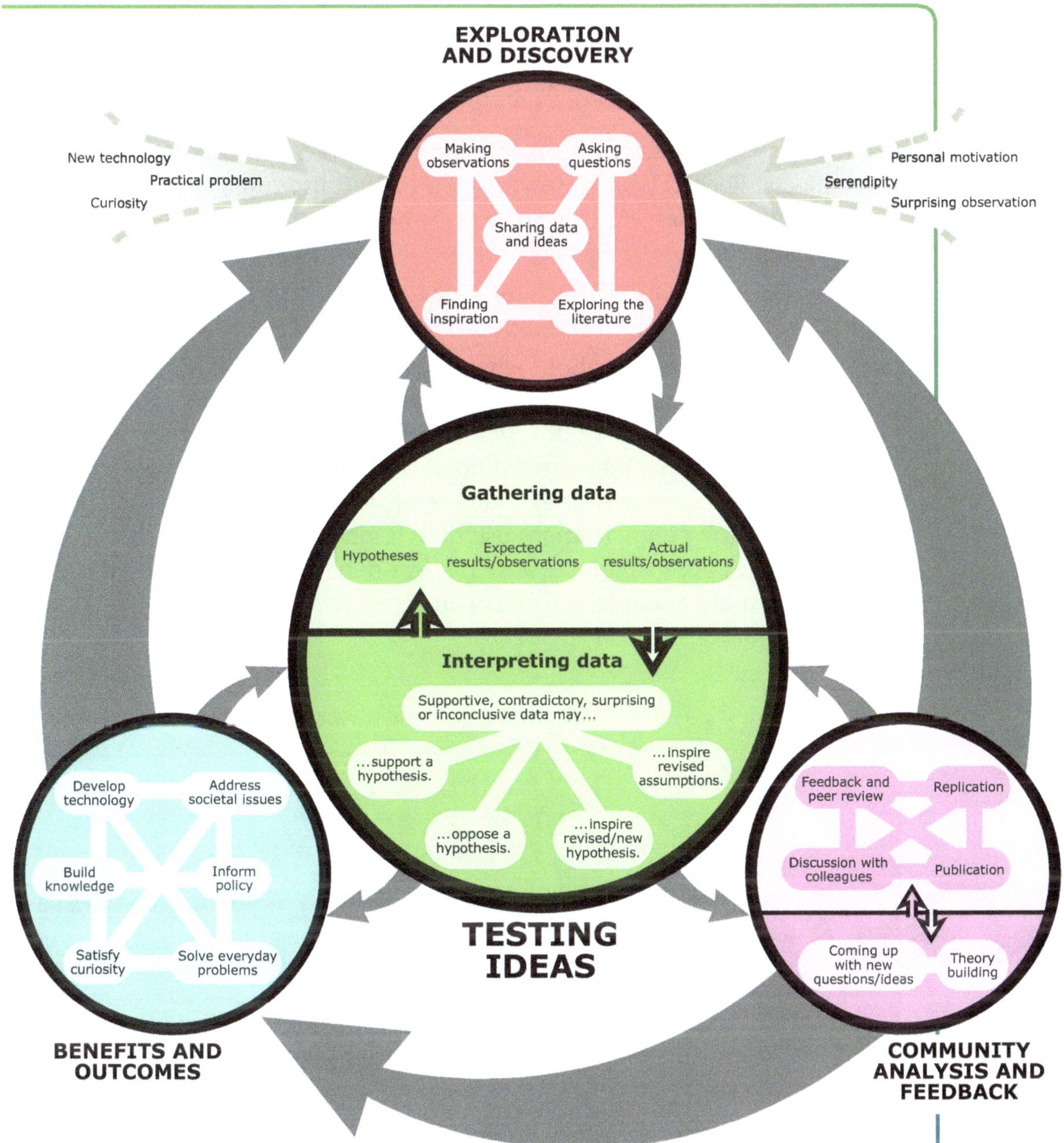

Image courtesy of the University of California Museum of Paleontology—Understanding Science www.understandingscience.org.

Figure 1.3 How Science Works. Science is a process that helps people understand the world around them. The characteristics shown here help to define the process of doing science.

"When humans jog, they take short steps quickly. Slower runners tend to have their heel strike first and then they push from their toes." In contrast, writing "People jog" is not as good an observation because it is not descriptive. What do you think about the comment "He is jogging because he wants to lose weight"? This is an interpretation because it attaches meaning to an action, rather than being an objective description. Try to make all your observations as objective as you can. Take the time to notice all the different types of strides that are possible and the differences in the way humans use their arms, legs, and feet.

3. Observe the different ways that humans hold and use the objects. Record your observations in detail in your science notebook under the title "observations of how humans hold and use objects."

CAUTION: Do not swing, throw, drop, or fool around with any of the objects in a way that might harm you or your classmates.

Make observations about what parts of the hand and arm your partner uses to hold objects. For example, after examining how your partner turns the pages of a book, you might write: "When humans are reading a book, they use the tip of the thumb and forefinger to pick up a page, and then they rotate the forearm to turn the page." In contrast, writing "Humans use their fingers to turn pages" is not as good an observation because it is less descriptive. If you would like to learn more about Jane Goodall and the observations she made, read the sidebar, *The Chimp Scientist*.

SIDEBAR

ENGLISH SPANISH

The Chimp Scientist

. . .I have often wondered exactly what it was I felt as I stared at the wild country that so soon I should be roaming. Vanne [Goodall's mother] admitted afterward to have been secretly horrified by the steepness of the slopes and the impenetrable appearance of the valley forests. And David Anstey [a game warden in Tanzania] told me several months later that he had guessed I would be packed up and gone within six weeks. . . What had I, the girl standing on the government launch [landing dock] in her jeans, to do with the girl who in a few days would be searching those very mountains for wild chimpanzees?

Jane Goodall

Jane Goodall began to observe and study the chimpanzees of Gombe Stream in Tanzania about 50 years ago (figure 1.4). At that time, some people thought that it was not scientific to talk about an animal's mind or personality. However, Goodall's careful and detailed observations and evidence changed many people's minds. Today, her work stands as one of the great contributions to science made during the 20th century.

In the past, some people thought of science only as experiments that men in white coats conducted in laboratories.

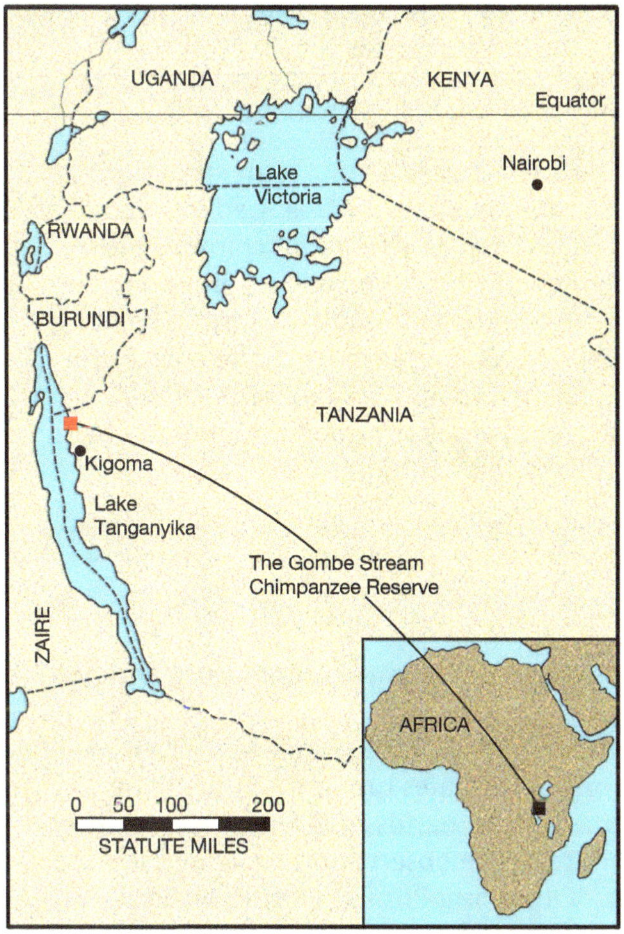

Figure 1.4 The Gombe Stream Chimpanzee Reserve. The Gombe Stream Chimpanzee Reserve is located near Kigoma, on Lake Tanganyika in Tanzania.

Indeed, some people still have that mistaken image today. As you have been discovering, however, science is a particular way of knowing and learning about the world. Each of us can learn the methods of science and put them into practice.

Goodall's work is a great example of the role of observation and reflection, two key aspects of science. The notes pictured in figure 1.2 are copied from a page of Goodall's field journal. Like your science notebook, her notebook documents her observations. The notes recorded on the page reveal the value of careful organization for keeping accurate records. Can you tell what Goodall's notes mean? She made detailed observations while she kept careful watch on her chimpanzee subjects. She designed her own abbreviations and note-taking style to make it easy to write quickly and understand later.

To learn about how chimpanzees behave in the wild, you have to go into nature and observe them across long periods of time. It took Goodall close to four years to collect a significant amount of information. The chimpanzees were hard to find and resisted her presence for the first several years. But eventually they came to trust her—this itself was a key observation. Goodall carefully recorded their every move, interaction, gesture, and grunt. She found that chimpanzees are caring, clever, and capable of lasting attachments. Through years of observations, she also learned that some individuals are capable of extreme aggression, even cannibalism.

Scientists used to think that humans were the only toolmaking animals and generally assumed that humans were the only organisms that could think or have emotions. Today, due to the work of Goodall and other scientists in related fields, we recognize that many intellectual abilities once thought to be unique to humans are present in other animals. For instance, Goodall observed the problem-solving ability of chimpanzees. They recognized a need for tools, planned ahead of time to prepare tools, made the tools, and used the tools. One such observation involved a chimp named Mike (see figure 1.5). He finally solved the problem of how to get a banana that Goodall held out to him. Apparently too nervous to take the banana from her hand, he finally used a stick to knock it to the ground. Only then did he grab it and run.

Figure 1.5 **Mike bangs kerosene cans together.** Although Mike was not the largest or strongest male, he made his way to the top by creating frightful noises with these empty cans.

Jane Goodall's long-term studies and careful observations resulted in important pieces of evidence about chimpanzee behavior. This knowledge is important because the chimpanzee is the closest living relative of the human species. Accurate information about chimpanzees helps us come to a better understanding of what it means to be human.

4. With your partner, discuss the different ways humans use and hold objects.
5. Insert a chart like the one in figure 1.6 into your science notebook. Title the chart "observations of primates moving." You will be assigned to observe either arms and hands or legs and feet. Write column headings appropriate to the observations you are assigned.
6. Watch the video segment, "Observing Primates," while recording your observations in your chart. The table is designed to show the

http://www Online Resource

Observations of Primates Moving

Primate	How arms are used	How hands are used
Gibbon		
Orangutan		
Gorilla		
Common chimpanzee		
Pygmy chimpanzee		
Human		

Figure 1.6 **Sample table of observations of primates moving.** A table similar to this one will help you organize your observations.

primates in the order they appear in the video. Be sure to make objective, descriptive observations for each primate in the same way you did for humans.

Record as much detail as you can in your table. As you observe humans in this video segment, add any new observations to those that you made in steps 1 and 2. If you are observing hands, pay particular attention to how each primate uses its fingers and thumbs so that you can compare those motions with the human grip observations you made earlier.

7. Switch science notebooks with your partner and review his or her observations. Ask your partner to clarify any observation that you do not fully understand. Use your partner's questions to refine the wording of the observations in your chart.
8. Discuss the following questions with your partner and record answers in your science notebook.
   a. How is human movement and posture different from that of other primates?
   b. In what ways do primates (other than humans) use their thumbs?
   c. In what ways does human thumb use differ from that of other primates?
9. Scientists ask questions based on their observations. Think about your careful observations of the primates. Based on that, brainstorm with your partner all the questions that come to mind about how primates move about and use their hands to grip and manipulate objects. Include at least 1 question that relates to how different primates could use a key to unlock a padlock. Record this list in your science notebook.

   Come up with at least five questions. If you have trouble coming up with questions, or if you would like to learn more about the ways that humans use their hands, read the sidebar, *Do You Have a Grip on That?*

10. Discuss with your partner which of the following 2 questions meets the criteria for a testable question, as described in the need to know, *Testable Questions*. In your science notebook, explain the reasons for your choice. This will help you gain a clearer sense of the nature of testable questions.
    a. What is the importance of an opposable thumb for the ways in which humans use their hands?
    b. How does using an opposable thumb affect the time it takes to use a key to open a lock?

11. Reexamine the questions you wrote with your partner in step 9 by using the think-share-advise-revise (TSAR) strategy. Then, place an asterisk by the questions that you and your partner agree are testable.

    As you read each of the criteria, *think* about whether or not the questions you generated in step 8 are testable or not testable. Then *share* your ideas with your partner and *advise* each other for possible corrections in your thinking. Finally, *revise* your ideas based on your discussion. Use a different-colored pen or pencil for your revisions.

    For more information about using the TSAR strategy, see appendix B3, *How to Use the Think-Share-Advise-Revise (TSAR) Strategy*.

    HOW TO

12. With your partner, choose the question you would like to test. It may be 1 of your own questions or 1 of the 2 questions presented in step 10.

NEED TO KNOW

ENGLISH SPANISH

Testable Questions

Science begins with observing and asking questions about the natural world around us. Then it moves to a stage in which we begin to answer those questions by using a combination of further observation and experimentation. Scientists often have many questions that they would like to answer, but science is limited to answering questions that are testable in the natural world.

1. A question is likely to be testable if it uses question words like "whether," "when," "where," "what," "how many," "how much," and "how often," rather than question words like "why."

   For example, the question, "How many fingers does a gorilla have on its right hand?" is easy to answer. Yet the question, "Why does a gorilla have four fingers on its right hand?" is not.

2. If the specific issue to be tested is stated in your question and you list the specific items that will be involved, then your question is likely to be testable.

   The question, "How much do gorillas eat?" is not easily tested. A more easily tested version of this question is, "How many pounds of bananas do adult male gorillas eat?"

3. A question is likely to be testable if it describes the conditions under which to conduct the test.

   An even better version of the gorilla and banana question would be, "How many pounds of bananas do the three adult male gorillas in the National Zoo eat in one week?" This question specifies the conditions that interest us; we are not asking how many pounds of bananas this gorilla would eat in the wild or during his lifetime.

4. A question is likely to be testable if it describes the criteria that will be used to judge the outcome of the test.

   Does a half-eaten banana count as having been eaten? How about a banana that is three-quarters eaten? Can you phrase this gorilla and banana question so that it is easier to test than the other questions listed above?

5. A question is likely to be testable if it can be tested using available resources and procedures.

   In the end, questions about gorillas and bananas, although they may be testable by some people in some parts of the world, are not testable questions for us in the classroom. All researchers (students and teachers alike) are limited by the resources that they have available.

SIDEBAR

ENGLISH SPANISH

Do You Have a Grip on That?

Tying a shoe, threading a needle, throwing a baseball, taking notes, or carrying a bucket. Think of all the tasks—large and small—that humans accomplish with their hands every day. Have you ever thought about the amazing number of ways in which humans can use their hands to grip things?

Scientists have studied the different grips of primates for some time. They have more than one way of classifying what they have observed. One school of thought divides grips into two major categories: power grips and precision grips. You use the power grip when you grasp an object with your palm and then curve your fingers around it. The power from the power grip comes as you apply pressure with your whole hand. This power is made possible by the muscles of your palm that are located at the base of your thumb. These muscles are responsible for moving your thumb. The power grip requires the full grip of the hand—when you use a hammer or open a tightly closed jar, for example. The precision grip is a more intricate grip that requires a specific alignment of the thumb with one or more fingers. The thumb applies pressure against another finger or fingers in order to accomplish a precise motion, such as picking up a coin. Sometimes it lets you make a precise motion that also requires strength, such as placing a key in a lock and turning it.

Another school of thought subdivides the power and precision grips into as many as 12 categories by defining the grip according to the exact placement of the palm, fingers, and thumb. Figure 1.7 shows seven of the 12 grips.

Cylindric grasp. The palm is used to grasp a cylindrical object; the thumb opposes all of the fingers.

Fist grasp. The fingers are wrapped around a narrow or small object; the thumb opposes the fingers, lying on top of them and securing the grip.

Figure 1.7 Types of grips. These photographs show seven of the 12 ways humans can grip objects. Which grips did you see the primates use in the video segment, "Observing Primates"? Which grip did you test in the Explore activity, *Primates Exploring Primates*?

Hook grasp. The fingers are used as a hook; the thumb is not used.

Spheric grasp. The palm is used to grasp a spherical object; the thumb opposes the fingers.

Tip prehension. The tip of the thumb and one or more of the fingers are used like tweezers to pick up small objects; the thumb opposes one or more fingers.

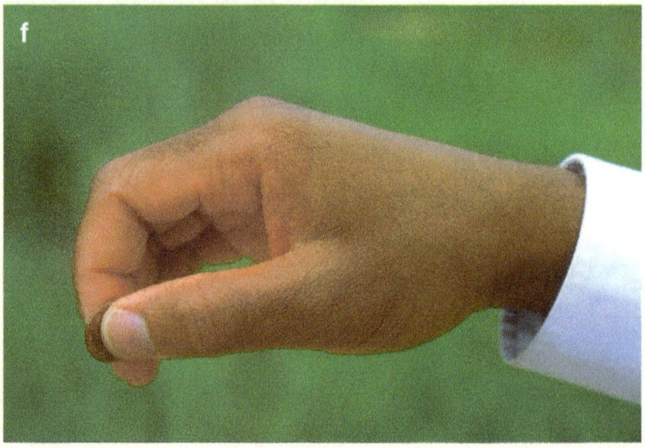

Palmar prehension. The flat surface of the thumb is pressed against the flat surface of the finger or fingers. This grip is similar to the way the tip of a pair of pliers holds an object.

Lateral prehension. The thumb presses against the side of the index finger with other fingers providing additional support. Another type of lateral grip involves the sides of two fingers pressing against each other.

13. Predict what you think the answer to your testable question will be. Record your prediction in your science notebook and explain why you have made this prediction.
14. Work with your partner to plan an investigation for the question you will test. Briefly outline the steps of your protocol for the question you will test. Then write it in your science notebook. You may adapt the steps in *Protocol for Testing the Question* to suit your question, if they apply.

PROTOCOL

Although you may be testing your own question during your experiment, this protocol will give you an idea of the detailed steps that are included in a protocol. It is important that protocol steps be followed carefully and in order.

If you are designing your own investigation, you may wish to review the guiding questions in *How to Plan and Carry Out an Investigation*, in appendix C2.

HOW TO

PROTOCOL

Protocol for Testing the Question

A protocol is a plan or procedure for a basic investigation that can answer a testable question. The protocol also serves as a standard for controlling variables. When you construct a more challenging or creative testable question, you can modify the standard protocol to collect the data you need to answer or better understand your question.

1. Obtain a lock, a key, and masking tape as your teacher directs.
2. Decide who will carry out the task first and who will keep track of the time and record observations.
3. Be certain that the lock is locked. Place the lock and the key on the table in front of you and your partner.
4. When the person responsible for keeping track of the time says "begin," the person performing the task will pick up the key and attempt to unlock the lock.

   This task lasts 30 seconds. The timer-recorder should keep track of the time and record whether or not the person was successful.
5. Switch places and repeat steps 3 and 4 of this protocol.
6. Use the masking tape to tape 1 partner's thumbs to the side of each hand, as shown in figure 1.8.
7. Repeat steps 3 and 4 with the hands taped.
8. Switch places and have the second person tape his or her thumbs. Repeat steps 3 and 4 once more.
9. In your science notebook, record any results from your team's experiment that you haven't yet documented.

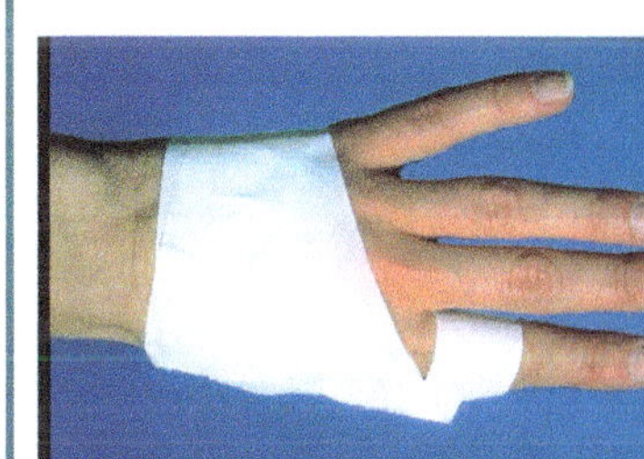

Figure 1.8 Ask your partner to help you tape your thumbs to the sides of your hands, as shown. You should not be able to use your thumb at all, but you should be able to move and use each of your other fingers.

15. As you carry out your investigation, be sure to record your observations and data in your science notebook.
16. Use your results to answer the question you tested. Record your answer in your science notebook.
17. Write a few sentences to explain the connection between the results and your answer to the question.
18. With your partner, discuss the limits, exceptions, or alternative interpretations of your results. Record your ideas in your science notebook.
19. Share your results with the rest of the class.
20. Reflect on all the steps of this activity, starting with your observations of humans moving. Write a statement in your science notebook about what practices of science you engaged in during this activity.

    Use the How Science Works diagram from step 1 of this lesson to review all the science practices and identify those you were engaged in during this activity.

Analysis

Discuss the following with your partner, and then record your own answers in your science notebook.

1. In this lesson, you focused on the similarities between different primates. You focused on the way hands and arms are used. What other similarities are there in the behavior of different primates?
2. A scientist might also compare the characteristics and behavior of other organisms. Imagine a scientist comparing a goldfish to a dog. Both can be pets, but they are very different. What do you think a scientist might be able to say about each animal by studying their characteristics and behaviors? What do you think the differences would tell a scientist?

Figure 1.9 Sea lions at Pier 39 in San Francisco.

Further Challenges

1. Visit the zoo, a park, or San Francisco Bay (figure 1.9) and make detailed observations of the way various primates or other animals move. You could record your observations in writing, drawings, or video.
2. Develop an experiment designed to answer another 1 of your testable questions from this activity.

EXPLAIN

More Evidence

In the last activity, you examined some behavior and characteristics of primates' hands, feet, and movement. But do these characteristics capture the essence of what it means to be human? Take a moment to review the class's brainstorming results of distinctive human features from the Engage activity, *How Different Are We?* How many of the features on that list are not clearly physical characteristics?

Perhaps the class's list included such notable human characteristics as using language to communicate, creating art and music, and participating in ceremonies. Humans have an exceptional ability to develop behaviors and technologies that permit us to cope with new situations. For example, we can minimize the effects of an extremely hot environment by using the technology of air-conditioning to modify part of that environment. This ability is called **cultural adaptation**, and it is one of the distinguishing characteristics of humans.

Did you know that some kinds of animals do not even care for their young? Contrast that to humans who live with their parents for many years before they are on their own. There are other animals that fall in between these two extremes. This development represents another distinguishing characteristic for organisms.

In the Explore, you focused on behavior and physical characteristics. However, scientists study more than these features to learn about animals. They also examine ancient artifacts and the way organisms develop. Observations about these features of humans and other animals can help scientists learn much more than simply observing behaviors.

In this lesson, you and your teammates will consider how scientists can learn from humans in the past and from the way living things develop.

Materials

Part A

online resource

different-colored pens or pencils two colors of highlighters

Part B

Resources on gestation and the development of animals after birth

PROCESS AND PROCEDURES

Part A: A Look at Ancient Humans

In part A of this Explain, you will have a chance to learn more about a human who died more than 5,000 years ago. You will discover the items that were found with him. Because he was buried in ice and snow high in the Italian Alps, his body and belongings were well preserved. Because of that, they have given scientists a lot of information about the man and his culture.

1. Consider the questions, "How do scientists study events that occurred 5,000 or more years ago?" and "Why might scientists be interested in studying events from that long ago?" Take notes as you participate in a class discussion of this question. Write your best answers in your science notebook.

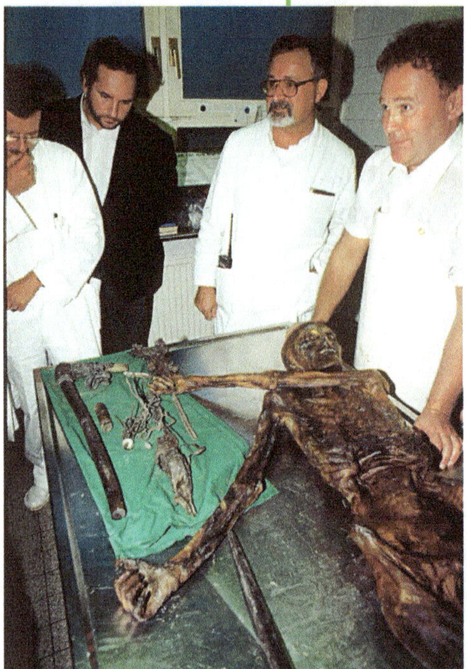

2. Read the need to know about an ancient human by yourself. As you read, use the Mark It Up! strategy to focus your attention. To do this, complete the steps in the following bullets.
   - Use one color of highlighter to mark key information about the man that was found.
   - Use a different color highlighter to mark actions that scientists did.
   - Underline words, phrases, or sentences that are unclear to you.
   - Circle words that you know but which are used in an unfamiliar way.

   If you need more information about the Mark It Up! strategy, see Appendix B10.

NEED TO KNOW

ENGLISH SPANISH

A man died high in the Alps (figure 1.10), above the tree line. His body was frozen into a glacier and remained there for more than 5,000 years. Then, in 1991, some hikers discovered the well-preserved corpse. They assumed that it was a modern hiker who was killed the previous winter, and they notified authorities. Because the authorities did not realize that

Figure 1.10 The Alps, where the body of the Iceman was found.

they were dealing with something very old, they removed the body from the ice carelessly. Unfortunately, this carelessness resulted in the loss of valuable evidence.

Imagine their surprise when it was discovered that the body was ancient, probably 5,300 years old. He was male, stood 165 cm tall, weighed about 38 kg, and was about 45 years old at the time of his death. He had a series of blue markings on his skin. This included several sets of blue parallel lines on his lower back, stripes on his right ankle, and a cross behind his left knee. His hair was medium length and had been cut. His physical features were similar to those of humans today. His final meal had been eaten about eight hours before he died. This meal included a crackerlike bread, some green plant material, and some meat. X-rays and CAT scans revealed that before his death, he received a blow to his head as well as an arrow wound through his back left shoulder, which punctured an artery.

The Iceman and his possessions are among the most valuable anthropological evidence ever found. Usually, the types of evidence that are preserved from humans who lived 5,000 years ago are hard substances, such as teeth, bones, or metal parts of tools. Soft tissues such as skin, hair, plant materials, and leather generally are destroyed by decay. In this case, however, many unusual material remains were recovered with the Iceman. These remains are called artifacts.

Scientists often draw logical conclusions about things they cannot directly observe. For example, scientists come up with ideas about the skin and outward appearance of a dinosaur from its fossil remains. In the same way, scientists come up with ideas about the appearance or culture of early humans from artifacts that are preserved.

3. Use the information found in the table in figure 1.11 to help you imagine what the Iceman was wearing and carrying when he died. A few artifacts are shown. You might also wish to do an Internet search to find out what additional items looked like. With your team, share the artifact that you think is most interesting. Describe what you think a scientist might be able to tell about the Iceman or other people that lived at that time based on that artifact.
4. Watch the video segment, "A Glimpse of the Iceman." As you watch, take notes on ideas that come to mind about what the Iceman's life and culture were like.

http://www Online Resource

Flint spearhead

Stone disk

Spelt wheat

Artifact	Scientists' Comments
Grass cape	The cape is carefully stitched together; isolated repairs were made with grass thread.
Arrows and leather quiver (carrying case for arrows)	This is the oldest leather quiver ever found; some arrows have flint arrowheads, other arrows are unfinished.
Very long bow	The bow is unstrung.
Pouch with worked bits of flint	Flint is a type of stone that flakes easily into sharp pieces.
Clothing and boots	The clothing and boots are made of leather; grass is stuffed inside boots.
Copper-bladed axe with wooden handle	This is one of the oldest axes of its type and the first with handle intact; it is attached by leather ties and glue; it has a copper blade.
Mushrooms strung on a leather strip	These mushrooms are a type of fungus with medicinal (antibiotic) properties.
Flint dagger and grass sheath	The worked flint blade is on a wooden handle; this is the first sheath of this age recovered.
Bone needle	
Grass rope	
Stone disk threaded with a leather strap	
Bits of a primitive wheat and wheat pollen	This type of wheat grew only at low altitudes.

Figure 1.11 **Artifacts recovered with the Iceman.**

5. To prepare for a team discussion, construct a table titled "evidence and ideas about the Iceman" in your science notebook. The table should look like figure 1.12.

Figure 1.12
Table of evidence and ideas about the Iceman. Write down evidence in the left column and your ideas in the other two columns.

Evidence	What it might indicate about the Iceman	What it might indicate about the Iceman's culture

6. With your team, discuss each piece of evidence from the Iceman's body and possessions. Contribute your ideas about what each observation may indicate about the Iceman and his culture. Record all ideas in the table you made in your science notebook.

7. Write an "E" next to statements that you think constitute scientific evidence.
   a. The Iceman carried arrows made with flint arrowheads.
   b. The Iceman used mushrooms as medicine.
   c. The Iceman spent some time at lower altitudes.
   d. The Iceman had bits of wheat from a lower altitude on his clothing.
   e. The Iceman wore clothing made from animal products.
   f. The Iceman knew how to sew.
   g. The Iceman was a hunter.
   h. The Iceman used grass for several purposes.
   i. The Iceman lived at a time when people knew how to work with metals.
8. Some of the statements in step 7 describe evidence and some describe inference. To learn more about what these terms mean, read the need to know, *Evidence and Inference*. The information will help you learn the difference between evidence (observations you can make directly) and inference (logical assumptions based on evidence, but that you have not directly observed). Think about each of the statements in step 7 as you read.

NEED TO KNOW

ENGLISH SPANISH

Evidence and Inference

When scientists find human remains and artifacts, they work carefully to gather as much information as they can from their findings. Often what they find is incomplete, but the remains are *evidence* that scientists can use to draw conclusions about the individual. Scientists try to be aware of their own beliefs, moods, and prior experiences when they look at evidence. That way, they will not let these factors affect their conclusions. Conclusions that follow logically from some form of direct evidence are known as *inferences*. (Conclusions that do not follow logically from evidence are just guesses. Guessing is not an acceptable way to draw scientific conclusions.)

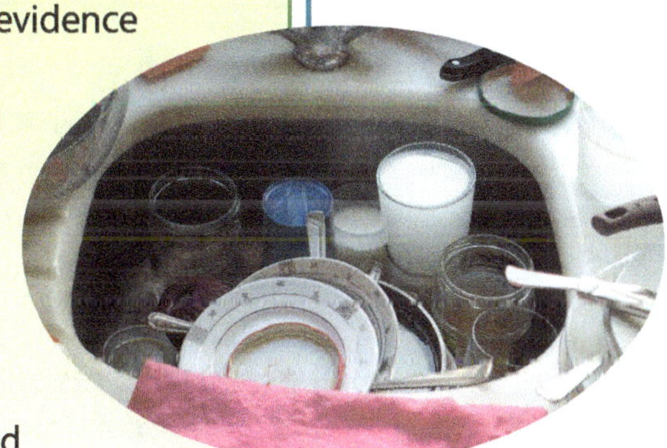

For example, if you came home and found some dirty dishes in the sink, you might infer that your older brother was in a rush to eat before leaving the house. An illogical assumption would be that a robber had stolen some things and then eaten something before leaving. There is no additional evidence that a robbery had taken place. Sometimes more than one inference could be drawn from the evidence. For example, another inference could be that your brother had been too lazy to wash the dishes before he left.

In science, the strongest inferences are the ones that are made with the most evidence and that provide the simplest explanation. If you saw that in addition to the dishes, his phone was on the table with unread text messages, a lamp was left on, and there was a note reading "Back at 6 after job interview—Dave," you might conclude that the strongest inference is that your brother was in too much of a rush to wash his dishes. With all of that evidence, the idea that Dave was just lazy is a weakly supported inference.

9. Decide with your team whether each of the statements in step 7 describes *evidence* or *inference*. Record your ideas next to the list, including why you think each statement is evidence or is inference.
10. Look back at the ideas you recorded about each artifact in step 6. Were your ideas about how the Iceman used the items based on evidence or inference? Write "evidence" or "inference" next to each statement, using a different-colored pen or pencil.
11. Discuss with your team the kinds of cultural artifacts that are worn and carried by people today. In your science notebook, make a table with 2 columns. List the Iceman's artifacts on one side and the modern-day equivalents of those artifacts on the other side.

Part B A Long Childhood

Have you noticed that puppies can be separated from their mother within two months of birth, and a newborn foal (horse) will be up and walking around within minutes of its birth? A human child, however, usually stays with its mother for years and does not begin to walk until it is close to one year old. Have you ever wondered why this is so?

The collection of physical characteristics that sets humans apart from other organisms results in some behavioral differences as well. You have explored evidence of an ancient human's culture and compared it with your own culture. In part B, you will have a chance to learn more about the development of different organisms. You will consider how that development provides more information about the organisms. You are also likely to come up with more intriguing questions of your own.

1. Discuss with your class what you know about the development of humans and other animals. Consider the following questions in your discussion.
   - How long is gestation (the time from conception to birth)?
   - What do parents do to care for the young?
   - What are the milestones of childhood?

2. In your science notebook, create a table similar to that shown in figure 1.13. The table will help you organize and record information about the gestation and development of humans and other animals.

Timing of important developmental events in different animals

Animal	Length of gestation	Method of obtaining food	Completely mobile	Sexually mature
Humans				

Figure 1.13 Sample table of information on animals. A table similar to this one will help you organize your information in a way that makes it easy to use.

3. Select an animal to research. Use the headings in the table to guide your research then record your answers.

   Use the resources that your teacher provides or resources from the library or the Internet.
4. Contribute your answers to a class data table. Record the class's data in the table in your science notebook.
5. Discuss with your classmates the patterns that emerge as you compare the data for humans with those of the other animals. Use the data to identify the animals that have long "childhoods." Discuss your ideas with the class to determine a definition for a "long" childhood.
6. Write "long" next to those animals that have a long childhood and circle the pieces of data that support that idea.
7. Briefly discuss the following questions with a partner. Then, as you read the essay, *The Importance of Being Children*, use the Mark It Up! strategy to mark information that can help you answer the questions. Highlight information that answers a question. In the right margin, write the letter of the question next to it and any comments or questions you have. Underline anything you are not sure about. Once you complete the reading, discuss your ideas with your partner again and then record your best ideas in your notebook.

   a. How does a long childhood help humans develop complex culture?
   b. How does our language ability help us develop complex culture?
   c. How does our ability to learn help us develop complex culture?
   d. Compare and contrast learned behavior and innate behavior.
   e. Give 2 examples of other animals that exhibit a certain capacity for learning, language, and culture.
   f. Is it possible to infer whether the Iceman would have had a long childhood? Use evidence to support your answer.
8. In your science notebook, write an answer to each of the questions in step 7.

The Importance of Being Children

ENGLISH SPANISH

"When will I be a grown-up?"

This is a question that four-year-olds all over the world ask in hundreds of different languages. The words may be different, but the question is the same—"When will I know all that I need to know to be me?"

Think about the following examples:

- A human newborn usually doesn't start to walk until he or she is at least 10 months old.
- On average, children and young adults in the United States are required to go to school until age 17.
- In the United States, people are not allowed to vote until they are at least 18 years old.

Compared with other animals, humans take a long time to become adults. Why do humans have such a long childhood?

A Long Childhood Allows Humans Time to Learn

One answer to the question about why humans have a long childhood is that humans need to learn a tremendous amount of information and concepts. It is amazing to stop and think about how much you really know. Humans and many other animals have incompletely formed brains at birth. However, a human's brain continues to develop for a much longer time than other animals' brains. As a result, our life experiences

influence us to a greater degree. The ability of the brain to be shaped and changed by experience is particularly true of human infants and young children. It is during these critical years that an incredible amount of information is learned. The long time it takes for our brains to develop, reflected in our long childhood, helps explain why humans have such a great capacity for learning.

Humans' ability to learn helps us develop the skills and knowledge needed to survive and contribute to our culture. Culture is the sum of all the knowledge that is passed from generation to generation. The skills and knowledge that are passed down often strongly depend on the places and times at which people are born. Think about the Iceman and the kinds of information and skills he needed to learn in his culture. For example, he needed to learn and practice hunting. He probably needed to learn to identify mushrooms to make sure he used the correct kind. The markings on his skin indicate that he also probably learned certain ceremonies and rituals. To obtain items he needed, he also probably had to understand a system of trade or exchange. What kinds of knowledge and skills are important for people to learn in your culture?

Humans are also very good at using the knowledge we learn to solve problems in new ways. Obtaining, organizing, and sharing knowledge through education allows human cultures to change rapidly. This type of change is called cultural adaptation. More than any other animal, humans make and use tools to solve problems. Compare the tools and technology used by the Iceman with the tools and technology available to you. This technological progress is an example of the power and speed of cultural adaptation. Cultural adaptation is distinct from biological evolution.

A Long Childhood Allows Humans Time to Acquire Language

Our long childhood and our great ability to learn also give us time to acquire language. Language is one of the most important things that humans learn. Most languages have numerous complicated rules of grammar and tens of thousands of vocabulary words. Once we learn language, we can communicate an infinite number of thoughts and ideas. Others who speak the same language can understand our thoughts and ideas.

Even though the process of learning a language appears to be tremendously complex, it occurs quite naturally. Children who grow up hearing one language at home and a different language among their peers can become fluent in both languages. Often, they appear to do this with little effort. Language, like many cultural behaviors, is learned, or *acquired*, as part of normal development. However, after age 10 or so, the capacity for learning a language decreases. This limitation explains why it is more difficult to learn a new foreign language in high school or college.

Language helps us record and communicate our culture efficiently. Imagine what our world would be like if we eliminated just one of our uses of language, the preservation of information in libraries. Passing on all this information by word of mouth would be difficult. Another benefit of language is that it allows people to get to know each other quickly. This helps us live closely with many other humans who are not our relatives. Language helps us maintain our complex social organizations. The interplay between language and culture helps to further distinguish humans from all other animals.

The long childhood of humans helps us develop our language and learning skills. Both of these help us live in complex, technologically advanced societies. But what about other animals? How do they learn and communicate? Do some other animals have complex societies?

How Do Other Animals Communicate?

Animals can communicate with each other through sounds, smells, displays, and movements (see figure 1.14). Some communication is purely visual, such as the colorful breeding plumage of many birds. Bees have a complex display known as the bee dance. Through this dance, illustrated in figure 1.15, a bee can communicate the location of food to other bees in the hive. Other social animals, like apes and monkeys, make different noises in different situations. For example, baboons can make and understand up to 22 different calls. They can also identify individuals by specific calls. But do we consider these activities language? Though animals have the capacity for language-like behavior, they lack precise words for things and do not have formal grammar. Thus, these communications are not considered language in the way that linguists or scientists would define language.

Figure 1.14 Communication between wolves. Animals communicate in a variety of ways with different levels of complexity. Wolves communicate with each other through vocalizations such as howling, barking, and growling, as well as through behaviors such as licking and submissive postures.

Do Other Animals Learn?

Many researchers study how animals learn. One broad category of behavior is called learned behavior. *Learned behaviors* are the responses an organism develops as a result of specific interactions with the environment. The other type of behavior is innate behaviors. Organisms are born with the ability to perform **innate behaviors.** An example of an innate behavior is a baby's ability to nurse. Babies are born knowing how to nurse and do not generally need to be taught this behavior. Scientists have discovered that all behaviors fall somewhere along a continuum between those that are primarily innate and those that are primarily learned.

In primate societies, flexible learned behaviors are important characteristics. One famous example of learned behavior occurs in some macaque monkeys, as shown in figure 1.16. In the early 1950s, on an island off Japan, scientists had begun feeding

a A foraging honeybee returns to the hive after finding a nectar source. Other bees soon leave the hive and fly to the same source. How does the first bee inform the other bees about the food source?

b The returning bee gives nectar to the other bees, then begins a dance on the vertical face of the hive.

c If the dance is a round dance, other bees search for plants nearby.

d If the dance is a figure 8 with wagging of the abdomen, bees search at a distance.

Figure 1.15 The bee dance. Honeybees communicate the location of food through a set of signals known as the bee dance.

Figure 1.16 Behavioral flexibility. When a feeding station was established, one macaque learned to wash sweet potatoes in the water before eating them. This behavior was adopted by most of the macaques in the community and even was passed on to the next generation of macaques.

macaque monkeys potatoes to get closer to the monkeys and to make observations. In 1953, a scientist saw a young female macaque monkey wash a sweet potato in a stream before eating it. Within a few months, the monkey's playmates and mother were also seen washing their potatoes. Within five years, more than three-fourths of the individuals in the population were washing their potatoes before eating them. Within 10 years, almost all of the monkeys in the population were washing their potatoes. They are still doing it today. This behavior demonstrates a creative response to a new situation and the transmission of that behavior between generations.

Social Organisms

The societies of humans and other primates are familiar to most people. However, biologists are also interested in the social organization of other animals. Termites are a very interesting example. Termites live in large families, called a colony because it is so big. In a termite family, the king and queen are actually the only father and mother in the colony. Termite offspring who stay within their home colony adopt one of two main "jobs." Most become workers who dig tunnels, find food, feed and care for the young, and fix the nest, as seen in figure 1.17. A small number of the offspring become soldiers. In some termite species, the soldiers have huge, muscle-filled heads and sharp jaws for fighting. Soldier termites are special because they also have *altruistic* behavior. In other words, soldier termites instinctually sacrifice their own lives to save their fellow nest mates from danger. Termites are just one of many fascinating examples of complex animal societies. You may wish to research the social organization of other animals on your own.

Figure 1.17 A queen termite with worker termites around her.

Analysis

Answer the following questions in your science notebook. Be prepared to share your ideas in a class discussion.

1. Review the table of evidence and ideas that you constructed in step 5 of part A, *A Look at Ancient Humans*. Consider how the evidence available to you limits your ideas about the Iceman. Then answer these questions.
   a. If you could have 3 more pieces of evidence to help you complete your understanding of the Iceman, what would they be?
   b. What information would these pieces of evidence give you that would help you understand how the Iceman lived or died?
2. Review the description of the Iceman's body from the reading in part A, step 2, and then answer these questions.
   a. Do you think there have been greater changes in humans physically or culturally in the last 5,300 years? Give 1 reason to support your answer.
   b. Propose a reason for why the Iceman's physical features are similar to those of humans today.
3. Review the distinctive human physical features that you investigated in the Explore activity, *Primates Exploring Primates*.
   a. How would the Iceman and his people have produced the kinds of artifacts that were found with him? In particular, how would they have used distinctive features of their hands and brains to make artifacts?
   b. How do people today use their distinctive physical features to make objects similar to the artifacts found with the Iceman?
4. Many human characteristics are shared by other animals. For example, imagine a guitar player in a concert. Other animals do not have the precision grip, flexibility, and coordination to play the guitar. But, that does not mean that they cannot grip items and use their hands. Examine the picture below. What do you think each animal might be able to do better than the others? You may wish to refer to the grasps in figure 1.7 and the hand shapes in figure 1.18 to think about different actions.

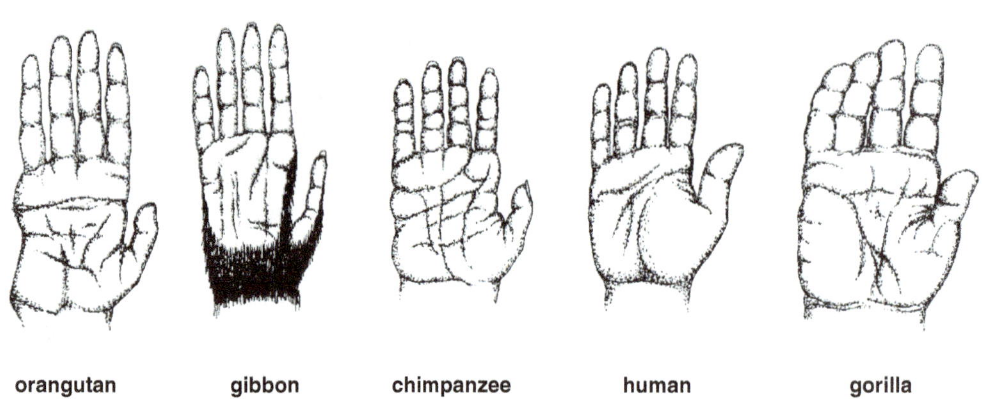

Figure 1.18 Ape and human hands. How do you think the shapes of hands and fingers affect what grips can be made?

5. Look back at figure 1.3, How Science Works. How do you think scientists' work analyzing artifacts and development can help with the process of science? Be specific about how these contribute to the elements within the circles.

Further Challenges

1. Together with other interested classmates, find out about the existence of culture and society in other animals.
2. Do further library research on the learning abilities of other primates or dolphins.
3. Learn more about your own learning style. Your teacher may have resources to get you started.

   Your learning style is the way you prefer to learn. There are questionnaires you can take to find out more about how your brain works best for learning new information. If you are interested in taking a learning style questionnaire, talk to your school counselor.

How Similar Are We?

ELABORATE

"She swims like a fish."
"He runs like a cheetah."
"She sings like a bird."

Have you ever been compared to another organism? Why do humans do this? Perhaps it is because we recognize that we share fundamental characteristics with other living things.

You have now thought more carefully about how scientists can learn more about humans and other animals using physical characteristics, ancient artifacts, and information about development. In this Elaborate activity, *How Similar Are We?*, you will consider the question of how scientists can describe the similarities humans share with other organisms. You will also develop your ideas around what it means to be alive.

Materials (per class)

unifying principle cards (1 card per team)
online resource
different-colored pens or pencils

PROCESS AND PROCEDURES

1. Brainstorm with your team a list of characteristics that humans share with other organisms, recording all ideas in your science notebook.

   Challenge one another to find fundamental similarities between humans and these organisms. For example, what do you have in common with a plant? Plants may not move like humans do, but like humans, they cannot survive for too long without water.

2. Identify any characteristics on the list from step 1 that you think humans share with *all* other organisms (such as the *Volvox* in figure 1.19). Mark these with the word "all" in a different-colored pen or pencil.
3. Choose a team member to report the characteristics that you identified as being shared among all organisms.
4. During the class discussion, record the class's list in your science notebook under the title "class list of characteristics shared by all organisms."
5. Read the essay, *Describing Life: An Impossible Challenge?,* as your teacher directs to learn about six unifying principles of biology. You will read one section of the essay. When you have finished your part, do the following in your science notebook.
   - Write a definition of the principle using your own words.
   - Describe how you think humans might fit with the principle.

   This portion of the activity will help you explore whether scientists recognize the similarities that the class identified. You will also learn whether there are additional similarities that the class didn't mention yet.

6. Discuss your ideas with others in the class who read the same section as you. Make sure you have all the information you need to share with your home team, who read other sections.

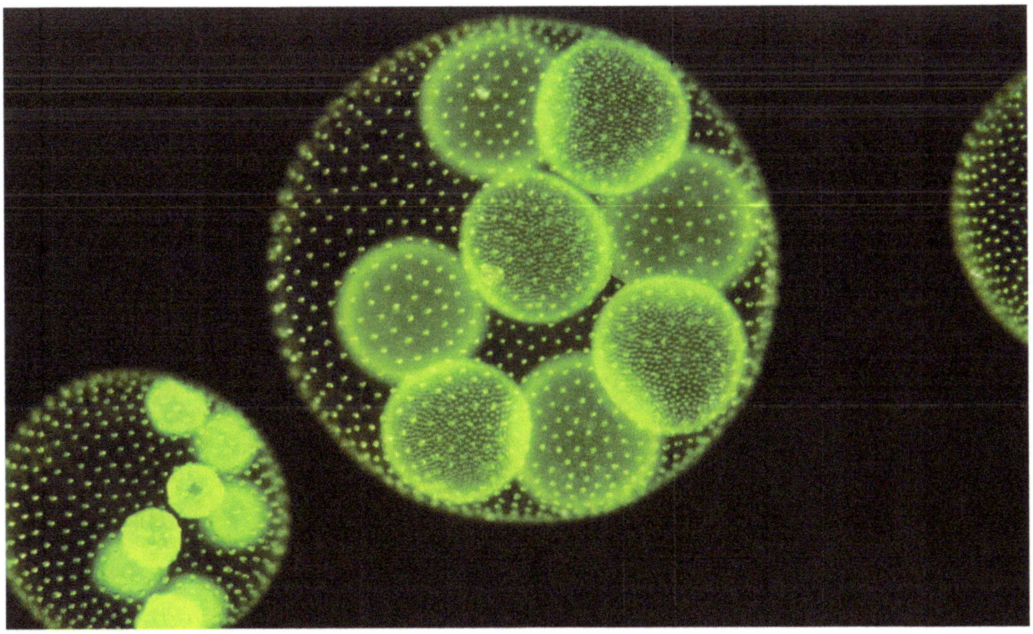

Figure 1.19 What characteristics do humans share with this organism, called a *Volvox*?

7. Return to your home team. Share about the section you read and how humans fit the principle. Make notes about the other principles as your teammates share their ideas.
8. Contribute ideas and questions to the class discussion of the following questions.
   a. Does the evidence support the idea that humans fit with the way scientists might describe all living things?
   b. Do the unifying principles leave out some characteristics that the class identified in step 4 as being shared among all living things? Does this change your ideas about those characteristics?

Describing Life: How Do Scientists Do It?

ENGLISH SPANISH

You probably already have some ideas about what makes living things unique. If you were to ask your classmates to identify a tree, a dog, and a rock as living or nonliving, all their answers would be the same. However, if you asked your classmates how they know what is living and what is not, their responses probably would vary. And suppose you asked them about a less familiar object, such as the scaly, grayish-green lichen on a boulder (like the one depicted in figure 1.20)? This time, some of your classmates might not be so sure if it is living or not.

All forms of life share many basic characteristics. This holds true even for very different life-forms such as humans, apple trees, spiders, and microscopic bacteria. Understanding these common characteristics is fundamental to understanding biology. In fact, these characteristics are so important that we have summarized them as six unifying

Figure 1.20 **Lichens are organisms that consist of a close association between a fungus and a photosynthetic organism such as an alga.**

principles of biology. The flow of topics in this course is organized around them. In this unit, you will study evolution, one of these principles. As you begin to achieve a deeper understanding of the remaining five principles, you will be developing a rich understanding of biology.

Evolution: Change Over Time

One significant characteristic of living systems is that they evolve, or change, over time. The fossils of organisms from a long time ago reveal many organisms that look different from organisms today. Changes in populations of living things have resulted in an amazing diversity of characteristics. Changes over time can also lead to the generation of new species. All species are linked because they share a common ancestor with all other living things. *Evolution* represents the first unifying principle considered in this course.

Homeostasis: Maintaining an Internal Balance

A second characteristic of life—and a second unifying principle of biology—involves a living system's ability to maintain an *internal balance*. This is called homeostasis. All organisms regulate their internal systems in response to changes in their surroundings. Have you ever noticed how your heart beats faster when you are startled? Quickly pumping blood through your body helps you respond quickly to a potentially dangerous situation. In fact, all organisms show interesting kinds of internal regulation. Bacteria can adjust their production of certain chemicals. Plants respond to changes in humidity by opening or closing tiny holes in the underside of their leaves. And some animals can change colors to match their environment (see figure 1.21).

Figure 1.21 **An anole lizard can change color in response to its environment.**

Energy, Matter, and Organization: Relationships Between Living Things

Another common characteristic of life is *organization*. All living systems are highly organized forms of *matter*. One of the most remarkable similarities among all living things is that they are mostly made up of only a few types of atoms. These are mainly carbon, nitrogen, oxygen, hydrogen, phosphorus, and sulfur. In organisms, atoms are often held together in large, complex molecules.

The molecules of living materials are organized into complex structures known as cells. Cells are the basic structural units of living matter, but they are too small to see with the unaided eye. As figure 1.22 illustrates, cells are baglike structures surrounded by a barrier called a membrane. Membranes allow cells to accumulate certain substances and exclude other substances. In this way, cells can maintain a different composition from their environment.

A related property of all living systems is that they require *energy* to build their structures and to carry out their activities. Organisms typically harness light energy from the Sun or chemical energy from other organisms. Think about what happens when an organism dies. A dead bird, for example, loses its ability to use matter and energy to keep its body repaired and functional. Over time, it will become indistinguishable from the matter around it. Together, the ideas of matter, energy, and organization represent the third unifying principle of biology.

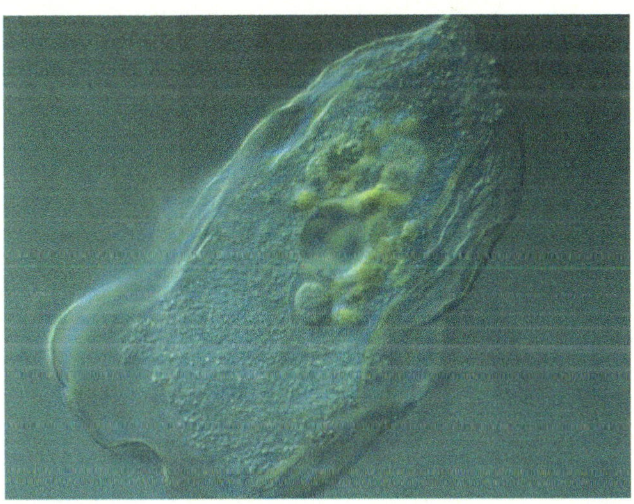

Figure 1.22 The cell is the basic unit of living matter. Cells typically range in size from 1 to 100 μm in diameter. Notice the organization of the interior of this *Thecamoeba* cell, which is 10 μm in diameter.

Continuity: Reproduction and Inheritance

All living systems depend on plans that are encoded in genetic material, or DNA. The specific information in your DNA has shaped how your body has grown and developed. DNA is a long and complex molecule that stores information. One of the most significant characteristics that unifies living systems is the universal nature of this DNA. All organisms, from bacteria to plants to humans, use DNA to provide instructions for building and maintaining their cells and bodies. In addition, all organisms have the ability to transfer those instructions—through DNA—to the next generation during reproduction. Together, DNA and reproduction make up a fourth important unifying principle of life: *continuity* (see figure 1.23).

Figure 1.23 DNA allows genetic information to be continually passed from one generation to the next.

Development: Growth and Specialization

The *ability to develop* represents the fifth unifying characteristic of living systems. Growth is an important activity for young organisms, as the body assembles new tissue. As an organism grows, the way in which the organism is organized also changes. For example, many plants begin life as small seedlings that push up through the soil. They grow into mature plants that look very different from the seedlings. Differentiation—the way in which cells "specialize"—occurs in organisms with more than one cell. This allows the body to "divide the labor" that it must carry out. Differentiated cells, like skin cells and neurons, develop specific shapes and carry out specific functions. Development is the sum of all the growth and differentiation that occurs over the lifetime of an organism.

Ecology: Interactions Between Living Things

Finally, all living systems on Earth interact and depend on other organisms in the web of life (refer to figure 1.24). Imagine a wooded area on an early summer day.

Figure 1.24 **Organisms along and in this stream interact and depend on one another.**

Plants provide shelter and food for a variety of birds. A rabbit has created a burrow nearby and is feeding on wild berries. Not far away, a fox just left her den in search of food for her young. This community of different, but connected, living systems illustrates the sixth unifying principle of biology: the *interactive and interdependent nature of life*.

We might say that in this short list of characteristics we have described life—as it exists on Earth. Can we say that any one of these principles *defines* life, in the sense that it alone is an indicator of life? Probably not. Just as a *combination* of characteristics identifies you as a human, a *combination* of these principles indicates the presence of life.

Analysis

1. In your science notebook, record your answer to this question: "What characteristics unify all living things?" In other words, what makes something alive?
2. Earlier in this chapter, you recorded your initial ideas about similarities between humans and other organisms. In your science notebook, review and revise the answers that you wrote for the following questions. Use a different-colored pen or pencil to make any revisions you would like. Put a single line through your previous thinking to indicate you have revised it.
   a. Engage, *How Different Are We?, Analysis* question 3
   b. Explain, *More Evidence, Analysis* question 2b
3. In your science notebook, provide an explanation for why living things share fundamental characteristics.

Further Challenges

Read these observations about a virus, called a bacteriophage (see figure 1.25), that infects bacteria.

A bacteriophage

- contains genetic material,
- reproduces only when inside another organism,
- has an outer case made of protein,
- injects genetic material into a bacterial cell, and
- uses the energy and the structure of the bacterial cell to make parts that assemble into copies of itself (and often kills the bacterial cell).

SEP Use your knowledge of the unifying principles of life to construct a well-reasoned argument to answer the question, "Is the bacteriophage alive?"

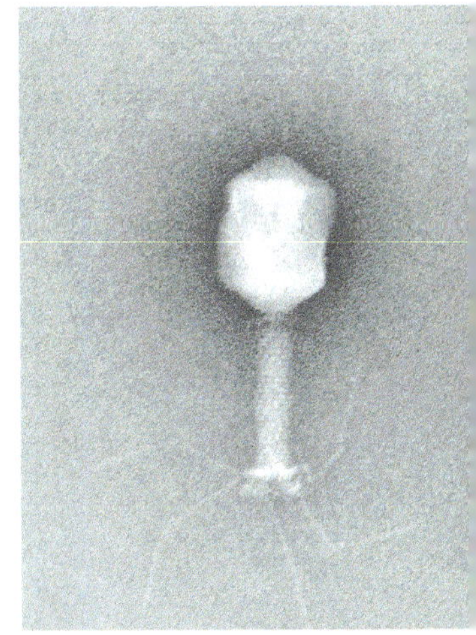

Figure 1.25 The bacteriophage (photographed at 240,000x) is a virus that attacks bacteria.

EVALUATE

First Encounter with the Critter

In this Evaluate activity, *First Encounter with the Critter*, you will demonstrate what you have learned in chapter 1. You will use your project to show how this new knowledge helped you answer the question, "What does it mean to be alive?" from a scientific perspective. You will have an opportunity to use your imagination as you evaluate your understanding of the science ideas in this chapter. In *First Encounter with the Critter*, you will "adopt" a critter, describe it, and show how it is similar to and different from humans. You will revisit your critter throughout the year. This will help you develop your description and understanding of the critter as you learn more about science.

Materials (per person)

1 habitat card

PROCESS AND PROCEDURES

1. Read the scenario, *Keep Your Eyes Open*. Then study the habitat card that describes where you are going. Draw a sketch of what you imagine the habitat looks like and label several important features. Write down a detailed description of the habitat you will be visiting.

   Since this will be an ongoing project for the year, you will be recording facts about your "critter" in a folder separate from your science notebook.

SCENARIO

Keep Your Eyes Open

It was a thrill to hear that the research funds came through. Now you can embark on your long-awaited trip into the wilderness to observe and record data about several endangered species.

The conservation organization that is sponsoring your trip has assigned you to a team that will spend several weeks in one location. The organization has given you a list of organisms that you should observe. As a born naturalist, however, you are determined to keep your eyes open for whatever you find of interest.

2. Choose 1 of the multicellular organisms that lives in this habitat and adopt it. Then determine its scientific name. Write down its scientific name in large lettering. If it has a common name, write that down as well.

   Your teacher may suggest some resources to help you consider the diversity of life in your habitat as you decide which critter to adopt. He or she will also help you research an organism's scientific name. You may be asked to adopt a critter from a particular group, in order to allow the class projects to best represent the diversity of life.

3. Describe your organism by drawing 1 or more diagrams of it. Include labels with brief descriptions of its distinctive structures.

   It is okay to use your imagination along with logic to describe your critter. For example, if your critter lives in an extremely cold environment, you should examine your critter carefully and think about what features it may have to help it live there.

   Remember, this is a time to show what you have learned about the unifying principles. Write about each of the 6 principles with at least one sentence per principle. Use the types of evidence that you have learned about in this chapter to support your ideas. As you continue to add information about your critter later in the year, you will have a chance to revise it, but you should write your best ideas at this point.

4. Create a Venn diagram to compare your critter and humans. Use your descriptions and evidence from step 3 to fill out your diagram.
5. Prepare to present your work using the Chapter 1 Critter Rubric. Your teacher will direct you on what you will turn in. Be prepared to present your work to the class.

6. Take notes while other students are presenting. Pay particular attention to the ideas that seem to be included in more than 1 presentation. Note aspects of presentations that you think are particularly strong. Ask questions at the end of each presentation to clarify any information that you don't understand.
7. Using the criteria in the Chapter 1 *Critter Rubric*, identify what you think are the 3 strongest projects in your class. List these in your science notebook and justify your choices.

Analysis

On your own, write the following questions in your science notebook and take notes to answer them. Use your notes to participate in a class discussion. When you are through, write the best answers to the questions in your science notebook.

1. Use the rubric's criteria to rate your own project and presentation. How would you explain your rating?
2. How would you revise your own project to better match the criteria in the rubric?
3. What common characteristics did the class generally identify as distinguishing humans?
4. Scientists have changed their views of the features that distinguish humans from other primates. How does this fact illustrate the process of science? Refer to figure 1.3, How Science Works diagram, to support your ideas.

What happens when someone cuts down the trees along a riverbank? The trees are gone, you say. But what effect does the loss of these trees have on the river's ecosystem? What about the birds that depend on the trees for nesting or the insects that use the leaves for food? How will the loss of shade affect water temperature and the fish that live in the river? Many organisms depend on these trees in different ways. So, how has this complex web of interactions been altered?

In the *Launch into Biology* unit, you saw how humans share some characteristics with other organisms, but saw some characteristics are different. In this unit you'll think about the ways humans and other organisms interact with each other. What impacts do those interactions have? How are different organisms dependent on each other? As you continue your study of biology in this unit, you will learn about concepts of interaction and interdependence in living systems. You will also begin to investigate the issue of human influence on ecosystems.

UNIT 1

ECOLOGY: Interdependent Relationships

By the end of unit 1, you should understand that

- a community of organisms interacts with the abiotic environment to form ecosystems;
- ecosystems are complex, but it is possible to analyze them;
- populations are limited in size by the amount of available resources and
- ecosystems can be modified by human actions.

UNIT CONTENTS

CHAPTER 2

Interdependence among Organisms

We do not inherit the land from our ancestors;
we borrow it from our children.

Haida Tribe Saying

Life in a densely populated city like San Francisco. Life in a forest. Life in the oceans. Life in a refugee camp in Africa. Life on a farm in rural Ohio. Think of an organism that lives in one of these settings. In what ways does this organism depend on other organisms for survival? How can such interdependence among these organisms be described? What factors influence their interdependence?

An **ecosystem** is made of all the living (biotic) and nonliving (abiotic) components in an environment. In this chapter, you will learn about the interdependence of organisms in various ecosystems. First, you will seek evidence of interactions in your own school yard or neighborhood. Next, you will study interactions in ecosystems throughout the world. You will further explore ecosystems by considering the types of resources that exist in them. You will then analyze the influences those resources have on communities of organisms. You will apply your understanding of communities and resources to different types of "island" ecosystems. During your study of island ecosystems, you will analyze patterns of population growth in a real group of people who lived on an island off the coast of South America. You will also examine how global climate change is affecting organisms living in specialized habitats, in this case, the tops of mountains, or so-called islands in the sky. Finally, you will evaluate your understanding by describing how your critter will use resources and interact with other organisms in a particular ecosystem.

GOALS FOR THE CHAPTER

By the end of chapter 2, *Interdependence among Organisms*, you should understand the following concepts:

- ✔ Organisms interact and are interdependent.
- ✔ Interactions occur among biotic and abiotic parts of an ecosystem.
- ✔ Factors such as climate and carrying capacity affect the interdependence of organisms.
- ✔ Humans influence the biosphere in many different ways.

To help you develop these understandings, you will complete the following activities.

ENGAGE / EXPLORE	Hidden Interactions?
EXPLORE	Interactions in the World around Us
EXPLAIN	The Pasture Story
ELABORATE	Mystery on Easter Island
ELABORATE	Islands in the Sky
EVALUATE	Critters and Interdependence

The chapter organizer can help you keep track of your learning and the flow of ideas across the chapter. Refer to it frequently as you complete the chapter's activities.

ENGAGE

EXPLORE

Hidden Interactions?

Key Idea: Organisms interact with other organisms.

Linking Question:
What are the different ways organisms interact with each other and with the environment they live in?

CHAPTER 2

Interdependence among Organisms

EXPLORE

Interactions in the World around Us

Key Idea: Organisms interact with their environment in complex ways.

Linking Question:
How do interactions that involve the use of shared resources influence the number of organisms an environment can support?

EXPLAIN

The Pasture Story

Key Idea: The use of an environment's resources influences carrying capacity.

Linking Question:
Does the concept of carrying capacity apply to humans?

EVALUATE

Critters and Interdependence

Part A: Resources and Ecosystems
Part B: Critters in Ecosystems
Key Idea: Every organism depends upon abiotic factors in the environment and interacts with other organisms.

Linking Question:
How is my critter limited by abiotic and biotic factors in its ecosystem?

MAJOR CONCEPTS

- ✔ Organisms interact and are interdependent.
- ✔ Interactions occur among the biotic and abiotic parts of an ecosystem.
- ✔ Factors such as climate and carrying capacity affect the interdependence of organisms.
- ✔ Humans influence the biosphere in many different ways.

ELABORATE

Islands in the Sky

Key Idea: Multiple limiting factors can operate to affect population levels.

Linking Question:
How do limiting factors affect nonhuman organisms?

ELABORATE

Mystery on Easter Island

Key Idea: Human population growth is restricted by limiting factors.

ENGAGE
EXPLORE

Hidden Interactions?

Imagine crawling out of bed in the morning. You flip on a light, begin streaming your favorite music on your phone, and stroll to the bathroom as music plays in the background. For breakfast, you eat a bowl of cereal. It's an especially cold morning, so you fix yourself some hot chocolate. The television is blaring with news of an overnight fire in the nearby national forest. It's getting late. So you pack your lunch, give your dog a quick pat on the head, and rush off to catch the bus.

This scenario describes a typical morning for many teenagers. Think about how you interact with your environment during your morning routine. All organisms, including humans, take part in many interactions in their environment. In this Engage-Explore activity, *Hidden Interactions?*, you will make observations of organisms in their environment and you will consider what interactions you see in the natural world.

Materials

pencil

PROCESS AND PROCEDURES

Go outside as your teacher directs. In your science notebook, record the following observations and answers to the questions.

1. Look carefully at all the organisms around you. Find 2 or more different organisms that are interacting with each other.

   Remember to consider all the living things you observe, not just animal life.

2. Create a quick drawing of the interacting organisms from step 1. Add labels or a brief explanation for the interactions.

   When you make observations, only write those things you see. Do not suggest a meaning for what you see. For example, a good observation is "the organism moved quickly behind a plant when another organism approached." Writing "the organism hides to avoid a predator" is not a good observation. That statement assigns meaning to what you see. Remember what you learned about evidence and inference in chapter 1? In this case, a good observation is your *evidence*. Assigning meaning to this evidence is an *inference*.

3. Listen for sounds made by living organisms. Write a brief description for how 1 organism is using sound to interact with other organisms.

4. Focus on the smells of the natural world. Write a brief description for how 1 odor you smell may be evidence for an organism interacting with other organisms in its community.
5. Name and describe or draw 2 different organisms near you that may be prey for other organisms.

   It's okay if you do not know the exact names of the organisms you see. You can ask your teacher or name the organisms descriptively. For example, you could write "small brown toad" or "tall fern-like plant."

6. Name and describe or draw 2 different organisms that have a distinguishing texture.
7. Write a brief description for how each texture may encourage or discourage other organisms to touch it.
8. Name and describe or draw 1 organism that appears to be living in isolation from all other organisms.

Analysis

Discuss the *Analysis* questions with a partner. Write answers to the questions in your science notebook.

1. Does the organism observed in step 8 require any interactions with other organisms to survive? Why do you think so?
2. During this activity, do you think the class was able to observe all of the different types of organisms present in the study area? Explain your answer.
3. Write an inference for one of the observations you made. What additional observations (evidence) could you make that would support your inference?
4. The title of this Engage-Explore activity is *Hidden Interactions?* Why do you think the writers of your textbook want you to think about this question?

Interactions in the World around Us

EXPLORE

Think about the different organisms you observed in the Engage-Explore activity, *Hidden Interactions?*. How were the various organisms interacting? Do both organisms benefit from an interaction? Is one organism harmed while the other benefits? Are both organisms harmed by the interaction? These are just some of the questions an ecologist might ask when studying the interactions among organisms in a certain area.

Ecologists, like the ones shown in figure 2.1, attempt to describe and understand the interrelationships that exist between organisms and their environments. First, they gather information about organisms and their

Figure 2.1 Some ecologists analyze ecosystems by studying the changes in the quantity and the quality of the ecosystem's components.

environment in a particular area. They might compare this area with different regions to look for similarities, differences, and patterns. Finally, they try to explain the patterns they observe. They might run experiments to test their hypotheses.

In this Explore activity, *Interactions in the World around Us*, you will consider the questions posed above as you practice thinking like an ecologist. You will look for patterns and analyze the interactions you observed in the Engage-Explore activity, *Hidden Interactions?*. You also will have the opportunity to compare your observations with images from around the world. The images you will see on the video represent a wide variety of interactions in many different settings.

Materials (per team of 2)

online resource (watch as a team)
1 task card

PROCESS AND PROCEDURES

1. With your partner, share the observations you recorded during the Engage-Explore activity, *Hidden Interactions?*, and discuss the following questions. Be prepared to share your ideas with the rest of your class.
   a. What types of interactions did you see between organisms in which only 1 organism appeared to benefit from the interaction?
   b. What types of interactions seemed to benefit all organisms involved?
   c. Give examples of how odor, taste, and sound can lead to interactions that either attract or repel other organisms.
   d. Name at least 5 living and 3 nonliving components that are part of the area where you made your observations.

   The living components of an area are called biotic factors. Nonliving components are called abiotic factors.

2. Participate in a class discussion about your observations and answers, as your teacher directs.

   When sharing your ideas, consider how you might describe the ecosystem where you made your observations. The opening pages of chapter 2 include a definition for ecosystem.

3. In your science notebook, *use your own words* to write a definition for the term "ecosystem."
4. Obtain a task card with a set of 4 tasks from your teacher. These tasks will help guide your observations as you watch the video segment, "Images from around the World."
5. In your science notebook, make an organizing table like the one in figure 2.2 to help you take brief notes about the examples you will see in the video.

   If you wish, you may split up the tasks with your partner. For help setting up your table, you may want to refer to *How to Use and Create Organizing Tables* in appendix B4.

HOW TO

Example	Positive influence of humans on other organisms	Positive influence of humans on the abiotic environment	Positive influence of humans on other humans	Negative influence of humans on other organisms
1				
2				
3				

Figure 2.2 Sample table of observations of images from around the world. A table similar to this one will help you organize your observations.

6. Watch the video segment, "Images from around the World." Provide examples of the influences assigned on your task card.

   Be sure to record examples in your table.

http://www Online Resource

7. Create a new 3-column table in your science notebook and title it "interactions among organisms."
   a. Label the first column "one organism benefits and the other organism is harmed."
   b. Label the second column "both organisms benefit."
   c. Label the third column "both organisms are harmed."
8. Join another pair of students who had a different set of tasks. Discuss your video observations as a team. Choose at least 2 observations from your discussion to record in each column of your *Interactions among Organisms* table.
9. Study the table in 2.3 to learn more about how ecologists classify interactions among organisms. Add the terms from figure 2.3 to appropriate locations on your *Interactions among Organisms* table.

Interaction	Effect of the interaction	Description
Mutualism 	+ / +	In this interaction among organisms, both organisms benefit. For example, lichens consist of an alga and a fungus that live in close association. The alga produces food through photosynthesis, and the fungus provides moisture and nutrients. Another example is the interaction between flowering plant species and insects that pollinate them.
Predator-prey 	+ / –	One organism (predator) eats another (prey). This interaction harms the prey (often killing it) but benefits the predator.
Parasitism 	+ / –	One organism (the parasite) lives on or in another organism (the host), using it as a habitat and a food source (figure 15.3). The interaction may or may not kill the host species over time.
Competition 	– / –	Organisms may compete for such things as food, space, sunlight, nutrients, or water. Competition often limits the growth of populations of organisms. In competition, both organisms may be harmed through the effort necessary to obtain a limited resource.
Commensalism 	+ / 0	One organism benefits from the interaction, and the other organism is unaffected. For example, many orchids use trees as surfaces upon which to grow. The trees are not harmed or helped by the orchids. The orchids, however, benefit by growing near the top of the tree, where they can collect more sunlight and rain. True commensalism is relatively rare in nature.

Figure 2.3 **Interactions in ecosystems classified by the effects of the interaction.** The survival or reproduction of a species may benefit from the presence of another species (+) or be harmed by it (–). In the table, "+ / +" indicates that both species benefit from an interaction; "+ / –" indicates that one species benefits and the other is harmed; benefits; "– / –" indicates that both species are harmed; and "+ / 0" indicates that one organism benefits and the other is unaffected.

10. In your science notebook, write 1 or 2 general statements about the interactions of organisms (see figure 2.4) with each other and with the environment.

    Think about the interactions you recorded in your *Interactions among Organisms* table, your observations from the Engage-Explore activity, *Hidden Interactions?*, and the video images.

Figure 2.4 Ticks are parasites on humans as well as many other animals.

Analysis

1. Think of two non-human organisms that interact in the natural world. Then, look back at Table 2.3. What type of interaction do these organisms have? Provide reasoning for your answer.
2. The interaction between plants and the animals that eat them is called **herbivory**. Consider the following two examples of herbivory:
   a. Mountain pine beetles lay their eggs under the bark of pine trees such as ponderosa and lodgepole pines. The eggs hatch and the larvae feed on the trees, killing them. Millions of acres of pine forest have been destroyed by the tiny pine beetle.
   b. Hummingbirds drink nectar produced by plants with tube-like flowers, such as honeysuckle. When they do this, they transfer pollen from one plant to another. This pollination helps the plant to reproduce.

   If you added herbivory to a new row in Figure 2.3, how would you complete the second column, "Effect of the interaction?" Explain your reasoning.
3. Do you think that humans have more or less influence on the environment than other organisms have? Provide reasoning for your answer.
4. Do you think that humans have a responsibility to monitor how we influence the environment? Provide reasoning for your answer.

The Pasture Story

EXPLAIN

In the Explore activity, *Interactions in the World around Us*, you studied some of the complex interactions that take place in a variety of ecosystems. One organism's actions influence those of another. Populations influence one another as well. Humans influence other humans, other species, and the nonliving components of the ecosystems where we live. In some situations, we feel that we can control the influence that other humans have on our lives. In most situations, the decisions made by other people in our communities have significant influence on our lives. In this Explain activity, *The Pasture Story*, you will study a historical dramatization of farmers grazing cows on a shared pasture. This will help you think about the factors that influence how individuals share resources.

Materials (per team of 2–4)

online resource
computer with Internet access

PROCESS AND PROCEDURES

1. View the video, "The Pasture Story." Work in teams as your teacher directs.
2. What happened to the pasture that the farmers shared?
3. In your science notebook, write your ideas about why this happened to the pasture. Discuss these ideas with your teammates.
4. Divide your team into 2 groups. Each group will make the decisions for 1 of the dairy farmers (Kiana or Jason) in the "Pasture Profits" simulation.
5. Go to the "Pasture Profits Simulation" and run your first simulation.

   Follow your teacher's instructions to simulate managing the pasture's resources while maximizing your profits. If time permits, you may be able to run the simulation again to try a different management strategy.

6. With your teammates, graph the 4 sets of data found on the summary reports onto the *Pasture Profits* handout.

   The 4 sets of data are: total number of cows present; percentage of pasture area covered by grass (pasture health); total amount of milk produced; and number of units of milk produced per cow. To graph the latter data set you will use the scale on the right side of the graph.

7. Look over your completed *Pasture Profits* graph. With your partner, circle the areas of the graph you think are the most important.
   a. Write the phrase "What I see" near the areas you circled. Describe the pattern you notice.
   b. Write the phrase "What it means" under your "What I see" phrase. Then write 1 or 2 short phrases that summarize what you think caused the pattern you observed.

   These two phrases together are called *highlight comments* because they represent the essence, or the highlights, of the data in the graph. For more information, see *How to Write Highlight Comments and Captions* in appendix B6.

8. After you have completed step 7, compare your graph and *highlight comments* with the other partners in your team. Then discuss the following questions as a team.
   a. Which management scenario best protected the pasture?
   b. What is the relationship between the percentage of pasture area covered by grass (amount of food available) and the number of cows?
   c. What is the relationship between the percentage of pasture area covered by grass and the production of milk per cow?
   d. Using your graphs, determine the maximum number of cows the pasture can support and still have the ability to renew itself.

      You may want to look at other teams' graphs for additional information that may help with this task. After discussing these questions with your teammates, review your "What it means" phrases from step 7. Change or add to these phrases based on your discussion.

e. Read the following paragraph for information that will help you answer this question: Do you think different management strategies affect the carrying capacity for the pasture? Record your answer in your science notebook, citing evidence from your simulations.

Carrying capacity is the maximum number of individuals of a certain species that a habitat can support. The carrying capacity for a habitat can change over time if the habitat changes in size or if resources within the habitat change in abundance or quality. In the pasture example, carrying capacity was the maximum number of cows that the pasture could support without destroying its ability to renew itself.

9. Read the background information provided in "The Abundant Earth" segment. On the basis of your reading, use your own words to define the following concepts. Record your definitions in your science notebook.
    a. Natural resource
    b. Renewable resource
    c. Nonrenewable resource
10. What is the best management option for the commonly owned pasture in the "Pasture Profits" simulation? Why do you think so? In your science notebook, write your claim and support it with evidence and reasoning.
11. Contribute your ideas about steps 8 and 10 to a class discussion.

Analysis

With your team, consider the following questions. Record your answers individually in your science notebook.

1. What specific types of interactions among organisms were present in the pasture simulation?

    You might find it helpful to refer to the *Interactions among Organisms* table in your science notebook, which you completed in the Explore activity, *Interactions in the World around Us*.
2. List 3 examples of resources in the modern world that humans use that are like the commonly owned pasture in "The Pasture Story" video. Describe how people manage each example.
3. Are renewable resources more likely to be biotic or abiotic resources? Explain your answer.
4. Write a paragraph that summarizes the challenges you and your teammates faced in maximizing personal profits on a pasture owned in common. In your summary, include a response to this

question: What kinds of information made it easier for your team to make decisions about how to best manage the pasture?

5. Like every model, the simulation "Pasture Profits" has realistic and unrealistic features when compared to an actual pasture. Explain at least 2 realistic and 2 unrealistic aspects of the "Pasture Profits" model.

ELABORATE

Mystery on Easter Island

Islands are good places to study ecosystems because they are somewhat isolated. Because they are isolated, their resources are much easier to measure. It is especially interesting to study the patterns of population growth for animals on islands. Such is the case for a human population that lived on Easter Island from the year 500 to the present. This Elaborate activity, *Mystery on Easter Island*, will give you an opportunity to examine the mystery that surrounds Easter Island. You will study data about the history of its human population. You will then reflect on some of the interactions that took place in that ecosystem.

Materials (per team)

online resource
computer with Internet access
different-colored pens or pencils

PROCESS AND PROCEDURES

1. Read the need to know, *The Mystery*, to learn more about Easter Island.

NEED TO KNOW

ENGLISH SPANISH

The Mystery

Noted writer Jared Diamond called Easter Island "the world's most isolated scrap of habitable land." Easter Island is located in the Pacific Ocean, more than 3,218 km (2,000 mi) west of South America. The island has an area of about 166 km^2 (64 mi^2). Its subtropical location gives it a mild climate. And its volcanic origin provides fertile soil.

Easter Island in 1722

Despite its mild climate and fertile soil, when Europeans first reached the island in 1722, it was without a single tree or bush more than 3 m (10 ft) tall.

There were no native animals larger than insects. The people had no real source of firewood to warm themselves. Chickens were their only domestic animals. Yet the evidence indicates that at one time in its history the island offered abundant food and building materials. This was when the first Polynesian colonists arrived, some 1,400 years earlier. What happened?

Easter Island When the First Colonists Arrived

The first colonists traveled to the roughly triangular island (see figure 2.5) in approximately CE 450. When they came to the island, it was covered by a forest dominated by a now-extinct species of giant palm tree (see figure 2.6). These abundant palm trees were an important resource for the colonizers. In fact, ancient Easter Islanders created rock art showing images of palm trees and other important island resources such as birds, fish, turtles, and dolphins (see figure 2.7).

The Easter Island Palm

The giant palm is similar in appearance to the Chilean wine palm. We can look at the Chilean wine palm and make inferences about the

Figure 2.5
Easter Island is located more than 3,218 km (2,000 mi) west of South America. It is roughly triangular, with sides of 18, 18, and 24 km (11, 11, and 15 mi). How would you calculate the surface area of Easter Island?

Figure 2.6 This Chilean wine palm is similar to the now-extinct species of giant palm that was abundant when humans first arrived on Easter Island.

Figure 2.7 These petroglyphs were carved by ancient Easter Islanders.

characteristics of the Easter Island palm. (Remember from chapter 1 that inferences are conclusions that follow logically from evidence. The Easter Island Palm appears similar to the Chilean wine palm, so scientists use evidence from it to infer characteristics of the Easter Island palm.) The Chilean wine palm is one of the largest palms in the world. It stands more than 20 m (about 65 ft) tall and has a trunk diameter of 1 m (about 3 ft) or more.

Questions and Clues

Why did the giant palms that once covered Easter Island become extinct? What happened to the population of colonists on Easter Island? Visitors to Easter Island have been asking these and many other questions about the mysterious island for hundreds of years. For some questions, we can only speculate answers. However, evidence from archaeological digs and inferences based on population studies provide important clues for uncovering Easter Island's history.

2. In your science notebook, write down what you think caused the giant Easter Island palm to become extinct.
3. To learn more about factors that may have affected the palm, study figure 2.8. Draw the graph in figure 2.8 in your science notebook.
4. Make highlight comments on your graph following steps a through e.

HOW TO

For additional help, refer to appendix B6, *How to Write Highlight Comments and Captions.*

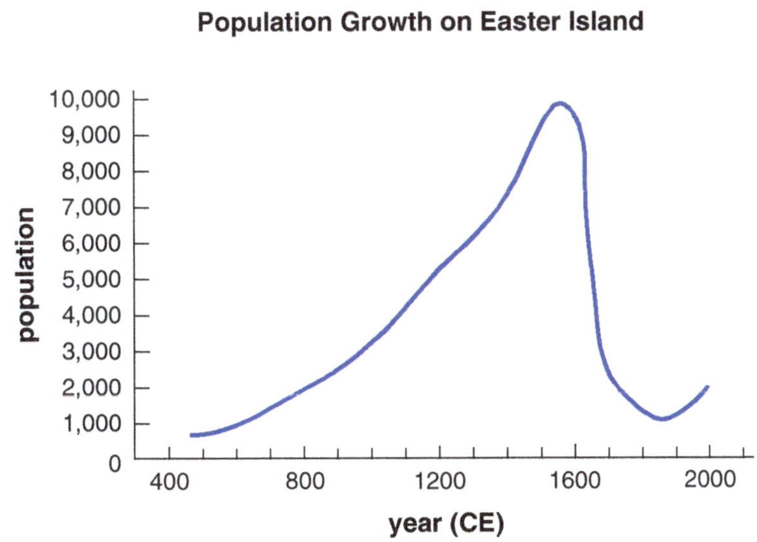

Figure 2.8 **Population growth for the first Easter Island inhabitants.** Researchers Paul Bahn and John Flenley used historical studies and population formulas to infer population levels on Easter Island over time.

a. Circle the graph for the years between 400 to 850. Draw an arrow from the area you circled to the margin and write "What I see." Below that, write a phrase that describes what you see in that area of the graph.
b. Repeat step 4a for the years between 850 and 1250.
c. Repeat step 4a for the years between 1250 and 1600.
d. Repeat step 4a for the years between 1600 and 2000.
e. Below each "What I see" phrase, write "What it means." Then write 2 or 3 phrases that describe what your observations about the graph mean.

5. Share your answers for step 4e with your teammates. What was similar among all your ideas? What was different?
6. With your teammates, study the vegetation charts and information in figure 2.9. Write a brief description of the resources on Easter Island that likely influenced the human population that lived there. Include at least 3 examples of biotic (living) and 3 examples of abiotic (non-living) resources.
7. View the "Easter Island" segment. You will learn more about the islanders, the mysterious stone statues (figure 2.10), the land, and the vegetation. As you view these segments, think about this question: How did the presence of humans affect the Easter Island ecosystem?
8. Discuss with your partner the changes that took place on Easter Island after the first humans arrived.
9. Write 1 or 2 paragraphs that summarize the changes in vegetation on Easter Island between 950 and 1980. Include your ideas about what you think caused those changes.

   Consider the interactions and interdependence that took place among the people as well as among the people and the shared biotic and abiotic resources.

10. Participate in a class discussion of steps 5–9.

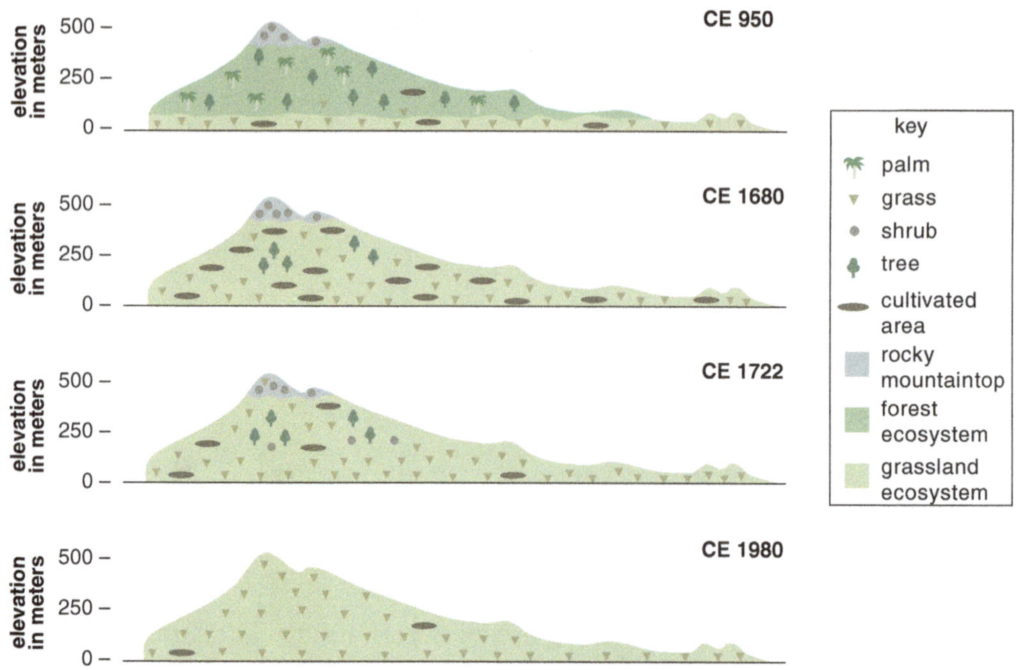

Figure 2.9 Easter Island vegetation distribution between CE 950 and CE 1980. Biologists study ancient pollen and spore samples to determine the historical vegetation distribution.

Figure 2.10 These monoliths (called *moai*) were carved from compacted and hardened volcanic ash by using stone tools. They stand up to 11 m (about 36 ft) tall and weigh up to 77 metric tons (85 short tons).

11. To get additional ideas to help you explain the mystery on Easter Island, read the essays, *Interdependence Involves Limiting Factors and Carrying Capacity* and *Growing, Growing, Grown*. Use the Mark it Up! strategy as you read these essays. Complete the steps in the following bullets.
    - Use a highlighter to mark key information about limiting factors, carrying capacity, and population growth.
    - Underline words, phrases, or sentences that are unclear to you.
    - Circle words that you know but which are used in an unfamiliar way.

    If you need more information about the Mark It Up! strategy, see Appendix B10. You may want to preview the questions in steps 12–17 before you read.
12. What limiting factor or factors slowed the growth of the human population on Easter Island? What other factors might limit the growth of a population of organisms? Give at least 3 specific examples.
13. Look back at the population growth graph for Easter Island (figure 2.8) that you drew in your science notebook in step 3. Color-code this graph so that each significant trend in the rate of growth is a different color.

    Hint: Look for areas of the graph that show slow increase in population size, rapid increase in population size, and rapid decrease in population size.

Interdependence Involves Limiting Factors and Carrying Capacity

ENGLISH SPANISH

Newspapers report continued famines in Africa. New animals are being added to the endangered species list. What are the causes of these events? Are they caused by human mismanagement, or do some happen naturally?

We can find some of the answers by studying the concepts of carrying capacity and limiting factors. **Carrying capacity** is the maximum number of individuals of a certain species that a habitat can support. It is influenced by biotic (living) and abiotic (nonliving) factors. A **limiting factor** is anything that slows, or limits, population growth. Limiting factors can be biotic or abiotic. They influence the carrying capacity of an environment. Take the Easter Island ecosystem as an example. Biotic factors include the giant palms, rats, and humans. Abiotic factors include space, sunlight, freshwater, nutrients, and temperature. In this essay we will discuss abiotic factors that affect the carrying capacity of an environment. As you read, consider whether each limiting factor could have affected the Easter Island ecosystem.

Climate

Climate is the average weather conditions in a given area over long periods of time. Climatic conditions are abiotic and include temperature; sunlight intensity; precipitation in the form of rain, snow, and fog; wind; and humidity (the amount of moisture in the air). The combination of climatic factors, along with individual weather events, partly determines the kinds of organisms that can live in a particular region. Along with individual weather events, climate also influences the carrying capacity for those organisms.

Temperature

How does something like temperature limit population size? One example involves the mosquito and the organisms that feed on it. A cold day in the fall in temperate regions will kill almost all mosquitoes. The decrease in temperature directly affects the ability of the mosquito to survive. However, the effects don't stop with the mosquitoes. The organisms that feed on the mosquito, such as bats, now have a smaller food supply. The decline of the mosquito population due to cooler temperatures now lowers the carrying capacity for other populations until temperatures warm in the spring.

The same type of interactions takes place between various wildflower populations and the insects that pollinate them. As temperatures drop in the autumn, wildflowers die, leaving the seeds of the next generation in the soil. The insects then either die, become dormant, or migrate. In both examples, temperature causes limited population growth for a period of time.

Water

What about water? How can it affect population size? Because water is needed by all living organisms, it is a limiting factor that influences the carrying capacity of many ecosystems. Look at figure 2.11. Water limits the carrying capacity of a savanna in East Africa. Also, because different organisms have different water needs, the location of water sources influences how populations are distributed throughout an ecosystem.

Light

Light also can be a limiting factor. The penetration of light into the ocean determines the depth to which organisms that photosynthesize

Figure 2.11 Water is a limiting factor in the savanna in East Africa. Here, only grasses and scattered trees can grow. These plants in turn support grazing animals. The grazing animals move on when the rainfall in their location decreases.

can grow. This is generally not beyond the depth of 180 m (590 ft). If silt reduces the water clarity in a pond or lake, plant growth is limited. In dense rain forests, the tallest trees spread their leaves and take most of the light. The ground below them is shaded. The shade prevents other plants from reaching great heights. Light affects animals indirectly. The amount of light influences the number of plants that can grow. This then influences the carrying capacity of the environment for organisms that feed on the plants.

Space

Space is another abiotic limiting factor. Every individual needs living space, but some need more than others. For example, individual corn plants grow well when they are planted close together. A mountain lion requires many square kilometers of space to find enough food. However, the availability of this space is affected by population density. **Population density** is the number of individuals in relation to the space a population occupies. An increase in density can have a drastic effect on a population. For example, think about the experiment described in figure 2.12. Researchers gave caged mice more food than they needed each day. As the mice reproduced, the cages became very crowded. Because of this, some female mice stopped caring for their young. Mice continued to be born, but many newborn mice died from neglect. Eventually, the death rate of young mice reached 100 percent. Space, as a limiting factor, affects all living populations and the carrying capacity of particular ecosystems for those populations.

This essay described four abiotic limiting factors: temperature, water, light, and space. What abiotic limiting factors affected Easter Island? Why do you think so?

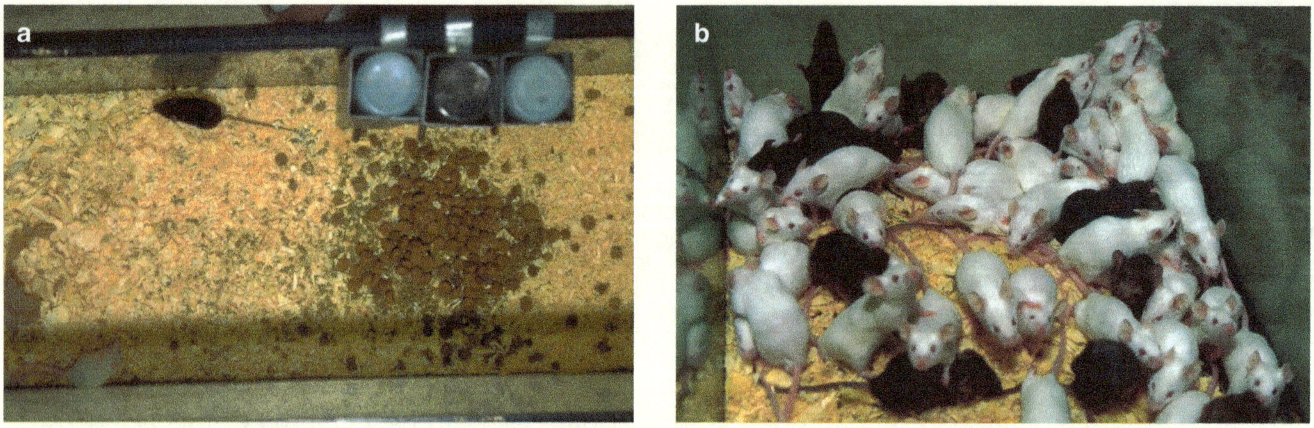

Figure 2.12 **Experiments involving population growth of mice.** (*a*) In one population experiment, researchers provided mice with more than enough food. (*b*) As a result, the population grew dramatically. Space then became a limiting factor that resulted in a high death rate among young mice.

Growing, Growing, Grown

ENGLISH SPANISH

Why did the human population on Easter Island grow and grow and suddenly crash? Do all populations of organisms grow in this way? Under ideal conditions the total *size* of a population rises steadily across time. The growth *rate* of the population also increases. This is called **exponential growth**. In other words, the larger a population gets, the faster it grows. A graph of an exponentially growing population looks like a J (figure 2.13). However, a population must have an unlimited supply of resources to grow this way.

This growth principle applies to all species. In just one year, one adult female housefly could lead to nearly 5.6 *trillion* flies. (figure 2.14). If all species can grow exponentially, why isn't the entire Earth buzzing with flies or trampled by other species? The answer is limiting factors such as food supply. Populations tend to increase in size by a factor of 2: 1, 2, 4, 8, 16, 32, 64, 128. This is exponential growth. However, food supply tends to increase only by an *addition* of 1: 1, 2, 3, 4, 5, 6, 7). This is arithmetic growth. This means there is not enough food for exponential growth to continue. As the number of individuals increases, competition for resources such as food, water, and space also increases. Then the death rate increases and exponential growth ends. Because all ecosystems have limiting factors, exponential growth cannot continue forever. What type of population growth occurs in the natural world?

In nature, conditions are not ideal and resources are limited. Because of this, populations have a pattern of growth called **logistic growth**. The logistic growth curve

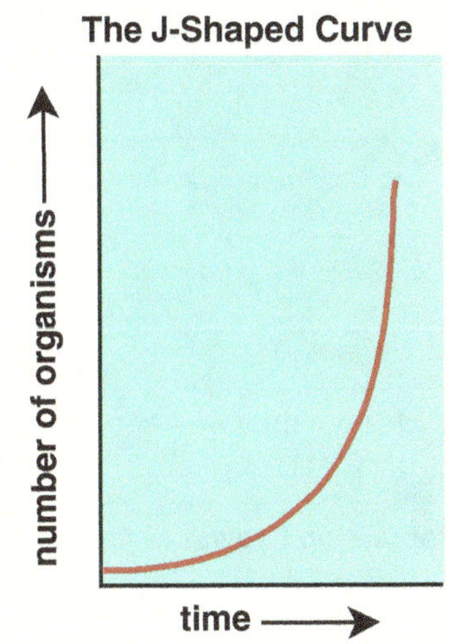

Figure 2.13 The J-shaped curve is characteristic of populations that are growing exponentially.

Generation	Population size
1	120
2	7,200
3	432,000
4	25,920,000
5	1,555,200,000
6	93,312,000,000
7	5,598,720,000,000

Source: Data from E. J. Komondy (1984). *Concepts of Ecology,* 3rd ed. Englewood Cliffs, NJ: Prentice Hall Publishers.

Figure 2.14 Predicted population growth of the common housefly in one year. This prediction is based on the following observations and assumptions. An average female fly lays 120 eggs at a time. About half the eggs develop into females. There are seven generations in one year. Individual flies live for one generation.

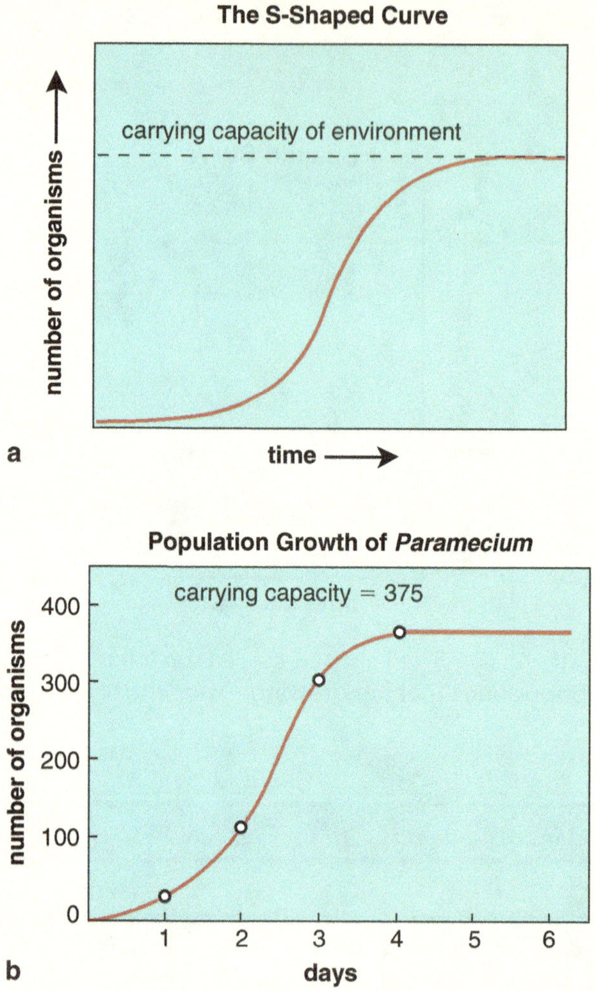

Source: From G.F. Gause (1934). *The Struggle for Existence.* Baltimore, MD: Williams and Wilkins.

Figure 2.15 **(*a*) An idealized S-shaped growth curve. (*b*) A population growth curve for the single-celled organism *Paramecium*.** Limiting factors force this population of organisms to level off.

looks like a flattened S (see figure 2.15a). For example, populations of *Paramecium* growing in nature (figure 2.16) show logistic growth. The population levels off at the carrying capacity of the environment (see figure 2.15b).

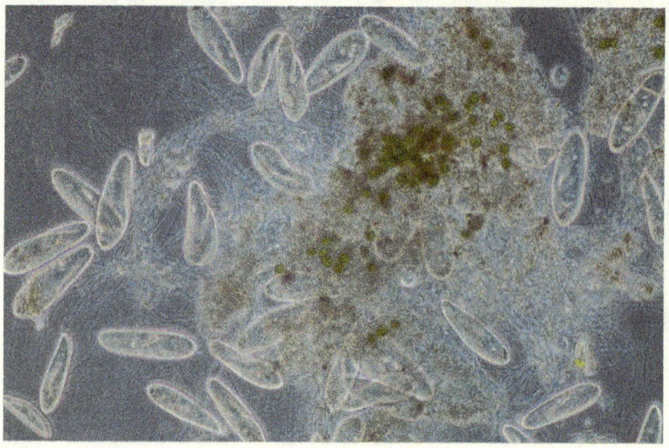

Figure 2.16. Growing population of *Paramecia*. *Paramecia* are single-celled organisms that are widespread in freshwater, brackish, and marine environments. They are often abundant in stagnant basins and ponds like those in Golden Gate Park.

How is the human population growing? The world's human population was at 1 billion people in 1820. By 1930, population had doubled to 2 billion people. It had taken 110 years to double in size. The world's population was over 3 billion people in the 1960s. The population doubled to 6 billion people in 1999. It had doubled in size in less than 40 years. The number of humans is expected to increase to 9 billion people by 2040.

The human population is growing at an exponential rate at this time. Think about this as you consider the following questions. How will continued rates of human population growth affect global resources? What happens to the humans when the population becomes larger than Earth's carrying capacity? What will happen to the quality of life for future generations? None of these questions has an easy answer. Yet future generations depend on us not only answering these questions but taking appropriate action as well.

14. Add a line to this graph that shows where you think the island's carrying capacity for people was when the colonizers first landed on Easter Island. Label this line "initial carrying capacity." Why did you draw your line there?

    You might find it helpful to compare the vegetation distribution graphic in figure 2.9 with the population growth graph in figure 2.8 as you think about where this line should go.

15. What evidence can you point to on the population growth graph (figure 2.8) that suggests that the number of people on the island exceeded its carrying capacity?
16. Add another line to your graph. Show where you think the island's carrying capacity for people was around the year 1690. Label this line "later carrying capacity." Why did you draw your line there?
17. Using terminology from the essay, *Growing, Growing, Grown*, identify which type of population growth the human population on Easter Island experienced between the years 450 and 1250.

Analysis

With your partner, discuss and answer the following questions. Record your answers in your science notebook.

1. Think about the growth in the human population on Easter Island between 1000 and 1600. Compare that with the rate of growth during the first years that the colonizers were on the island. What might account for this large increase?
2. What does the example of Easter Island tell you about unchecked population growth?
3. Think about the relationship between Easter Island's population size and the resources available per individual. How is that relationship similar to the relationship between the number of cows and the

common pasture from the previous activity (Explain, *The Pasture Story*)? What does each relationship demonstrate?

4. In the essay *Growing, Growing, Grown* we noted that the Earth's human population is experiencing exponential growth. Which type of growth curve is this? How does the graph show this?
5. In what ways is the Earth system as a whole similar to an island ecosystem such as Easter Island? Do you think the Earth has a carrying capacity for humans? Use specific examples of resource consumption and interactions to explain your answer.

ELABORATE

Islands in the Sky

In the previous Elaborate activity, *Mystery on Easter Island*, you studied the ecosystems on Easter Island. Organisms that live on Easter Island are surrounded by many miles of ocean. The situation causes the majority of plants and animals to be isolated from mainland populations. This Elaborate activity is titled *Islands in the Sky* because plants and animals living on the tops of mountains are nearly as isolated as those living on islands in the ocean. They are adapted to cold and sometimes arid conditions. These organisms may not have an easy way to move from one mountaintop to another. This activity will focus on one sky island resident from the mountains of the western United States: a small mammal called the American pika. You will use your understanding of the interactions among organisms, limiting factors, and scientific inquiry to gain a better understanding of why pika populations have been changing.

PROCESS AND PROCEDURES

1. Read the scenario, *Mystery Mountain Mammal* to become familiar with the star of this activity, the American pika.
2. Copy figure 2.19 from the scenario into your science notebook and study it. Write highlight comments on your data table. Share your ideas with your partner.

   HOW TO

   Remember, highlight comments include "What you see" as well as "What it means." Compare the numbers between the time periods, as well as between the two elevation categories. For additional help, refer to appendix B6, *How to Write Highlight Comments and Captions.*
3. Write a caption for the data table, based on your highlight comments. Follow the steps below to do this.
   a. First, write a topic sentence that describes what is in the data table.

      For example, a topic sentence for the table in figure 2.14 (in the *Growing, Growing, Grown* essay) could be "This table shows the increase in housefly population over 7 generations."

SCENARIO

ENGLISH SPANISH

Mystery Mountain Mammal

This past fall, Malcolm and his family moved from a large city to a small town in the Rocky Mountains. It was difficult for Malcolm to adjust to small town living at first. He was so used to the lights and sounds of the city that the cool, quiet nights were a bit unnerving. However, by early summer, the snow was almost done melting, and Malcolm began to feel that he was starting to fit in. To celebrate the start of summer vacation, Malcolm and his new friend, Dylan, decided to hike up Mount Princeton and enjoy the warm summer day.

During the hike, the boys passed through dense forests of aspen, spruce, and fir trees, crossed an alpine meadow with a small stream, and scrabbled up rocky slopes. As the sights and smells of a warm mountain day filled their senses, they began to feel calm and peaceful. They decided to take a break on some sunny boulders by the trail. All of a sudden, a loud sound startled them. It sounded like a raspy chirp: "chee, chee, chee." Malcolm could not actually see the animal making the noise and wondered what it could be. Dylan had never heard an animal make that sound either. It didn't sound like a bird. They thought it might be a small mammal, like a squirrel or a chipmunk. As the boys continued climbing higher up the mountain, over loose rocky slopes, the sound became more frequent.

Back at Dylan's house late that afternoon, Dylan called his aunt Sandy, who works nearby as a ranger for the National Park Service. She suggested that they might have heard a pika, but she was not certain. She invited the boys to come to her office in the morning to talk more about the pika populations in the area. Before hanging up the phone, she suggested the boys look up information about the pika on the Internet.

Back at home, Malcolm decided to do a search about pikas. He came across an interesting article about the pika's mountain lifestyle.

http://www.whats_a_pika.com

What's a Pika?

The American pika (*Ochotona princeps*) is a small mammal that looks like a cross between a large hamster and a rabbit with short ears (figure 2.17).

The best place to catch a glimpse of a pika is on rocky slopes or in boulder fields. However, you would be more likely to hear the American pika than to see one. The pika uses a series of calls to communicate with others nearby. To avoid predation and the heat of summer, pikas retreat to the cracks and spaces between the rocks. They also use these spaces to store food for winter. During the summer, small piles of grass, flowers, and other plant parts drying in the Sun provide evidence of pika activity. Pikas consume these dried piles of vegetation during the winter months. This interesting behavior has earned the pika the nickname "haymaker."

The American pika is most closely related to rabbits and hares. Its most common predator is the weasel. Unlike other small animals, pikas do not hibernate during the winter. To withstand the cold, pikas have a thick coat of fur. Although fur insulates them from the cold, it inhibits cooling. Pikas can overheat and have been observed to die when trapped in temperatures over 25°C (77°F) for more than a few hours. Pikas thrive at high elevations where temperatures never get too high, rocks provide shade in the summer, and there is enough snow cover to provide insulation over the winter.

Figure 2.17 The American pika is related to rabbits and hares.

Pikas live in many mountaintop regions around the western United States. Each mountaintop may be home to one or several pika populations living on rocky slopes. Pikas from different populations may not encounter one another very frequently, because they tend to stay close to the rocks where they are born. The maximum distance that pikas are known to travel from their place of birth is about 3 km (1.86 mi).

In the mountains surrounding the Great Basin region of the United States (figure 2.18), which is made up of the state of Nevada and parts of the states of Utah, Idaho, Oregon, and California, a number of pika populations have been discovered by hikers and scientists over the years. Records of pika discoveries date back to 1898 (figure 2.19). More recently, a survey of pika populations in this area was carried out by scientist Erik Beever and his colleagues. They went to locations where pikas had been known to live. In each location, they looked for pikas and also searched for indirect evidence of pika activity such as calls, feces, and hay piles.

Figure 2.18 Pikas can be found in the mountains of the Great Basin region of the western United States.

Beever compared their observations with records of pikas dating back to the 1800s. His results show how many pika populations were found historically, compared with 1998 (figure 2.19). Beever and other scientists say that these numbers cause them concern about the American pika's long-term outlook in its sky island habitat.

Dates of population records	Higher-altitude populations	Lower-altitude populations
1898–1956	18	6
1990–1998	16	2

Figure 2.19 **Pika populations discovered between 1898 and 1956, and re-observed between 1990 and 1998.** Higher-altitude populations are defined as those living above 2,000 m (6,560 ft). Populations found below 2,000 m are called lower-altitude populations.

Malcom stopped reading to ponder. It must have been a pika that he and Dylan heard. Not only did they hear the calls, but Malcom also remembers seeing small hay piles around some of the rocks. Malcom looked at the table of population numbers in the article and thought, "What do these numbers mean for the pika's future?"

b. Next, join each "What I see" to its "What it means" to form a sentence.

An example of one sentence in a caption for the table in figure 2.14 might be "Each generation has 60 times more houseflies than the previous generation, which means that the population is showing exponential growth."

c. Organize the sentences to form a coherent paragraph that summarizes the information in the table.

The caption should describe whether the number of pika populations is stable, increasing, or decreasing over time.

4. With your partner, brainstorm a list of all of the limiting factors (both biotic and abiotic) that you think might affect pika populations. Create a 2-column table in your science notebook to organize the list into abiotic and biotic factors.
5. With your partner, choose 2 or 3 factors on your list that you think are most likely to lead to the extinction of pika populations. Write a sentence explaining how changes in each factor could negatively affect pika survival or reproduction.
6. With your partner, think about how changes to multiple limiting factors at the same time could affect pika populations. Describe 1 way that a changing abiotic factor could cause a change in a biotic factor, which could, in turn, negatively affect pika survival or reproduction. Write your description in your science notebook.
7. Imagine you and your partner are scientists investigating the cause of pika declines. Brainstorm your ideas about the following questions with your partner, then record your ideas in your science notebook. Be prepared to share your ideas in a class discussion.
   a. What data would you want to collect about pikas?
   b. How would you go about collecting the data?
   c. On the basis of your ideas about limiting factors, what patterns do you predict you would see in the data?
8. On your own, read the scenario, *Investigating Pika Declines*. The reading will help you learn how some factors might affect pikas.

SCENARIO

ENGLISH SPANISH

Investigating Pika Declines

The next morning, Malcolm headed off to meet Sandy, the park ranger. He couldn't wait to tell her what he learned about the pika and ask questions. Sandy agreed that he probably heard pikas calling to one another. She also told Malcolm that the pika declines he read about led the U.S. Fish and Wildlife Service to look into whether the pika qualifies for protection under the Endangered Species Act. Then she gave Malcolm a news brief summarizing recent research into pika declines in the United States.

http://www.whats_a_pika.com/pikaextinction

Factors Related to Pika Extinctions in the Great Basin

How might scientists determine which factors are causing pika declines? In 2003, Erik Beever's team tested several different hypotheses concerning pika populations in the Great Basin region. The team thought many different factors could logically contribute to pika population extinctions. They hypothesized that abiotic climate factors such as average summer temperature might affect how hot pikas get in the summer. Abiotic factors could also affect biotic factors, such as the abundance or water content of the vegetation that the pikas rely on. They also thought biotic factors including human activity and grazing cattle could affect pikas by disturbing the pika's habitat. Beever's team went to each location where pikas had lived or were living and collected data on as many potentially limiting factors as possible.

Why did some populations survive while others went extinct? To test their hypotheses, the team compared living and extinct populations for each biotic and abiotic factor of interest. Then Beever's team ran a statistical analysis comparing all factors at the same time. This model allowed the team to determine which factors most likely impact Great Basin pika population extinctions. A sample of their results is found in figure 2.20.

Abiotic factors	Is this factor related to pika extinction?	Biotic factors	Is this factor related to pika extinction?
Elevation of habitat	Yes, extinctions are more likely to occur at lower elevations.	How heavily livestock graze the location	Yes, extinctions are more likely to occur in heavily grazed areas.
Maximum summer temperature	Yes, extinctions are more likely to occur in locations with higher summer temperatures.	Distance from the habitat to the nearest major road	Yes, extinctions are more likely to occur in habitat that is close to main roads.
—	—	Distance to the nearest neighboring pika population	No, more-isolated pika populations are not more likely to go extinct.

Figure 2.20 This table provides a sample of the results of Erik Beever's study of abiotic and biotic factors and their effect on pika extinctions.

9. Reflect on the list of factors that you made in step 5, which you thought might contribute to pika population extinctions. Did Beever's team investigate any of the factors you listed? Write an entry in your science notebook about 1 of the following:
   - If Beever's team studied a factor you listed, describe your reaction at learning the result.

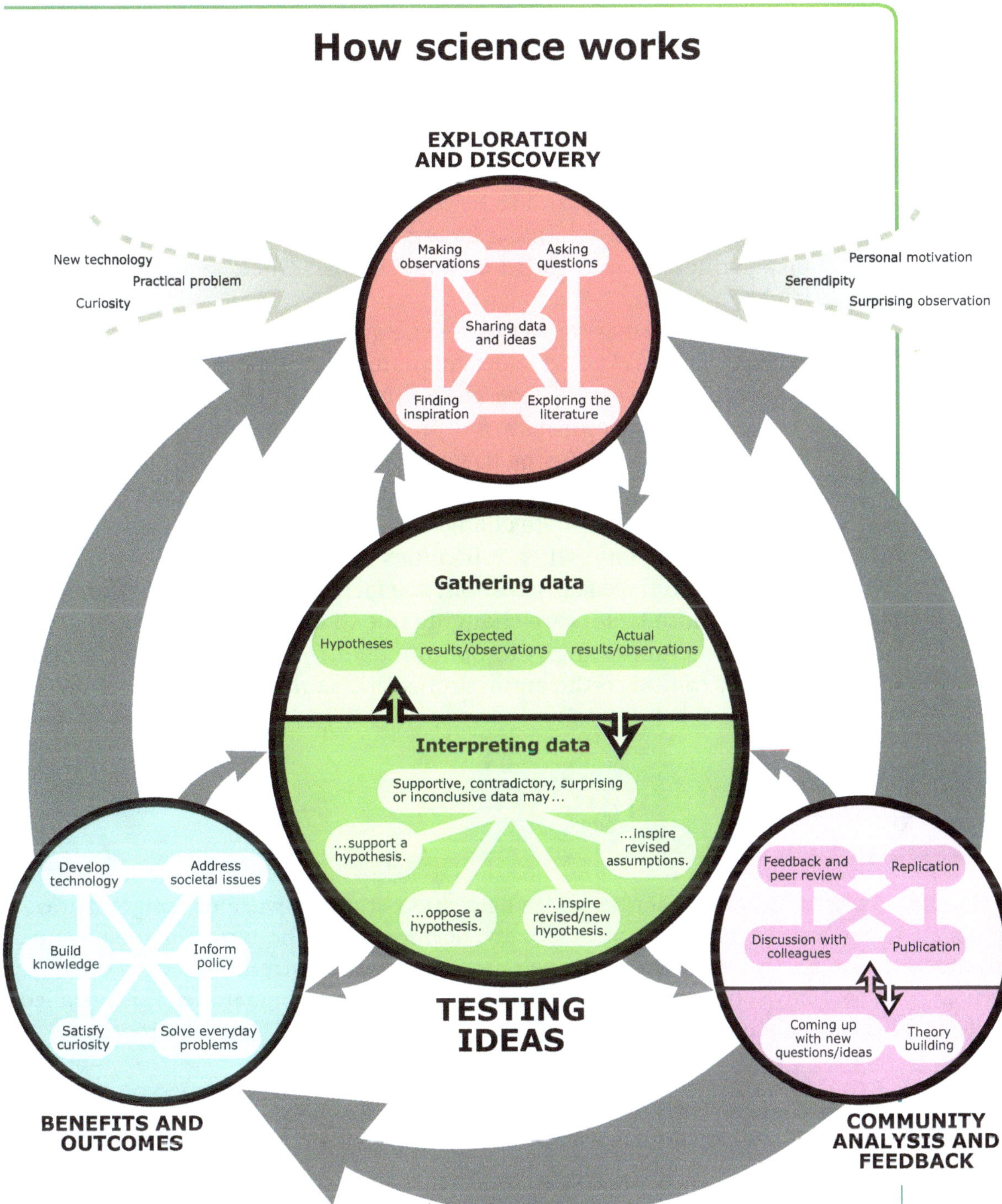

Image courtesy of the University of California Museum of Paleontology--Understanding Science www.understandingscience.org.

Figure 2.21 **The practices that are part of how science works.** Which practices do you think Beever's team engaged in?

- If Beever's team did not study a factor you listed, describe how you think scientists should study the factor that interests you.
- Describe another factor that you think Beever's team could have studied, and how.

10. To deepen your understanding of the process of science, create a 2-column table in your science notebook. In the first column, use

figure 2.21, *How Science Works*, to make a list of the scientific practices that Beever's team used while carrying out their study. In the second column, give examples of how Beever's team used each scientific practice you listed.

11. In your science notebook, make a list of the factors from figure 2.19 that Beever's team found to be related to pika extinctions. Do you think any of these changing factors suggest a connection between pika extinctions and global climate change? Circle factors you think may be related.
12. To gather information to answer the question of whether pika extinctions could be related to climate change, read the essay, *Climate Change and Its Impacts*. Use the Mark It Up! strategy to help you identify key information in the essay and to note any parts of the text that are confusing to you.
13. Write a paragraph addressing the question of whether pika declines may be related to climate change. You will know you have written a complete paragraph when it includes the following.
    a. What you have learned about the factors affecting pika extinctions
    b. How those factors are changing due to climate change
    c. Reasons supporting your opinion about whether climate change is likely to lead to the extinction of pikas on sky islands

Climate Change and Its Impacts

ENGLISH SPANISH

"Climate change" is a term used to describe long-term changes in the average temperature and precipitation over the Earth. Throughout geologic time, climates have changed in a regular pattern. There have been periods of ice ages and periods of warmer temperatures. Today, scientists use the term climate change to describe a warming trend observed over just the past century. Sometimes this is called global warming (figure 2.22). This recent

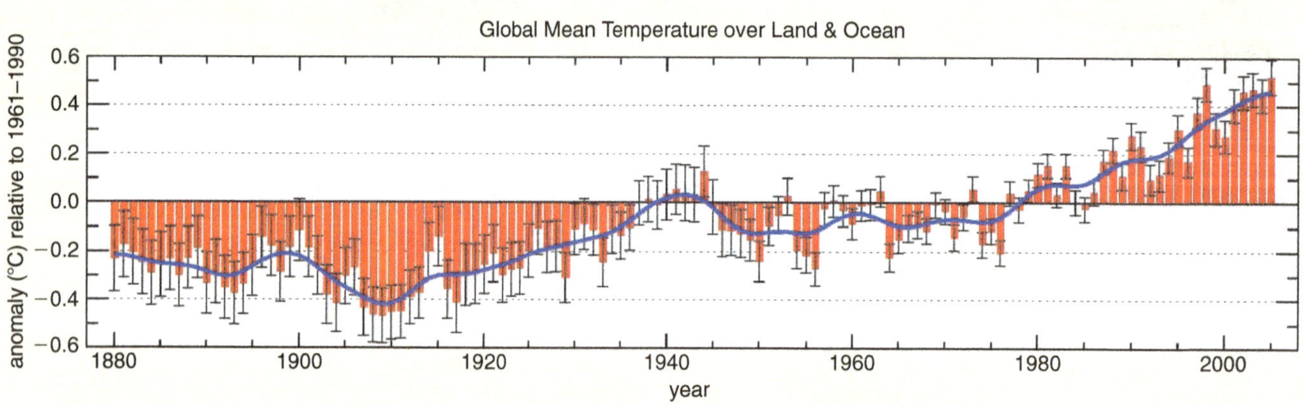

Figure 2.22 Global mean temperature over land and ocean. The graph pictured here shows yearly average global temperature (red bars) and five-year averages (blue line) compared with the average temperature for the 30-year period from 1961 to 1990. Gray bars indicate the extent of possible error in the measurements. What trends do you see?

warming trend has been linked to changes such as the melting of glaciers, ice sheets, and permafrost; a rise in sea level; changes in precipitation patterns; and changes in ecosystems. These changes have especially affected arctic regions. Some of these changes have occurred more rapidly than ever seen before.

What could cause Earth's climate to become warmer?

There are three possible processes that could cause Earth's climate to become warmer: (1) a shift in Earth's orbit around the Sun; (2) an increase in the intensity of the Sun's light; and (3) an increase in the amount of greenhouse gases in Earth's atmosphere. The first process provides the best explanation for changes in Earth's climate over the billions of years of the planet's existence. Variations in Earth's orbit occur in regular patterns over long periods of time. These variations cause changes in climate over hundreds of thousands of years. Based on these variations, scientists have predicted that Earth's average temperature should remain about the same or begin to fall over the next 50,000 years. The second process is a gradual increase in the intensity of the Sun's light over the last century. But these increases are too small to explain the recent warming trend. So, can an increase in greenhouse gases in the atmosphere explain the warming trend of the past 100 years?

How do greenhouse gases work?

Greenhouse gases have a chemical structure that allows them to absorb energy. When energy from the Sun's light travels to Earth, the surface of the Earth absorbs this energy. It radiates back into the atmosphere as heat. As heat energy passes through the atmosphere toward space, greenhouse gases absorb a lot of it. Then they radiate this heat back to the Earth's surface, to another greenhouse gas molecule, or to space. The action of the greenhouse gas molecules spread throughout the atmosphere creates a heat barrier or "blanket" around Earth (figure 2.23). This greenhouse effect is extremely important because without this heat barrier, Earth would be too cold to support life as we know it. But increasing amounts of greenhouse gases in the atmosphere can cause additional warming.

Where do greenhouse gases come from?

Greenhouse gases include water vapor, carbon dioxide, and methane. Carbon dioxide is one greenhouse gas that has increased a lot in the atmosphere over the past 100 years (figure 2.24). Carbon dioxide is released by burning wood, oil, and gas fuels. Cellular respiration by Earth's organisms also releases carbon dioxide. (Cellular respiration is a body process you will learn about in chapter 4.) The recent warming trend on Earth began after the industrial revolution. This fact suggests that greenhouse gases released by burning fuels are driving the current warming trend.

Figure 2.23 Greenhouse effect on Earth. When energy from sunlight reaches the Earth, it is absorbed and then radiated back into the atmosphere as heat (yellow arrows). Greenhouse gases absorb much of this heat energy and radiate it back to Earth (red arrows).

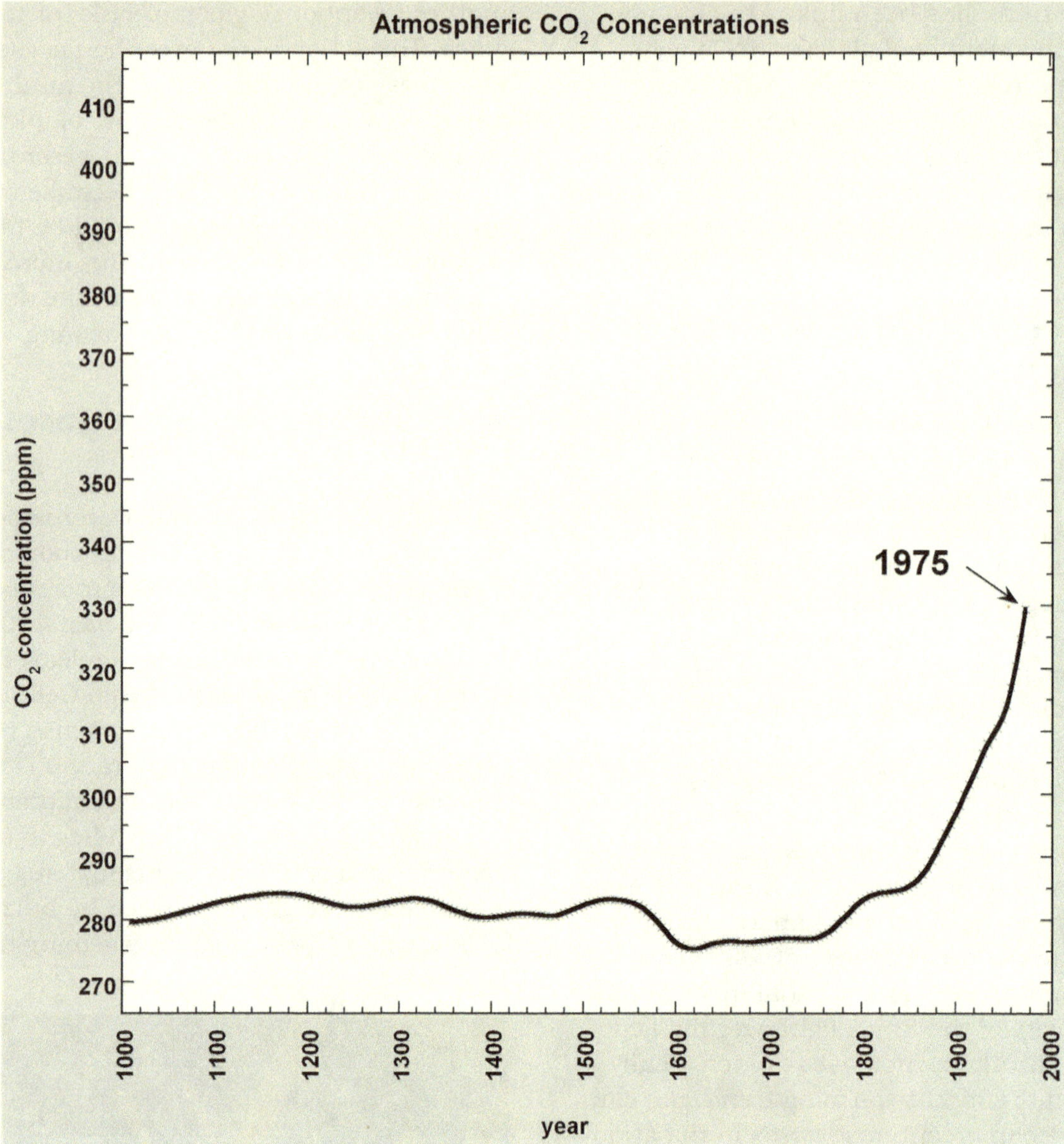

Source: D.M. Etheridge, L.P. Steele, R.L. Langenfelds, R.J. Francey, J.M. Barnola, and V.I. Morgan. 1998. Historical CO_2 records from the Law Dome DE08, DE08-2, and DSS ice cores. In *Trends: A Compendium of Data on Global Change.* Carbon Dioxide Information Analysis Center, Oak Ridge National Laboratory, U.S. Department of Energy, Oak Ridge, Tenn., U.S.A.

Figure 2.24 Changes in the concentration of the greenhouse gas carbon dioxide in the atmosphere between the years 1000 and 1975. The amount of carbon dioxide (CO_2) is given in parts per million (ppm) and was collected from bubbles trapped in ice cores from Antarctica. In 2015, the amount of carbon dioxide in the atmosphere reached 400 ppm. Based on that number, has carbon dioxide concentration continued to increase at the same rate or at a faster or slower rate between 1975 and 2015?

What does scientific evidence say about the causes of climate change?

Scientists have studied the three possible processes that could cause the Earth to become warmer. In 2007, a group of more than 2,000 scientists from across the world formed the Intergovernmental Panel on Climate Change (IPCC). The IPCC brought together the findings of thousands of other scientists. They analyzed all these findings to help them explain the recent climate changes on Earth. The IPCC concluded that greenhouse gas released by modern society is the most likely explanation for the recent warming trend. The entire team was awarded the Nobel Peace Prize in 2007 for its work (figure 2.25).

What are the impacts of climate change?

The most obvious impact of climate change is warmer winters and hotter summers across most of the world. But climate change has other impacts as well. Analysis question 2b in the Explore section described how pine bark beetles are killing millions of acres of trees in the western United States (figure 2.26). Two climate-related changes explain this result of the interaction between pine trees and beetles. Warmer temperatures lead to less rain and snow in an ecosystem. First, the stress that trees experience from drought makes them more likely to be infected by the beetles. Second, severely cold winters no longer limit beetle population growth as much as they did in the past.

Figure 2.25 Memorial plaque for Alfred Nobel. Alfred Nobel was a Swedish scientist who died in 1896. He left his fortune to begin the Nobel Prizes. Each year, Nobel prizes are awarded in physics, chemistry, medicine, literature, and peace.

What can you do about climate change?

Human activities have driven recent climate change, so we have the opportunity to tackle the problem. We can reduce our release of greenhouse gases by using fuels more efficiently. For example, you can ride the bus, bike, or walk to school. You can buy food that is grown locally in California instead of other parts of the US or other countries (figure 2.27). You can communicate with our government representatives at the local, state, and federal levels and let them know how climate change affects you in San Francisco.

Figure 2.26 Grove of dying pine trees. The brown trees in this picture are dying due to drought and pine bark beetles. This is one impact of climate change.

Figure 2.27 Market with locally grown food. Buying food that is grown locally helps reduce global warming because less fuel is burned to transport that food to grocery stores.

Analysis

1. Several lines of evidence collected by Erik Beever and other scientists suggest that pika populations went extinct without migrating elsewhere. Use what you know about pikas and sky island habitats to explain why pika populations did not migrate to another mountaintop habitat.
2. American pikas live in areas other than the Great Basin. In the Rocky Mountains, some pika populations live at elevations higher than they do in the mountains of the Great Basin. Populations living at higher elevations are subjected to much cooler summer and winter temperatures than the Great Basin pikas. These locations historically have much snowfall in the winter. A layer of snow keeps the ground temperature near freezing (0°C, or 32°F), even when air temperatures plunge far below freezing. In the Rocky Mountains, snowfall has been declining, and this trend is thought to result from climate change. What impact do you think decreased winter snow cover might have on pikas in the Rocky Mountain region?

   You may want to refer back to the pika characteristics described in the *Mystery Mountain Mammal* scenario to help answer these questions.

EVALUATE

Critters and Interdependence

It's time to evaluate what you have learned about interdependence among living organisms and the complex interactions that take place in communities and ecosystems. In this Evaluate activity, *Critters and Interdependence*, you will work in teams to create a story that describes interactions among various organisms and resources in a particular habitat. These organisms will include your critter, other classmates' critters, humans, and other native organisms.

Materials

Part A

online resource

Part B

descriptions and diagrams of your critter

PROCESS AND PROCEDURES

Part A Resources and Ecosystems

1. Identify the habitat card you had in chapter 1, *The Human Animal,* for the Evaluate activity, *First Encounter with the Critter.* Do this by referring to your science notebook.
2. Watch the video segment, "Ecosystems of the Earth." As you watch, perform the following tasks.

   Online Resource

   a. In your science notebook, record the name of each ecosystem.
   b. List at least 4 limiting factors presently found in each ecosystem.

      Remember, include both biotic and abiotic factors.
3. Participate in a class discussion that your teacher guides. You will summarize the interactions and resources that characterize each of the ecosystems depicted in the video segment.

   Think about the main ideas of this chapter and how they relate to those ecosystems.

Part B Critters in Ecosystems

1. Join with classmates who had the same habitat card as you in chapter 1.
2. Obtain a Chapter 2 *Critter Rubric* handout from your teacher. Discuss the criteria with your teammates.

   Ask your teacher to clarify any questions that come up during your discussion that your team cannot answer fully.
3. Introduce your new teammates to your critter and respond to any questions that your teammates might have.

   Be sure to describe all the features of your critter in detail.
4. Imagine that you, your teammates, and each teammate's critter are living together in the assigned habitat from chapter 1. Brainstorm ideas about the following.

   SEP

   a. Possible interactions among the organisms that inhabit this ecosystem
   b. The biotic and abiotic resources that might exist
   c. The ways the different critters might be interdependent
   d. How humans from a variety of cultures might interact with other organisms in this environment
5. Individually, write a story that features some of the ideas that your team suggested. Make sure it follows the criteria outlined in the rubric.

   Remember, your teacher will assess what you have learned from this chapter by reading your story.

Analysis

Reflect on the process you used to think about and write your story. Participate in a class discussion of the following questions.

1. What was the most challenging part of writing your story? Explain.
2. Which of the major concepts in this chapter did you have the most difficulty incorporating into the story? Explain.
3. You wrote about the interactions and interdependence of your critter with other organisms. You also wrote about the limiting factors at work and the carrying capacity of the environment. Which was easier to write about? Explain why.

All living things grow and repair injuries to their bodies. They also carry out functions. Whether you think of a plant or an animal, it needs to grow and carry out activities. In order to grow and repair its body, a living thing needs matter. Matter is the physical "stuff" that helps them grow. Living things use energy to carry out all their functions. That means that all living things need both matter and energy.

In unit 2, *Cycles of Energy and Matter*, you will explore matter, energy, and the relationship between them. You will see how the interdependence between the different parts of an ecosystem helps to ensure that living things get the matter and energy they need. In chapter 3, *The Cycling of Matter and the Flow of Energy in Ecosystems*, you will learn how matter and energy link all the organisms in a community. Then, in chapter 4, *The Cellular Basis of Activity*, you will see how processes in the body let living things get matter and energy from food. You will learn how the different molecules in food can affect the amount of energy you can derive from food.

UNIT

2

Cycles of Energy and Matter

By the end of unit 2, you should be able to understand that

- the conservation and transformation of energy and matter are found in all living systems;
- organisms release energy by breaking the chemical bonds of food molecules, forming different molecules that have lower amounts of energy;
- photosynthesis transforms light energy into chemical energy, with dramatic effects on all living systems;
- energy flows through ecosystems, and matter cycles in them; and
- well-designed investigations collect data, compare results to controls, draw conclusions, and report findings.

UNIT CONTENTS

CHAPTER 3

The Cycling of Matter and the Flow of Energy in Ecosystems

Climb the mountains and get their good tidings. Nature's peace will flow into you as sunshine flows into trees. The winds will blow their own freshness into you, and the storms their energy, while cares will drop off like autumn leaves.

John Muir

When you hear the word community, what do you think of? Do you think of your school or your neighborhood? Do you think of everyone in San Francisco or in California? In much the same way, energy and matter move through organized communities of living organisms. In biology, we speak of a **community** as the group of living organisms that inhabit a specific area and interact with one another. As you learned in Chapter 2, *Interdependence Among Organisms*, an **ecosystem** is a community of organisms and the nonliving part of the environment. Consider a bird flying over the mountains in this photograph. Now think about the mice it feeds on living in the rocks. Think about the trees and grass, the insects and bacteria, and the many other organisms that inhabit the soil and other nonliving parts of this area. Together these organisms make up a community. This community and the air, rocks, soil, and other nonliving parts of this environment make up an ecosystem. In the last chapter, you considered the relationships and interactions between the living and nonliving parts of an ecosystem. These relationships are important for many reasons, particularly when you think of the matter and energy that living things need to live. In chapter 3, *The Cycling of Matter and the Flow of Energy in Ecosystems*, you will use your experiences from a variety of activities and related essays to develop an understanding of how matter and energy move through ecosystems. You will also consider the role that organisms play in the movement of matter and energy.

GOALS FOR THE CHAPTER

Throughout chapter 3, you will work to gain an understanding of the following concepts:

- ✔ The particles that make up matter cycle repeatedly in ecosystems.
- ✔ As energy flows through biological communities, it eventually becomes unavailable for use by biological systems.
- ✔ We can investigate ecosystems by using models and controlled experiments.

To help you develop these understandings, you will complete the following activities.

ENGAGE / EXPLORE — A Matter of Trash

EXPLORE — Matter Goes Round and Round

EXPLAIN — Spinning the Web of Life

ELABORATE — Generating Some Heat

EVALUATE — Energy, Matter, and Disaster

Throughout chapter 3, remember to refer to the chapter organizer to help you remember what you have learned and where you are headed.

ENGAGE EXPLORE

A Matter of Trash

Key Idea: Waste is generated by every community and can show the way that organisms interact with others in the community.

Linking Question:
Is it possible to observe or measure the cycling of matter in an ecosystem?

CHAPTER 3

The Cycling of Matter and the Flow of Energy in Ecosystems

EXPLORE

Matter Goes Round and Round

Part A: Reflections on the Earthworm Habitats

Part B: Snails and Anacharis: *What Can I Learn from Them?*

Key Idea: The particles that make up matter cycle repeatedly between the abiotic and the biotic components of an ecosystem.

Linking Question:
How do matter and energy move through an ecosystem that contains many different organisms?

EXPLAIN

Spinning the Web of Life

Key Idea: Matter cycles, and energy flows, between organisms interacting in an ecosystem.

EVALUATE

Energy, Matter, and Disaster

Key Idea: Disruptions to the movement of energy and matter can affect every organism in a community.

MAJOR CONCEPTS

- ✔ The particles that make up matter cycle repeatedly in ecosystems.
- ✔ As energy flows through biological communities, it eventually becomes unavailable for use by biological systems.
- ✔ We can investigate ecosystems by using models and controlled experiments.

Linking Question: What happens when the flow of energy and the cycling of matter are disrupted?

ELABORATE

Generating Some Heat

Key Idea: Decomposers play an important role in the cycling of matter in an ecosystem.

Linking Question: When an organism dies, what happens to its matter and energy?

A Matter of Trash

Think about the last object you threw away. Did you put it into a bin for trash, recycling, or compost? Where did it go after the container was emptied? Do other organisms that live in your neighborhood use or create trash? Throughout this Engage-Explore activity, *A Matter of Trash,* think about the various forms of matter that you consider waste and what happens to this matter after you throw it out. You will also consider how the matter can connect the organisms in an ecosystem.

Figure 3.1 **Trash.** Where will the matter and energy in this trash end up?

PROCESS AND PROCEDURES

1. Write down at least 3 objects you have thrown away recently.
   a. Where do you think each object went after it was thrown away?
   b. Do you think any other organisms might be affected by the objects in the future?
   c. Do you think the objects are made of matter? Why or why not?
   d. Do you think there is energy in the objects? Why or why not?
2. Examine the discarded items in the trash demonstration that your teacher presents. Create a table or other visual diagram that includes the following information.

   Remember that this is the Engage section of the chapter and you are not expected to know all the answers. Use what you already know and logical reasoning to complete your table the best that you can.

   a. A list of the trash in the demonstration
   b. A list of the trash's origin (for example, newspaper originates from trees, wood, and pulp)

Figure 3.2 When plastics and paper are recycled, what becomes of the particles that make them up? When items are put into a compost pile, what happens to the particles that make them up? Based on the picture on the right and what you know about compost, is there energy in the pile?

c. An indication of which trash items match trash or waste in your household. (write "yes" or "no")
d. A list of at least 2 possible ways another living thing might be able to use each item. Consider ways the items could be used as they are and ways the items could be used if they are broken down. Look at figure 3.2 if you need some ideas to get started.

3. In your science notebook, write your best answer to the following question: Do you think that the trash contains energy? Describe evidence you could collect to back up your claim.
4. In your science notebook, write your best ideas about whether other organisms besides humans make trash. If so, what happens to it?

Analysis

Working individually, answer these questions in your science notebook. Be prepared to share your responses in a class discussion. You can read the sidebar, *Garbage Among Us—From Then Until Now!*, to learn additional information about garbage in different communities. Remember that in biology, the word "community" describes all the living things that interact in an area.

1. In what ways might particles in the waste matter of one organism be useful to another organism? Give 2 or 3 examples to support your answer.

   In the earlier questions, you considered the waste on a bigger level. Now think at the level of microscopic particles.

2. How does your answer to question 1 support the idea that organisms in communities depend on one another for matter and energy?
3. Think about how matter may be recycled by organisms in the following 3 communities. Remember that matter is the physical "stuff" that makes up everything. How do these communities differ in the types of matter they interact with? Does matter take different lengths of time to be recycled in these different communities?
   a. A community of soil bacteria and fungi on a forest floor
   b. An isolated village of humans in the Amazon rain forest
   c. A neighborhood of humans in San Francisco

4. Write about ideas you have in reaction to this question: What problems or concerns do modern human societies have related to the kinds of matter they use or to how long it takes matter to be recycled within their communities (figure 3.3)?

Figure 3.3 In many cities, the trash that is thrown away goes to a landfill like this one.

SIDEBAR

ENGLISH SPANISH

Garbage Among Us—From Then Until Now!

Matter—lots of it. Wood. Paper. Glass. Metals. Plastics. Rubber. Cloth. Food. Yard waste. All of these types of matter end up in landfills across the country. In fact, did you know the following?

- In the United States, each person throws away an average of 1,600 lbs of garbage each year. This is more than any other nation in the world.
- In 2013, Americans recycled about 34 percent of the waste they generated (254 million tons). This is up from only 6 percent in the 1960s. However, the total amount of waste generated by each person has also doubled over this time.
- A convoy of garbage trucks long enough to encircle Earth six times would be required to carry all the municipal waste generated in the United States in one year. (Municipal means "related to a town or city.")
- More than half of the states in the United States are having problems finding places to dump trash. Sometimes trash travels great distances before it is dumped (figure 3.4).
- Lettuce buried in a landfill may take more than seven years to decompose completely.
- A hot dog can last more than 10 years in a landfill.
- A steak buried in a landfill can retain its fat 15 years after burial.
- In spite of recycling efforts, waste paper is still a major component of landfills nationwide. It makes up about 15 percent of the waste in landfills.

You throw away things every day. But did you ever wonder what people could tell about you if they sorted through your trash? Garbage and waste can tell us quite a bit about how organisms acquire and use the matter and energy in their community. Let's consider three different communities: a bat community, an ancestral human community, and a modern-day human community. Reflect on the cycling of matter through them.

- How is this process similar and different in these communities?
- How does matter move from one organism to another?
- How much and what type of waste does each community produce?
- What happens to the waste in each community?

Figure 3.4 This barge is loaded with waste material that will travel to a final dumping ground.

Bats in Mammoth Cave National Park, Kentucky

Two bats flit about in the sky at twilight, barely visible in the last light of evening. They dart about, catching insects. One bat descends briefly, flying just above the surface of a pond to take a drink. Another nips a moth in midair. Soon, the bats dart into the entrance of a cave that is partially hidden by evergreens (see figure 3.5).

Figure 3.5 Brown bats live in caves throughout Mammoth Cave National Park, Kentucky.

Several fleas, ticks, and mites have latched onto the bats' coats during the evening. As the bats sit on a cave ledge, resting and cleaning themselves, some of the insects fall to the cave floor. They become food for the organisms that dwell there. During the night, many bats excrete their wastes—a substance called "guano." Guano is not solid like human feces, but rather thick and mudlike. It covers the floor of the cave like carpeting. Bacteria, fungi, and other single-celled organisms grow on the guano and use it for food—their source of energy. In turn, many of these small organisms are destined to become the food for insects, and some of these insects may become the food for bats.

Figure 3.6 **Cliff dwellings.** The Ancestral Puebloans lived in cliff dwellings. These structures can be found throughout the Four Corners region of the American Southwest.

The Earliest Ancestral Puebloans

The year is CE 950. As a hint of daylight appears in the east, a young Ancestral Puebloan woman awakens and checks on her two young children still asleep beside her in their cliff dwelling (figure 3.6). They live along a basin in the high-plateau region of the American Southwest. As the woman rises and starts a fire, her mate begins to stir. She retrieves some kernels of corn and begins to grind them into a coarse meal. It is late spring, and the couple hopes to finish planting the corn today. They want to work in the morning before the day becomes hot. To plant, they use a stick to make holes in the ground, drop kernels of corn into each hole, and then cover them with soil.

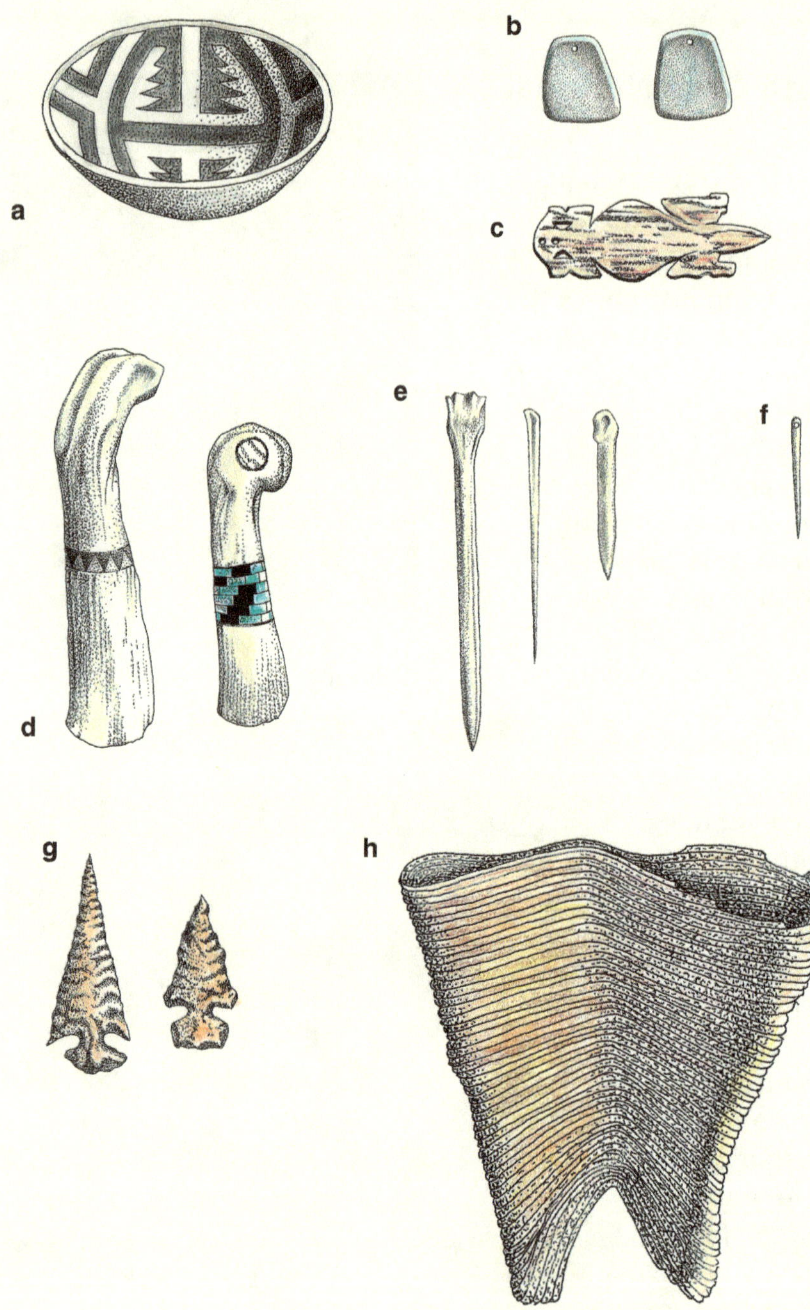

Figure 3.7 Artifacts from Pueblo Bonito. Archaeologists found these artifacts and more at the Pueblo Bonito site in the Chaco River Valley: (*a*) black-on-white bowl, 7 cm high; (*b*) turquoise pendants, 2.5 cm high; (*c*) shell ornament, 5 cm long; (*d*) bone scrapers, 16.5 and 12.7 cm long; (*e*) bone awls, 11.4–19.7 cm long; (*f*) bone needle, 7 cm long; (*g*) projectile points, 5 and 3.75 cm long; (*h*) basket, 39.4 cm high. They are dated at about CE 950–1000.

Later, the woman and her children join other women and children to gather yucca from the plateau. The Ancestral Puebloan families will not only eat the fruit and seeds of the yucca, but they will use the roots for soap and shampoo. They will use the strong, sturdy fibers from the leaves for making intricate baskets, sandals, aprons, mats, and cradle boards (figure 3.7). Small pieces of the yucca and other plant material are discarded in a pile along with corn husks and cobs, worn-out mats, and broken tools made out of bone.

Some mornings, the woman's mate joins other men in the community to hunt for rabbit and deer. In addition to preparing and eating the meat, the Ancestral Puebloans make clothing for the winter months from the rabbit pelts, and they make tools and utensils such as needles from the bones. Unused bones are discarded in the small but growing pile of waste. At the bottom of this pile, the organic material has decayed enough so that it is almost indistinguishable from the soil beneath it.

Family in San Francisco, California

A single mother looks in on her eight-year-old daughter, Sonya, who has the flu. Sonya has a big box of tissues beside her bed and a pile of used ones in the wastebasket nearby. She also has finished one juice box and has begun another. One of the family's three cats is sleeping at the foot of the bed. (Someone needs to clean the litter box today.)

This family lives in a two-bedroom, one-bathroom near the corner of 14th Street and Guerrero (see figure 3.8). Sonya has a 15-year-old brother, Stephen, who is keeping his distance because he doesn't want to get sick. He has a science presentation due Monday. He works away on his device most of the afternoon and prints the presentation four times before he is satisfied. He places the rejected copies in the recycle bin in the kitchen.

Their mother works at an architecture firm and has been working at home all day on a balsa-wood model of a San Francisco General Hospital remodel. (Scrap wood seems to be everywhere.) In the evening on the way home from the corner store (with 2 bags of groceries), the mother stops at a fast-food restaurant and picks up some chicken dinners, which everyone enjoys—even Sonya, who is beginning to feel better. As Stephen tosses the last dinner carton into the garbage, he notices that it's full again. Whose week is it to take out the garbage, anyway?

Figure 3.8 Sonya and Stephen's Mission District neighborhood in San Francisco.

Further Challenges

On a separate piece of paper, write 10–15 things you have thrown away in the past week. Exchange lists with your partner. Write 4 or 5 things that you might infer about your partner from his or her trash. Use information from the sidebar, *Garbage Among Us—From Then Until Now!*, to help you with your answer. Support your inferences.

EXPLORE

Matter Goes Round and Round

Have you ever watched ants in an ant farm? They always appear to be busy modifying their environment in some way. In this Explore activity, *Matter Goes Round and Round*, you will complete your observations of earthworm habitats. Think about what your observations tell you about how these organisms interact with their environment. You also will reflect on your investigation's design. You will use this understanding to plan and carry out another investigation to explore how other organisms interact with one another and with their environment.

Materials (per team of 4)

Part A

- 6 slides
- stereomicroscope
- hand lens
- earthworm habitats
- control habitats
- 2 paper towels
- 2 spoons
- small tray
- metric ruler
- balance (optional)

Part B

- 250-mL flask
- 4 25 × 200-mm test tubes
- test-tube rack
- thermistor or thermometer
- 2 pH probes or pH strips sensitive to pH 1–12 and pH strips sensitive to narrower ranges
- jar of tap water for storing pH probes (optional; if using probeware)
- 1-mL pipet (optional)
- petri dish halves (optional; if pH paper is used)
- 2.5-mL of bromothymol blue (optional)
- distilled water
- 1- to 1.5-cm freshwater snails
- 15-cm sprigs of *Anacharis* (elodea)
- light source
- aluminum foil
- graph paper
- different-colored pens or pencils

PROCESS AND PROCEDURES

Part A Reflections on the Earthworm Habitats

1. With your teammates, look at the earthworm habitats that you or your teacher set up several weeks ago to model natural earthworm habitats (figure 3.9). In your science notebook, record new observations and review the observations you made previously.

Figure 3.9 In what ways do you think earthworms interact with their environment?

The purpose of these model habitats is to provide evidence of interactions between living systems and the physical environment in nature, as well as evidence about the nature of these interactions.

2. Discuss the following with your teammates. Record your ideas in your science notebook.
   a. What evidence did you collect that supports the idea that earthworms interact with and change their environment?
   b. Describe any interactions that you notice between earthworms and their environment. What do you think happened in the earthworm habitats?
   c. What was the specific purpose of the control containers in helping you identify and describe the interaction of earthworms and their environment?

      Compare the containers that had earthworms with each other. (One contained soil and organic matter, and one contained only soil.) Why was it important to observe both types of containers? What did your comparisons tell you about the interactions of earthworms with their environment?
3. Draw a diagram in your science notebook that shows different components of an earthworm habitat. Add a plant and some microorganisms to the diagram. Then, as you read the essay, *Worms, Insects, Bacteria, and Fungi—Who Needs Them?*, draw arrows to indicate how you think the particles in a piece of banana might move through the different components of the habitat.

   Check with your teacher or teammates if you have questions about the concepts in the essay and how the concepts relate to your findings.
4. Discuss the following question with your partner: Does the diagram you drew in step 3 point to evidence that some matter could *cycle* (move through a series of events that returns to the starting point) through an ecosystem containing soil, organic matter, earthworms, microorganisms, and plants? Write down your ideas in your science notebook.

Worms, Insects, Bacteria, and Fungi—Who Needs Them?

ENGLISH SPANISH

Did you know that approximately 100 million bacteria can live in a single gram of fertile soil (figure 3.10)? Did you also know that some 250,000 earthworms can live in a 1.25-acre field of rich topsoil? That is 250,000 earthworms living in an area the size of a football field!

This is one reason why earthworms are considered a farmer's best friend. Both earthworms and bacteria are organisms that decompose decaying organic matter. They can work through 10 tons of topsoil a year, aerating it and increasing its fertility.

The Importance of Earthworms

The quality of the topsoil is important because topsoil serves as one important link between the biotic, or living, world, and the abiotic, or nonliving, world. All matter in the universe—both living and nonliving—is composed of tiny particles called **atoms**. When atoms of the same kind come together they make an **element**. Atoms of different types form **molecules**. Molecules and elements such as carbon dioxide, nitrogen, and phosphorus can be part of living things or nonliving things. This abiotic matter enters the living world when molecules and elements are absorbed by plants. These molecules and elements are returned to the nonliving world when they are excreted by plants, animals, fungi, or microorganisms as waste, and when organisms die. The waste ends up in the topsoil or in the atmosphere. In the soil, the molecules are broken down to simple molecules or atoms by the soil's inhabitants. At this point, the cycle can begin again as plants reuse these molecules.

Earthworms play a vital role in keeping the nutrient levels in the soil high because they consume partially decomposed organic matter such as dead leaves and roots. They then excrete nutrient-rich waste. This mixes with the soil, as in figure 3.11. Other organisms that perform a similar function in the soil are insects, bacteria, and fungi. In fact, bacteria perform most of the decomposition in soils. Together these organisms help return vital

Figure 3.10 A gram of soil contains a surprising number and diversity of bacteria.

Figure 3.11 Earthworms eat dead leaves and roots. Then they excrete matter that is high in nutrients. In this picture, earthworms have been eating and excreting, which has caused a mound of soil.

elements, such as phosphorus, calcium, and nitrogen, to the environment, where other organisms can use them. Thus, each atom can cycle through a community of organisms multiple times.

Matter Is Recycled in Ecosystems

Recyclable elements are important to the health of all communities, and they emphasize the role of organisms that decompose organic matter. This role is particularly apparent in the tropics. There, the layer of topsoil is thin, the temperature is high, days are long throughout the year, and rain is frequent. Under such conditions, organic matter is produced rapidly, and it also decomposes rapidly. Those nutrients that are not returned quickly to the living part of the ecosystem flow away in groundwater and are lost to the local community of organisms. Figure 3.12 illustrates this situation in a tropical rain forest. The community appears extremely lush because most of the nutrients are held almost continuously in living material. The matter recycles rapidly between organisms, the atmosphere, and the soil. The critical location for this rapid recycling is the relatively thin layer of dead plant material that forms the topsoil on the forest floor. In the constant heat and moisture of the tropics, the insects and microorganisms in this layer quickly break down much of the fallen leaves and other dead organic matter. The nutrients they release into the groundwater then move downward toward the sterile clay, beneath where the rich network of tree roots quickly reabsorbs them. The nutrients move up through the trees' roots, trunks, and branches to their leaves. It is here that the nutrients from the soil are incorporated into living matter through a variety of biosynthesis reactions. Simultaneously, carbon dioxide from the air is incorporated into living matter through photosynthesis.

In environments with cooler conditions, decomposition is much slower and virtually stops in the winter. In these environments, dead organic material can remain in the topsoil much longer without having nutrients transported away to other communities by rain or flooding. Though slower, these processes of recycling eventually return the nutrients in fallen leaves and other dead organic matter to the plants.

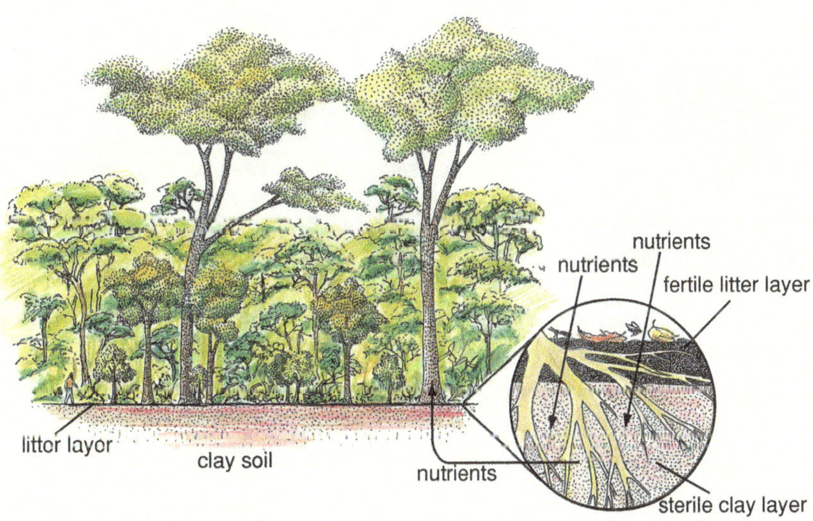

Figure 3.12 Tropical rain forest. In a tropical rain forest, most of the nutrients are held in the tissues of living organisms. As a result, the thin, nutrient-poor topsoils of tropical rain forests experience rapid recycling of any matter, including fallen fruits and leaves, that reaches the forest floor. What might happen to the nutrient levels of rain forest ecosystems if the number of living organisms were greatly reduced?

Part B Snails and Anacharis*: What Can I Learn from Them?*

In this part of the activity, you will have the opportunity to plan and carry out an experiment aimed at collecting evidence for the cyclical movement of carbon atoms in a model aquatic ecosystem. Your model ecosystem will contain snails and a plant, *Anacharis*. But first, you will need to find out about the cycles of several different atoms and molecules in ecosystems.

1. You may already know something about how matter cycles. In your science notebook, draw a picture to show what you know about the water cycle.

   One part of the water cycle is rain forming and falling. Try to add both biotic and abiotic parts to your diagram.

2. Share your ideas with your partners. Discuss whether your diagram shows matter cycling.
3. Read *The Cycling of Matter in Ecosystems*. As you read, use the Mark It Up! Strategy to annotate the reading.
   - Underline information that discusses water moving through abiotic parts of the ecosystem.
   - Circle information that describes water moving through biotic parts of the ecosystem.
   - Draw a star next to information you included in the diagram you drew in step 1.

The Cycling of Matter in Ecosystems

ENGLISH SPANISH

Figure 3.13
A beaver busy at work. What effect might this beaver have on the movement of matter in its ecosystem?

Imagine a riverbed ecosystem. In such an ecosystem, matter is always on the move from one place to another. Water runs through it. Nutrients move from the soil into plants rooted along the river. Beavers gnaw through branches and twigs and use them to build their dams (figure 3.13). Matter is also moving between organisms. A great blue heron nabs an unsuspecting fish from the river. When the types and amounts of matter in such an ecosystem remain essentially the same across time, the ecosystem remains in balance. When the types or amounts of matter in an ecosystem change, the ecosystem is forced to change as well.

Let's look more closely at the biotic and abiotic components of the riverbed ecosystem. Within this ecosystem, matter cycles continuously between the biotic and the abiotic components. Beavers are part of the biotic matter along this riverbed. In the spring, two to four young beavers are born to each female beaver. As the young beavers grow and become more self-sufficient, the local beaver community can withstand the temporary increase in its population. A few of last year's young may remain, replacing the beavers that die or disperse. But many of last year's young will move out of the community to establish other dams. The total population of beavers in this area will remain relatively stable. If the number of beavers in this area increased significantly, it would stress the available resources. These resources include the food supply and the sites for building dams. Eventually, the resources might become so depleted that the environment would be unable to support any beavers. They then would die off or move elsewhere. Such a scenario would cause the community to change.

Not only does matter move through an ecosystem, but some of it makes a complete cycle within it. For example, when beavers die along the river, their bodies gradually decay. The nutrients derived from their bodies eventually mix with the soil. Plants then acquire some of these nutrients from the soil. In turn, various animals (including beavers) feed on these plants. In this way, nutrients that once were part of an animal can cycle through various types of matter and be taken up by another animal.

Water is an example of abiotic matter that moves through and cycles within an ecosystem. Water is constantly moving through the river shown in figure 3.14. The water flowing through the river at a certain spot today is different from the water that flowed through this same spot yesterday. Water also is part of a cycle. Some is added to the river by rain or snow and from tributaries and groundwater reserves in the plateau. Some water leaves the river through evaporation. Much water also leaves as plants photosynthesize, through a process called transpiration.

If the net inflow of water is the same as the net outflow, the river remains essentially the same. A major increase in the water flow could cause the banks to flood, destroying many plants along the riverbed. A prolonged increase in the amount of water, however, may change the kinds of plants and animals that can live in the community. Similarly, a short-term decrease in the water flow may cause certain plants and animals to die as competition for water increased. Again, if such a change were sustained, the makeup of the community would change as well.

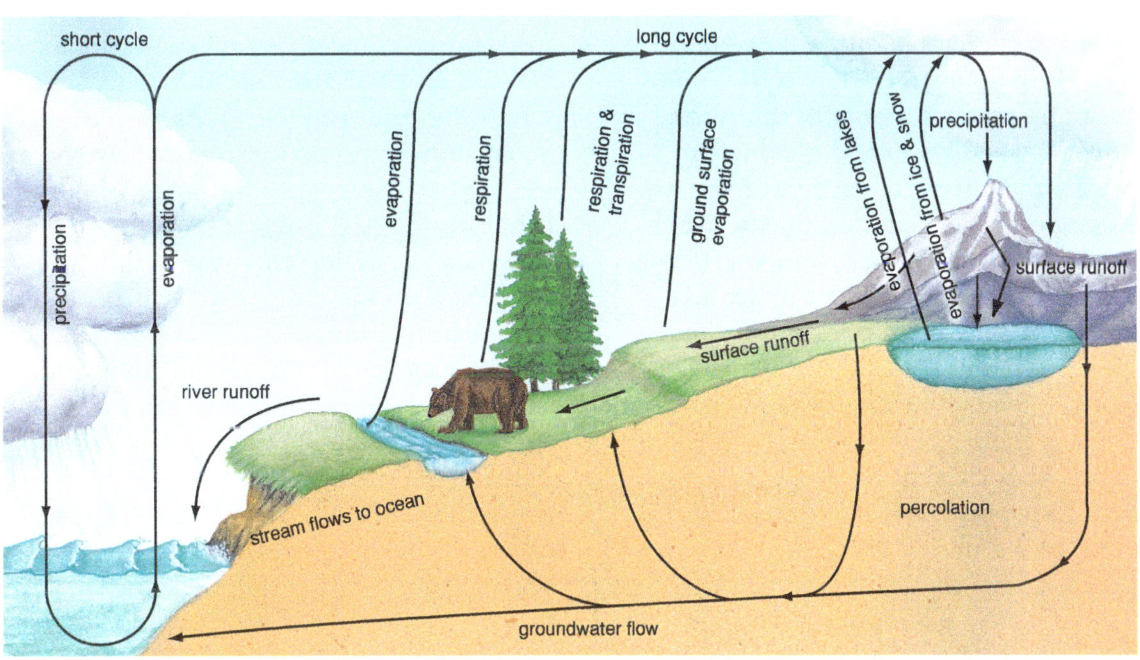

Figure 3.14 **The water cycle.** The water cycle collects and redistributes Earth's water supply.

4. Add to your diagram of the water cycle to show any new information you learned.
5. Carbon atoms are another type of matter that cycles in ecosystems. All living things are made up of molecules that include carbon. There are gases in the atmosphere that use carbon, such as carbon dioxide. Using what you know about water cycling in an ecosystem, try to describe how carbon might cycle in an ecosystem. Discuss your ideas with your team.
6. Read *The Carbon Cycle* to learn more about how carbon moves through the ecosystem. As you read, use the Mark It Up! Strategy to highlight the places carbon is found in the atmosphere. Draw an arrow next to any information that shows how carbon moves in the atmosphere.
7. In this activity, you will plan and carry out an investigation to collect evidence about how carbon cycles in an ecosystem. First, work with your team to put together what you know about carbon cycling for your investigation. With your team, carry out the following tasks to plan your experiment.
   a. Briefly summarize what you have learned about how carbon cycles in an ecosystem.

The Carbon Cycle

These same principles of the movement and the cycling of matter apply to the many other, less obvious components that make up the riverbed ecosystem. Consider, for example, a single atom of carbon. If you could follow a single atom of carbon in the riverbed ecosystem through time, you would see it cycle through many different molecules. At one point, it may form part of a protein molecule in a floating leaf of duckweed. At another time, it may become one of the atoms within a DNA molecule in the genetic material of a fish or a frog that ate the duckweed. At still another point, the carbon atom may remain for a long period within dead plant or animal material in the mud of the river until it is finally used by a bacterium or a fungus and rejoins the biotic community. Figure 3.15 illustrates some of these interactions and relationships in the carbon cycle. Again, when the total number of carbon atoms in an ecosystem remains approximately the same along with the proportion of carbon atoms to other atoms, the ecosystem remains essentially stable. When the number or proportion of carbon atoms changes significantly, the community changes.

Other cycles of matter in ecosystems depend on the existing environmental conditions. Essential elements, such as calcium, potassium, nitrogen, and phosphorus, are available to the biological community only after they dissolve in the groundwater and are taken up by the roots of plants. Through experience, people have discovered that many desert lands will produce large crops, at least

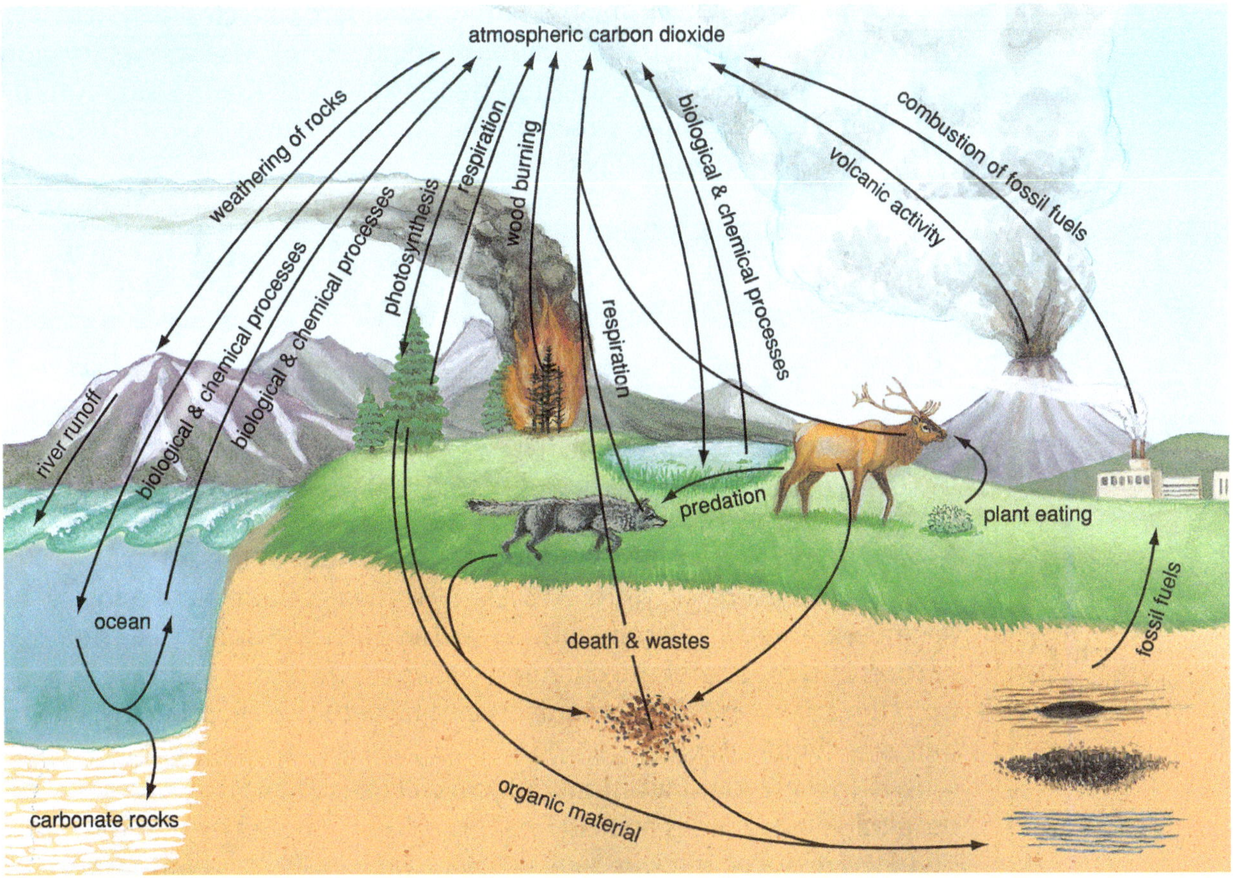

Figure 3.15 **The carbon cycle.** In a stable ecosystem, the total number of carbon atoms will remain approximately the same.

for a time, if they are irrigated. Often, essential elements are abundant in these soils, but they are not readily available to plants due to the shortage of water.

There is another issue that is related to the cycling of matter. Various organisms in a community often cannot efficiently metabolize foreign compounds such as pesticides. As a result, these compounds may build up to toxic levels in of some organisms. They also may build up in the environment and stay there for long periods of time before they are returned to the cycle.

Taken as a whole, when a community is stable within an ecosystem, such as the riverbed, the community exhibits a large-scale balance with matter continually cycling. The same is true in other communities such as those found in a desert, a temperate pond, a pine forest, or the arctic tundra. On a larger scale, the same also is true for the entire biosphere. The biosphere includes all the organisms as well as the soil, water, and air that surround and support them. As the environment changes, communities may change as well.

b. In your science notebook, draw a diagram of a closed system containing snails and *Anacharis* (figure 3.16). Use arrows to show the movement of carbon in this system. Use the information in the need to know, *Observing the Carbon Cycle in a Closed System,* to help you with this task.

Figure 3.16 **Snails and *Anacharis*.** How do these organisms interact to cycle matter in their environment?

NEED TO KNOW

ENGLISH SPANISH

Observing the Carbon Cycle in a Closed System

Plants and animals interact in a variety of ways. To limit the interactions that are possible in an experiment, you can conduct a study in a closed system. A **closed system** exchanges energy with its environment, but matter cannot move in or out of the system. For example, by setting up a closed system with an aquatic plant and an aquatic animal, you can study a portion of the carbon cycle. The **carbon cycle** is one of the chemical cycles in which matter moves from the environment to organisms and back to the environment in different chemical forms.

To help with your investigation, you will use pH. pH is a way to determine if a liquid is an acid or a base. When the gas carbon dioxide dissolves in water, it forms a weak acid, which lowers the water's pH. A decrease in pH indicates an increase in the concentration of carbon dioxide in your experimental system. Conversely, an increase in pH indicates a decrease in the concentration of carbon dioxide.

PROTOCOL

8. Carry out the following tasks to plan your investigation.
   a. Discuss with your team the materials available for you to use in your investigation, as well as the *Protocol for Monitoring pH in a Closed System.* Make additions to the diagram you drew in step 7b to capture ideas of how you might use your materials in an experimental setup.
   b. With your team, generate and choose a testable question and make a hypothesis about the cyclical movement of carbon in an aquatic ecosystem. Record your team's question and hypothesis in your science notebook.

      You must be able to test the question using the materials available.
   c. Plan an experiment to test your question. Outline your experiment in your science notebook. Your plan should include the following features.
      - Use of snails and sprigs of *Anacharis*
      - Use of appropriate controls

PROTOCOL

Protocol for Monitoring pH in a Closed System

1. Put 125 mL of distilled water in a 250-mL flask.
2. Use a pH probe or pH test strips to test the initial pH of the water. Record this starting pH.
3. Add water and organisms to your control and experimental test tubes. Handle organisms safely, according to your teacher's directions.
4. Perform the following steps to set up your test tubes to collect pH data.
   a. If you are using pH probes, calibrate them to a pH 4–11 range. Choose appropriate settings to collect data using 2 probes. Save your data and graph your results.
   b. If you are using pH probes, carefully position them in each test tube so that they do not interfere with your closed system.
   c. Adjust the pH probe (if used) and thermistor or thermometer (if used). Secure them in place by crimping a small piece of aluminum foil around the top of the test tubes.

      The aluminum foil also can serve to stop the exchange of matter between your closed system and its environment.

   d. If you will be using pH paper, set up your test tube with enough room to slide a pipet into the test tube and withdraw a water sample.
5. Set your test-tube rack next to a light source, as agreed upon by the class or directed by your teacher.
6. Use a pH probe or pH test strips to test the initial pH of the water. Record this starting pH.
7. Take pH and temperature readings for at least 2 hours. In your science notebook, record the pH readings every 5 minutes.
8. In your science notebook, monitor and record any changes you see in the test tubes. Note the times when they occur.
9. ***Wash your hands thoroughly with soap and water each time you leave the laboratory.***
10. Clean up your lab area as your teacher directs. Be careful to follow all directions for cleaning and storing the pH probes.

- Use of the *Protocol for Monitoring pH in a Closed System*
- Collection of data every 5 minutes, for at least 2 hours
- Inclusion of a plan for recording and graphing data

As you consider different designs, you may want to review the design of the earthworm habitats. Apply your understanding of that design to the one you develop here. Consider the controls you will need for your investigation.

d. Have your teacher approve your plan.

9. Conduct your experiment.
10. Generate a line graph showing how pH changed in your control and experimental test tubes over the time you tested. Color-code your graph. Add a legend and a caption so that anyone reading the graph can interpret your results.
11. Prepare a lab report of your findings. Include the following sections.
    a. A statement of the testable question and hypothesis
    b The procedure for carrying out the experiment, including the purpose of each test tube you set up
    c. A descriptive summary of your results, which refers to your graphed data
    d. A concluding paragraph or 2

    In your concluding paragraph, include a summary of the evidence from each test tube and what you learned from each test tube. Also include a paragraph with your best explanation that answers your question. Remember, explanations should include a claim, supporting evidence, and reasoning. Finally, include 1 or 2 sentences about questions you still have about the carbon cycle that you would like to answer.

HOW TO

See appendix C4, *How to Construct an Explanation,* for information that may help you write your concluding paragraph.

12. Work with your team to prepare a short presentation about your experimental design, results, and explanations. Make sure everyone has a role in the presentation.
13. With your team, present your investigation, results, and explanation to the class.

Analysis

1. Participate in a class discussion of the various investigations conducted and the results that emerged.
2. With your teammates, create a visual diagram, such as a concept map, that represents your current understanding of the cycling of matter through an ecosystem. Include specifics about how carbon, nitrogen, and water each cycle.

You might begin with the diagram you drew in step 3 of part A, *Reflections on the Earthworm Habitats,* and add to it a representation of the aquatic ecosystem that you studied in part B, *Snails and* Anacharis: *What Can I Learn from Them?*

Spinning the Web of Life

EXPLAIN

The next time you eat a hamburger, think about what it takes to make a pound of beef. You can see what goes into a hamburger by reading the scenario, *Recipe for 1 Pound of Beef.*

You have already explored how matter cycles through various ecosystems. You may also be aware that energy is stored in matter. When you eat various forms of matter, some of that stored energy is transformed. The transformed energy becomes available to you. It allows you to carry out the daily activities of life and the distinctive activities that make you who you are. Not all foods contain the same amount of energy. Not all organisms require the same amount of energy to live. So why does it take so much matter and energy to make 1 lb (0.45 kg) of beef? In terms of energy and matter, is there a difference between eating plants and eating animals?

Let's examine the larger picture of matter and energy in your world during this Explain activity, *Spinning the Web of Life*. By relating your food intake for one day to the plants and animals it came from (figures 3.17 and 3.18),

SCENARIO

ENGLISH SPANISH

Recipe for 1 Pound of Beef

Begin with 1 calf. Add the following ingredients over a period of about 2 years:

- 36 m^2 of grazing land
- 22.7 g of phosphorus
- 6.7 m^2 of farmland
- 45.4 g of potassium
- 5.9 kg of forage
- antibiotics
- 1.4 kg of grain
- hormones
- 181 g of soybeans

- pesticides
- 0.5 L (liters) of petroleum products
- herbicides
- 85 g of nitrogen
- 4,900 L of water, added regularly
- 43 million W (watts) of solar energy

Figure 3.17 Think of the many steps involved in putting bread or tortillas on your plate.

you will explore how the energy that is stored in matter flows through your community and fuels the activities of its organisms. You will consider if energy moves through an ecosystem in a way that is similar to or different from matter.

Materials (per person)

- large sheet of paper
- set of different-colored pens or pencils, 5 colors

Figure 3.18 Consider the sources of matter and energy that go into raising the eggs and chicken that may be part of your diet.

PROCESS AND PROCEDURES

1. Following the steps outlined here, generate a food web. A **food web** is a visual diagram of the interactions of organisms in an ecosystem.
   a. In your science notebook, create your ideal menu for 1 day. Include snacks as well as meals.

      Choose your favorite foods and snacks.

   b. Make a 3-column chart on a new page in your science notebook. Label the first column "Foods," the middle column "Ingredients," and the third column "Sources for ingredients." List the foods from your ideal menu in the first column and the ingredients in each menu item in the second column.

      Remember, many foods are combinations of different plants, animals, fungi, protists, or eubacteria. Record the ingredients of each food separately. For example, if you have a piece of cake for dessert, you should list oil, flour, sugar, butter, eggs, and milk.

   c. In the third column of your chart, list the sources for the ingredients of each menu item.

      For example, write "wheat" next to "flour," "sugarcane" or "sugar beet" next to "sugar," and "chicken" next to "egg."

   d. Create a second 3-column chart in your science notebook, on a new page. Label the *second column* "Prey." List all of the different animals that appeared in your first chart.

   e. Label the third column "Foods prey eats." For every animal that you have listed, do the following.
      i. In the third column, list several foods that it eats.

         For example, next to "cow," you would list the grass and corn that cattle eat. If you need help with this step, check with your teacher or your teammates.

      ii. Label the first column "Predator." In the first column, next to each animal that you listed in the second column, record 2 animals that might eat that animal.

         For example, next to "cow," you might list "wolf," "coyote," "mountain lion," or "human." Next to "fish," you might list "raccoon," "otter," or "seagull," depending on the type of fish it is. Again, if you need help, check with your teacher or teammates.

   f. Create a food web from the information that you generated in your science notebook, as follows.
      i. Obtain a large sheet of paper from your teacher. Fold it like an accordion to create 5 equal-sized sections (figure 3.19). Gather 5 different-colored pens or pencils. Use a different color for each section.

Figure 3.19 Obtain a large sheet of paper. Fold it to create five equal-sized sections. You will start your food web diagram in the bottom section by listing all the names of the plants.

Figure 3.20 Classify the organisms in your charts by their diets. For example: (*a*) Herbivores primarily eat plants. (*b*) Carnivores primarily eat animals. (*c*) Omnivores eat both plants and animals. (*d*) Decomposers primarily feed off dead organisms and the wastes or cast-off fragments of living organisms.

ii. Turn your paper so the sections go from top to bottom. In the bottom section, list all the names of the plants that you listed in the third column of your chart from step 1c.

iii. Write the names of all the herbivores (plant eaters) in the next section, above the plants (figure 3.20).

iv. Write all the names of the omnivores (animals that eat plants and animals) in the next section, above the herbivores.

v. Write all the names of the carnivores (animals that eat only meat) in the section above the omnivore row.

vi. Write your name in the row that is appropriate for your diet.

g. After you have all the names of the plants and animals organized, draw arrows from the organisms that *provide* energy to the organisms that *receive* energy.

Pay particular attention to the direction that you draw your arrows. They should show the direction of energy flow, *not* who will be eaten. Say, for example, that your plant row includes grass and your herbivore row includes cattle. You would draw an arrow from grass to cattle with the point of the arrow aimed at the cattle because grass provides energy to cattle.

2. Deepen your understanding of food webs by reading the essay, *Let's Ask Drs. Ricardo and Rita*. As you read, enter the following terms into your personal glossary.
   - herbivore
   - carnivore
   - omnivore
   - producer
   - primary consumer
   - secondary consumer
   - decomposer
   - biomass

Let's Ask Drs. Ricardo and Rita

ENGLISH SPANISH

Drs. Ricardo and Rita, longtime colleagues at the same university, answer questions and concerns about relationships in communities.

Dear Dr. Ricardo,

I am a biology student who has just completed drawing a food web for class. As I was making it, I learned that a food web *is the sum of all feeding interactions. I've included a sample (figure 3.21). Does it look right to you? My teacher keeps talking about communities, but I'm not sure what they are. How do communities relate to this food web?*

Sincerely,

Juanita Perez

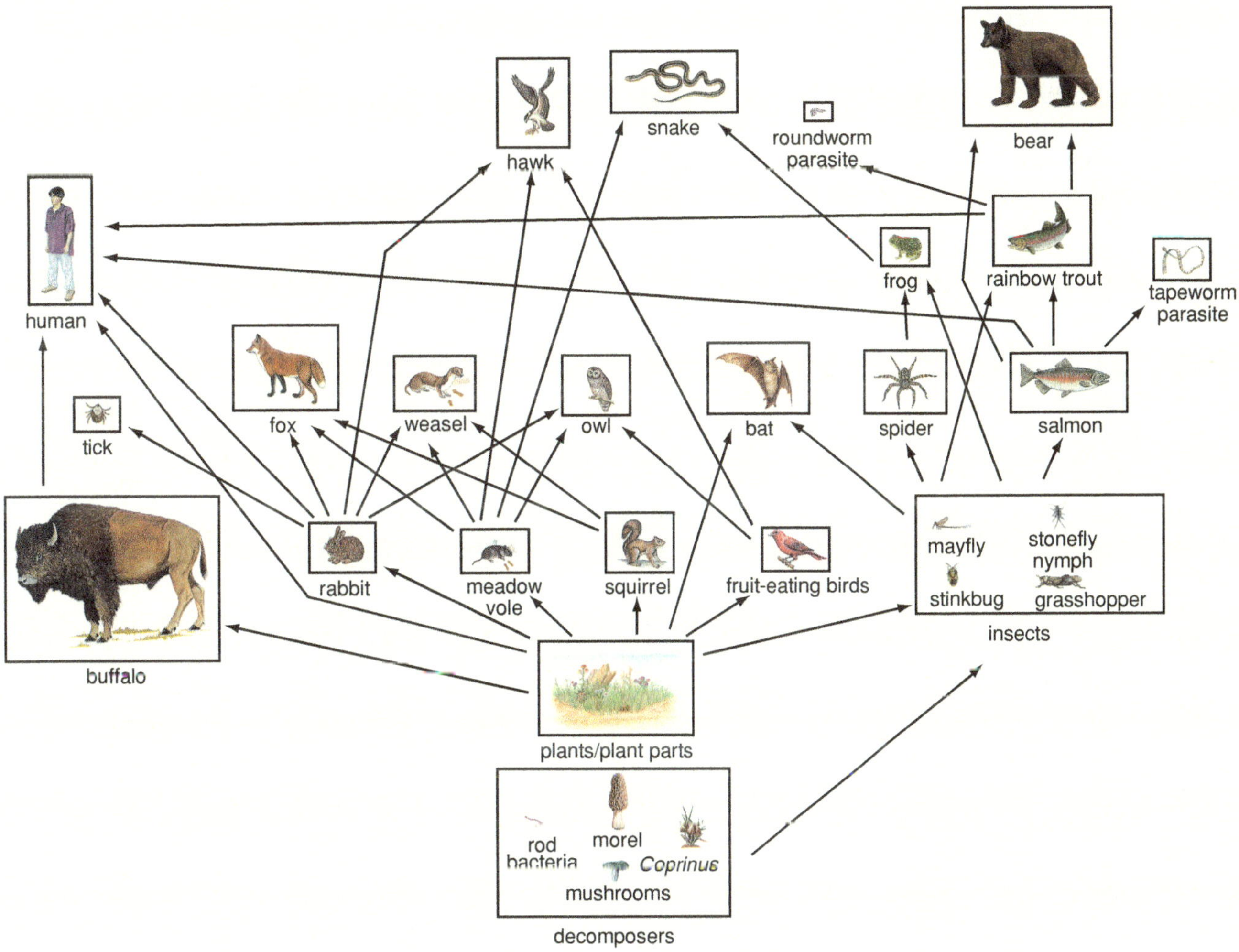

Figure 3.21 **Juanita's food web.** Food webs in a community can be very complex. Can you find any more relationships?

Dear Juanita,

Your food web is very detailed. It shows that you understand the relationships between different organisms. Because you included a large variety of organisms in your food web, you have also drawn a nice visual picture of a community. A *community* is a collection of organisms that live and interact with one another in a given area. You also described what biologists call biomass. **Biomass** refers to the mass of all living organisms in a given environment. As you probably know, some herbivores may not eat an entire plant. For example, a rabbit may eat only part of a violet. A carnivore, such as a mountain lion, may not eat the bones of the rabbit. What these creatures eat is considered the **consumable biomass**.

What happens to the remaining parts of the violet and the bones of the rabbit? These types of biomass, the leftovers, often represent a substantial portion of the total amount of energy and biological material in a community. This resource does not go to waste, however. Remember the **decomposers** in your food web? These organisms (for example, fungi and bacteria) specialize in using the matter that other organisms do not consume. Though generally not as visible as the other groups, these decomposers serve a vital function in communities. They reduce dead biomass, such as the partially consumed violet and rabbit, to molecules that the producers can reuse. So, you see, you could have drawn arrows from all the creatures to the decomposers. Because some small animals eat decomposers, you can also draw arrows from the decomposers to some of the insects.

Dear Dr. Rita,

I drew a food web in my class. But I still don't understand what it means in terms of the real world. Can you make this more realistic for me?

Sincerely,
Alfonso Washington

Dear Alfonso,

Thanks for your comment. I would love to help. Think of a river that may run near your community. Study the detailed drawing of such a riverbed (figure 3.22).

In the open water of the river, algae and microscopic water plants are the main **producers**. They make their own food through photosynthesis. These producers are consumed by insects and small aquatic animals such as the tiger salamander and various fishes. They are called **consumers** because they feed on other organisms. Higher-level consumers, such as the great blue heron, live by eating either first-level consumers (organisms that have eaten the producers) or smaller predators. Note that at every step in the process, some biological matter passes to the decomposers. The decaying matter they consume may be plant or animal parts, organic waste products, or whole dead organisms. The decomposers then break down this matter to simple molecules. In turn, the producers use some of these simple molecules.

The food web of the river ecosystem is based on solar energy that the producers convert into food. Not all ecosystems, however, support themselves in this manner. The seashore is a good example of a rich, natural ecosystem that is not self-supporting. Virtually all of the organisms that inhabit the zone between high and low tides are consumers. They ultimately depend on plant material and other living or dead organic matter that is brought in by each high tide. Many of the seashore creatures feed on this matter directly. For example, mussels and clams filter seawater for microscopic organisms and other tiny bits of food. Many other species are predators, such as the sea star, which preys on clams and mussels. The organisms in this community also depend on solar energy. But most of that energy is converted into food by producers that live in offshore ocean communities.

Figure 3.22 **Cross section through the edge of a riverbed in the southwestern United States.** The interactions among organisms in an ecosystem involve the cycling of matter and the flow of energy. What is the ultimate source of energy?

Dear Dr. Rita,

I am worried that the food web I drew is all wrong. When I drew arrows between organisms, I pointed the arrow toward the organism that does the eating. But my friend drew the arrow pointing to the thing being eaten. Who is right? What do these arrows really mean?

Sincerely,
Dwayne Robertson

Dear Dwayne,

Often, students who draw food webs for the first time want to draw the arrows pointing toward the organisms being eaten. But in this case, your food web is correct. Remember that the arrows represent energy flow, not the act of eating. For example, energy from grass is passed to a rabbit. The energy from the rabbit is passed to a mountain lion. Think of the arrows as meaning "the energy is passed to." A food web helps us understand the flow of energy through a community. Ultimately, the Sun is the source of energy for nearly all Earth communities.

3. Expand your food web in the following manner.
   a. Think of other organisms that might compete with you for your food. In your science notebook, make a list of organisms that eat some of the same foods that you eat.
   b. Add the organisms that you listed in step 3a to the appropriate level of your food web. Add the appropriate arrows.
   c. Add decomposers to the top section of your food web. Use arrows to indicate the relationship between these and the other organisms present.

      If you had mushrooms in your list of foods, they belong in this section.

4. In the next 3 steps, you will consider the flow of energy through ecosystems. Use figure 3.23 and your understanding of the concepts from this chapter to answer the questions in steps 4a–d.

   In the previous steps, you considered the cycling of matter in ecosystems. But energy flows through ecosystems. It is not cycled. This fact has important implications for ecosystems. The diagram in figure 3.23 shows the amount of energy available at different trophic levels in an ecosystem. Notice the pyramid shape, with the largest amount of energy available at the level of producers, less energy available to the first-level consumers, and even less energy available to the higher-level consumers.

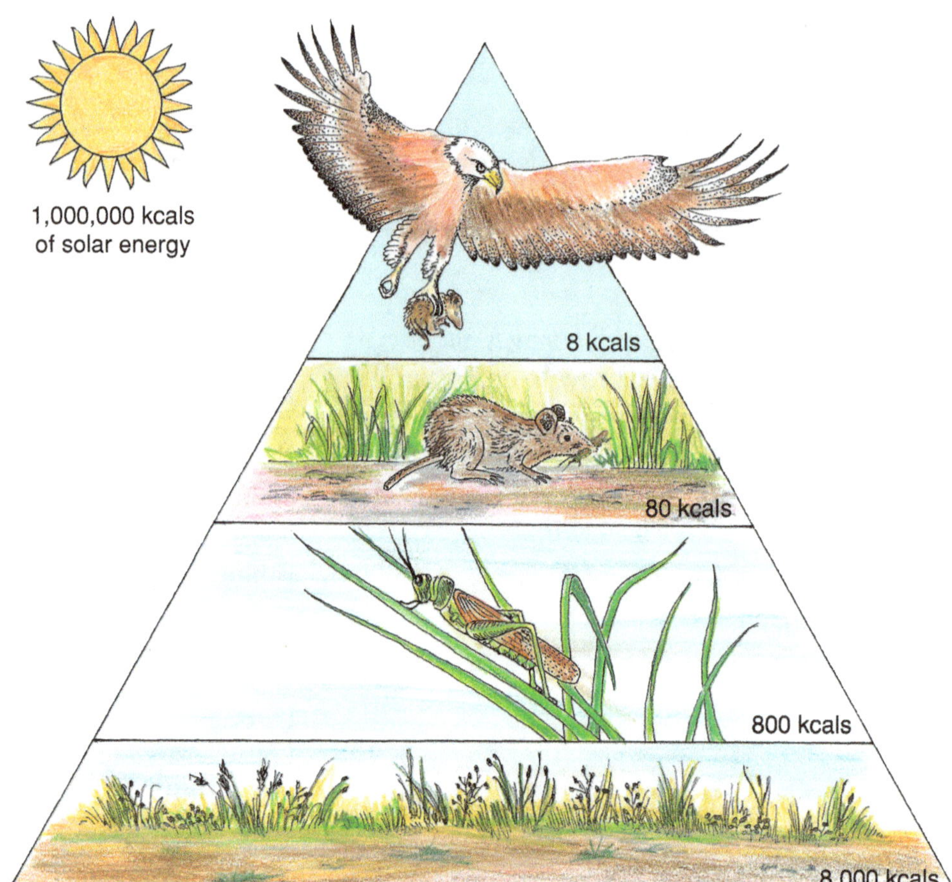

Figure 3.23 Energy pyramid. This idealized energy pyramid illustrates that only a portion of the energy available at one trophic level is available at the next-higher trophic level.

a. An examination of the diagram shows that energy seems to be missing in transitions among trophic levels in ecosystems. The first-level consumers have less energy available than the producers. But you know that energy is neither created nor destroyed, only changed in form. So the energy isn't really lost in the perspective of the universe. In what form or forms do you think the "lost" energy might be?

b. A famous ecologist named Paul Colinvaux wrote a book titled *Why Big Fierce Animals Are Rare*. If you asked a typical seventh grader why big animals are rare, what do you think the answer would be? How might you use figure 3.23 to help the student answer the question?

c. Explain to the same seventh grader why light from the Sun must be continuously supplied to most ecosystems.

d. How is figure 3.23 a scientific model for energy flow through an ecosystem? Explain why it is or is not a good model.

Do all organisms fit nicely into one trophic level?

5. To learn more about energy flow, read the essay, *Energy Flow in Ecosystems*. Study figure 3.24 carefully, as it may help you complete the task in step 6.

Energy Flow in Ecosystems

ENGLISH SPANISH

Each time that an organism uses energy, some energy is released to the external environment as heat. This means that only a portion of the solar energy that producers take up is stored in biomass that herbivores can eat. In turn, only a portion of the plant material that herbivores eat is converted into body parts that could become food for predators. Each transfer of energy from one organism to another results in a decrease in the amount of energy that is available (see figure 3.24).

Because some energy is released as heat with each transfer to a different organism, sometimes the energy in ecosystems is illustrated using a pyramid-shaped diagram. The producers are located at the bottom level of the pyramid and contain the most total energy. Consumers that eat those producers, also known as primary consumers or herbivores, contain a smaller amount of energy. Secondary consumers or predators are on the next-smallest level of the pyramid (figure 3.25). In any ecosystem, the number of levels in the pyramid is the same as the number of links in the longest possible food chain, with the top-level predators at the top. Each of these levels in the pyramid is called a **trophic level**. "Troph" comes from a Greek word that means "food" or "nourishment" (figure 3.23).

What about omnivores? What about decomposers? What trophic levels do they occupy? They can occupy more than one. A decomposer can exist in the second trophic level when it breaks down the remaining biomass of a plant that was partially eaten by an herbivore. The same

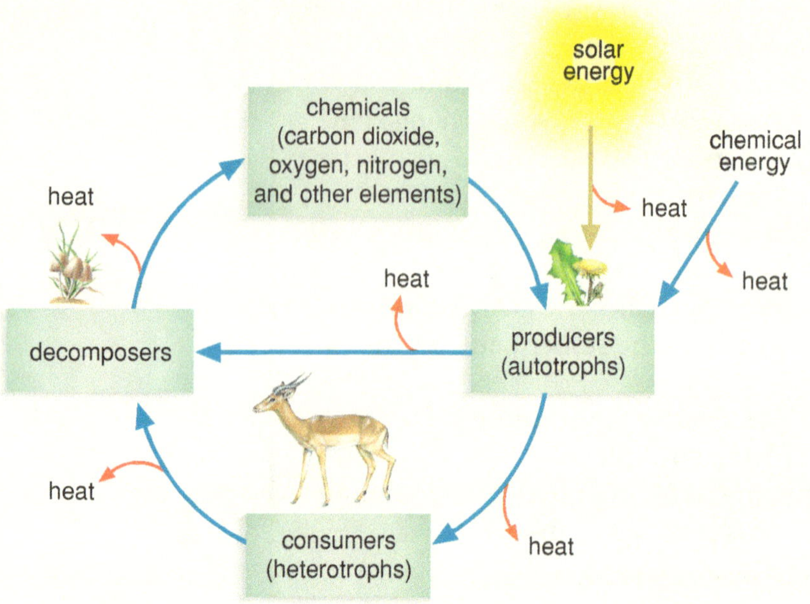

Figure 3.24 **Energy flow in a community.** As energy is transferred from one organism to another, some energy is always released to the environment as heat.

Figure 3.25 This mantis is a secondary consumer. The butterfly is a primary consumer because it eats from plants.

decomposer also might exist in the third trophic level when it breaks down the bones of a carnivore. Omnivores also occupy different trophic levels. When you eat a salad and a steak for a meal, you are occupying two trophic levels.

Scientists Can Estimate Energy in a Community

Ecologists can estimate the amount of energy that is stored in the biomass at each trophic level of a community. They do this by taking a sample of organisms and calculating the total biomass represented by the producers, herbivores, and higher-level consumers. They then estimate the amount of potential energy (measured in kilocalories) in this organic matter.

With these data, you can illustrate the amount of energy entering and leaving an ecosystem at each trophic level. Study figure 3.26 to see how energy is distributed from one trophic level to another. Notice that producers are the foundation. Each trophic level above the producers receives less energy. In some ecosystems, each trophic level receives about one-tenth of the energy of the level below it. Where does the energy go? Some is used to fuel activities needed to keep organisms alive, while some dissipates into the environment as heat.

Energy in the Food Humans Eat

The observation that some energy is transferred to heat when it flows from one trophic level to the next is the basis for the suggestion that people should eat at a lower trophic level when possible. For example, a person eating as an herbivore can create 1 kg of human biomass by eating 10 kg of plants. It takes about 100 kg of plants, however, to create 1 kg of human biomass if the person eats as a primary carnivore. You are eating as a primary carnivore if you eat the cow that ate the plants, instead of eating the plants directly (see figure 3.27).

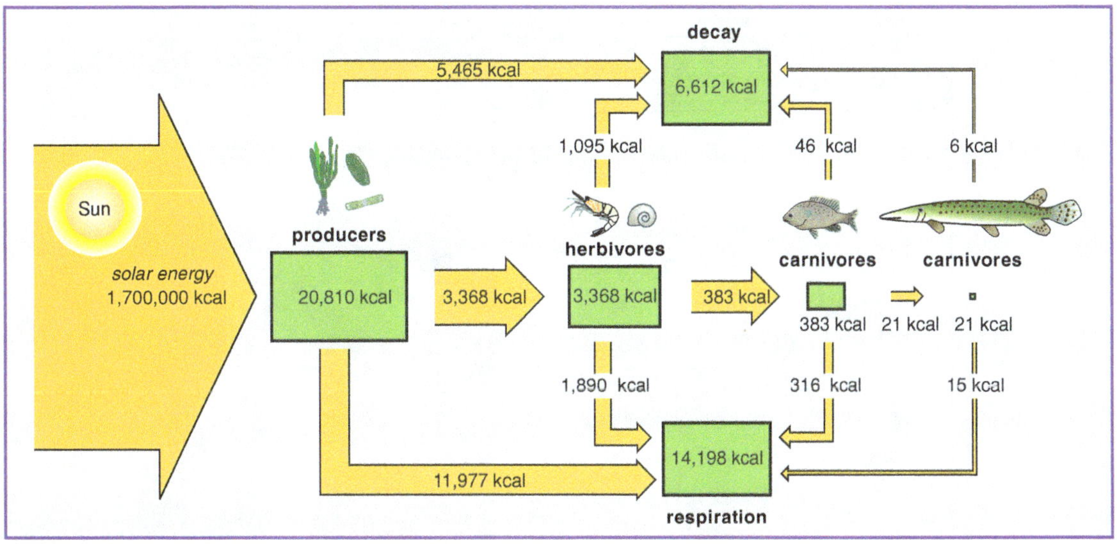

Source: Based on data gathered by H. T. Odum from a river ecosystem in Silver Springs, Florida.

Figure 3.26 Energy flow diagram for Silver Springs, Florida. How does the amount of energy (in kilocalories, or kcal) differ from one trophic level to the next? Can you calculate what percentage of energy is transferred from producers to herbivores?

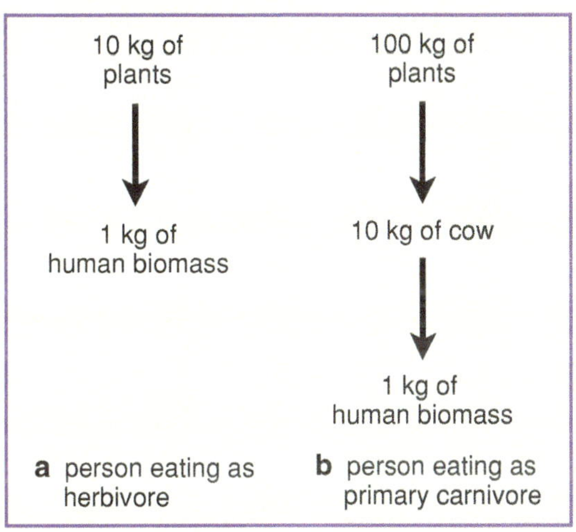

Figure 3.27 Human as herbivore and human as primary carnivore. The energy relationships between trophic levels are the basis for the suggestion that people should eat low on the food chain.

6. Work with a partner to generate a diagram that summarizes, in a way that makes sense to you, the data about the flow of energy in the Hubbard Brook ecosystem. Use information and data from the need to know, *Energy Flow in the Hubbard Brook Forest,* to complete the task. Your final diagram should include the following.
   a. Producers, primary consumers, and secondary consumers that are clearly labeled
   b. Boxes or levels that are scaled to represent relative amounts of energy

      In this context, "scaled" means that the boxes or levels are drawn in proportion to the amount of energy available at each trophic level. For example, if there is 10 times more energy available at the producer level than the primary consumer level, then the box you

NEED TO KNOW

ENGLISH SPANISH

Energy Flow in the Hubbard Brook Forest

The Hubbard Brook Experimental Forest in New Hampshire has been the subject of ecological studies since the 1950s (figure 3.28). This forest has been designated as a Long Term Ecological Research (LTER) site. In 1978, a researcher named James Gosz and many fellow scientists attempted to quantify energy flow throughout the site's different trophic levels. The data that they collected follow (figure 3.29). Note that all the data are in kilocalories per square meter per year ($kcal/m^2/yr$).

Figure 3.28 A hardwood forest like those studied at Hubbard Brook, White Mountain National Forest, New Hampshire. Scientists have carefully tried to account for the flow of energy and cycling of matter in this forest.

Measurement	Energy available to different trophic levels ($kcal/m^2/yr$)
Solar energy available	480,000
Solar energy reflected	72,000
Solar energy converted to heat	136,800
Solar energy used to drive evaporation of water from surfaces or through plants	201,600
Solar energy converted by plants into chemical energy	9,600
Energy in the form of producer biomass available to primary consumers	4,800
Amount eaten by primary consumers	3,500
Energy in the form of primary consumer biomass available to secondary consumers	200

Source: Gosz, J. R., Holmes, R. T., Likens, G. E., and Bormann, F. H. (1978). The flow of energy in a forest ecosystem. *Scientific American, 238*(3), 92-102.

Figure 3.29 Data from Hubbard Brook Experimental Forest. Use these data to trace the flow of energy in this ecosystem.

draw for the producer level should be 10 times bigger than the box for the primary consumer level.

c. Data about the energy that is available at each level
d. An indication of what supplies the system with energy and where "lost" energy goes

You might pattern your diagram after an energy pyramid like the one in figure 3.23 or the energy flow diagram like figure 3.26.

Analysis

Participate in a class discussion of the following. Record your ideas in your science notebook.

1. Where is the most energy available in your food web? Explain your answer.
2. Why does it take more biomass to produce beef, compared with chicken or eggs?

   It takes 16 lb of grain and soybeans to produce just 1 lb of beef. It takes 6 lb of grain and soybeans to produce a pound of pork, 4 lb for a pound of turkey, and 3 lb for a pound of eggs or a pound of chicken.
3. Compare the food web of a vegetarian in your class with the food web of someone who is not a vegetarian.
   a. What differences are evident?
   b. What do you think is significant about those differences?
   c. How does this relate to the recipe for 1 lb of beef?
4. Humans can eat organisms from multiple trophic levels. Why do you think top-level consumers (such as sharks) are rarely a part of a human diet?

Generating Some Heat

ELABORATE

We have learned that some energy leaves the food system as you move from one trophic level to the next. This energy is in the form of heat. Read about a surprising way to "see" heat leaving a system in the scenario, *Saturday Chores*.

SCENARIO

ENGLISH SPANISH

Saturday Chores

It's Saturday. You have a soccer game at noon. You promised your friends that you would be there a half hour early to practice. But you know you have to finish your Saturday chores before you leave. This is your week to cut the grass, so you get up early to begin the task. Because the morning is cool, you find yourself making quick progress. But time is passing, and practice is about to begin. In an effort to shave time off the job, you pile all of the clippings on some newly turned soil in a corner of the yard. You think, "I'll bag the clippings later." Saturday and Sunday come and go.

School begins again on Monday. The clippings still are sitting there on the soil in the yard (figure 3.30). Another Saturday rolls around, and you have another game. It's your sister's turn to cut the grass. Good! No grass today! Your game is early in the morning. On the way out of the house, your dad reminds you that last week's clippings are still sitting in the yard and that it is your responsibility to bag them before you head out. You run to the yard with the yard waste container and begin scooping up grass clippings (figure 3.31). Again, the morning is very cool. As you dig into the pile, you notice some-thing interesting. Steam is rising from the clippings. In fact, the center of the clippings pile is hot, especially down near the soil. You wonder why the inside of the pile is so warm on such a cool morning.

Figure 3.30. After a week, grass clippings are still sitting in the yard.

Figure 3.31 Fresh-cut grass that is piled after mowing gets hot after a few days.

Figure 3.32 Organic materials can be composted. This causes the matter to be cycled more quickly than if it were left to decompose on its own. Where does the energy from the composted matter go?

From your experiences in this chapter, what do you think generated the heat in the pile of grass clippings? In this Elaborate activity, *Generating Some Heat*, you may find an answer to this question as you elaborate on your knowledge of the cycling of matter and the flow of energy in ecosystems. You will look, primarily, at the role of decomposers. You and your teammates are about to participate in a compost design competition. Your goal is to design a compost system that generates the most heat (figure 3.32). As you complete the challenge, consider the question, "How is the flow of energy in an ecosystem affected by decomposers?"

Materials (per team of 4)

4 pairs of plastic gloves
1 measuring cup that holds approximately 250 mL
1 thermistor or nonmercury thermometer
1 compost column (made from 2-L bottles)
1 bag of potting soil
1 bag of shredded leaf or bark mulch
1 bag of grass clippings
1 box of compost starter inoculum (a mix of nutrients and microorganisms that is effective at decomposition)
other organic materials as needed
water
clear packing tape or masking tape
duct tape
insulation material (such as old towels)
1 trowel or spatula for mixing compost
1 plastic trash bag
1 pair of old nylons (optional)
rubber bands (optional)
graph paper
different-colored pens and pencils

PROCESS AND PROCEDURES

1. As a class, make the following decisions about your compost designs.
   a. How much compost starter inoculum all teams will use
   b. How much water all teams will use
   c. How much total organic matter (the mass) all teams will use

   Why is it important that all teams keep certain features constant?

2. Now you are ready to design your compost recipe. What ingredients will help you win the contest? Discuss the following questions with your team to help you decide on the best recipe. Record your recipe in your science notebook. Be sure to give reasons for your decisions. Have your teacher approve your recipe.
   a. What organic matter should you use?
   b. Should you use 1 source of organic matter or a combination of sources?
   c. Should you pack your compost column loosely or tightly?

   As you develop your recipe, practice the working-relationship skill of reaching consensus. Think about processes in organisms that give off heat. Can you use what you have learned to predict which kinds of organic matter would decay the most quickly?

3. Create your composting system with your team (figure 3.33).

SAFETY: Put on your plastic gloves.

   a. Use the recipe that your team agreed on. Using the measuring cup, measure your soil and your organic matter 1 at a time from the materials you have available. Pour them into the plastic bag.
   b. Mix the ingredients in the plastic bag thoroughly with gloved hands.
   c. Add the amount of compost starter inoculum that your class decided to include.
   d. Add the amount of water that your class decided to include.
   e. Use the trowel to mix the compost ingredients thoroughly.
   f. Remove the upper portion of the compost column and place your mixed compost into the lower portion. Place the top back on the compost column and tape the lid into place.
   g. Insulate your compost column, as instructed by your teacher. Make sure the airholes on the bottom are not covered.

Figure 3.33 Assembling a compost column. It is important to mix the materials your team uses.

4. Make a data table in your science notebook to record temperature data twice a day over 3–4 days. Record the initial temperature of the compost column in your table. Continue to record temperatures over the next few days. The winning team will be the one that has the greatest change in temperature in its compost column.
5. Graph the temperature changes that you observed in your composting system. Add highlight comments and a caption to your graph.

HOW TO

For help with this task, refer to appendix D3, *How to Create Graphs,* and appendix B6, *How to Write Highlight Comments and Captions.*

6. Construct an explanation of what happened in your compost column. Use the Explanation Tool, as seen in figure 3.34, to organize your ideas. Make a claim and use evidence from your graph to support the claim. Include relevant science ideas that you have learned in this chapter.

   If you need help constructing your explanation, refer to appendix C4, *How to Construct an Explanation.*

HOW TO

Question	
Evidence from data and observations	**Reasoning, including science ideas**
Claim ***(Your claim should answer the question.)***	
Explanation ***(Make sure to link your claim, evidence, and science ideas.)***	

Figure 3.34 **The Explanation Tool.** The Explanation Tool can help you organize your ideas.

Analysis

Complete the following tasks as a team.

1. Present your team's compost recipe and graph to the class. Use your highlight comments to help guide what is most important to present.
2. Based on all of the teams' reports, determine which compost recipes generated the most change in temperature.

   You and your classmates will judge the effectiveness of each composting recipe by examining each team's data.

   a. Why do you think the most effective recipes worked so well?
   b. What role did the microorganisms play?
3. The heat that you noticed is a form of energy. What form did this energy take before it was released as heat? What happens to this energy after it is released as heat?
4. The compost bottles and the heat they released help you think about the flow of energy in a community.
   a. What do your answers to the questions in number 3 reveal about the flow of energy in a community?
   b. How did the compost bottles serve as a model of a community to help you answer the question in 4a?
   c. What are the strengths and limitations of this model?

EVALUATE

Energy, Matter, and Disaster

By now you should be aware of how closely connected the flow of energy is to the cycling of matter in ecosystems. In this Evaluate activity, *Energy, Matter, and Disaster*, you and your teacher will evaluate what you have learned about these concepts. As you learned in the Explore activity, *Matter Goes Round and Round*, a major natural disaster can result in an unbalanced ecosystem that leads to dramatic changes. In this activity you will work individually to think about the impact that a natural disaster would have on various communities on Earth. Then you will respond to some questions about survival in different communities.

PROCESS AND PROCEDURES

1. Read about the following fictional catastrophe.

   Earth is entering a phase of instability that no one had predicted. Throughout both hemispheres, hundreds of volcanoes are erupting with great force (figure 3.35). Earth's atmosphere is thick with minute volcanic debris and dust. As much as 75 percent of the sunlight is now blocked from reaching Earth's surface. This period of eruptions is

Figure 3.35 The community around this volcano was drastically changed after the eruption.

expected to continue indefinitely. It is likely that soon virtually all sunlight will be blocked from reaching Earth's surface.

2. Review the *Energy, Matter, and Disaster Rubric* handout. Be sure to ask your teacher any questions you have about it.
3. Read the following questions and record your answers in your science notebook. Be sure to answer each part of each question.
   a. What might be the effect if 80–85 percent of the sunlight were blocked from Earth? What might be the short- and long-term effects on the following organisms: an earthworm, a shark, a maple tree, a saguaro cactus (figure 3.36), and a teenager?
   b. From the thousands that sprout, why will only 1 or 2 healthy trees grow into the available space between other existing trees?

      How is this pattern related to the availability of matter and energy on Earth's surface?
   c. Imagine that all sunlight is blocked from reaching Earth's surface.
      i. What might be the short- and long-term effects on the following organisms: the producers, the consumers, and the decomposers?
      ii. Describe how the cycling of matter through an ecosystem would be affected.

Figure 3.36 A saguaro cactus.

Analysis

Work with a partner to discuss the following questions about survival in different communities. Then record your responses in your science notebook.

1. Human societies that live by hunting and gathering usually have much smaller populations than groups in a similar setting that live primarily by growing crops. Why do you think that is so?

2. Suppose you found yourself snowed in for the winter in a remote mountain cabin with no way of contacting the outside world. You must survive for several months with only what is on hand to eat. Aside from a small supply of canned peaches, your only resources are 2 100-lb sacks of wheat and a flock of 8 hens. Discuss the relative merits of the following strategies.
   a. Feed the grain to the hens and eat their eggs until the wheat is gone. Then eat the hens.
   b. Kill the hens at once and freeze their carcasses in the snow. Live on a diet of wheat porridge and chicken.
   c. Eat a mixture of wheat porridge, eggs, and 1 hen a week. Feed the hens well to keep the eggs coming until all of the hens are killed.

3. Every breeding pair of bullfrogs produces hundreds of eggs each spring. During the time they are growing up in the pond, the small tadpoles feed entirely on microscopic water plants. Predators living in the pond eat a large fraction of the tadpoles before they transform into frogs (figure 3.37). As adults, however, bullfrogs themselves are predators. Discuss why this strategy is more advantageous than one in which the tadpoles would be predators and the adults would be herbivores.

Figure 3.37
Bullfrog. Although adult bullfrogs are predators, the tadpoles are prey. Many tadpoles are eaten before they reach adulthood.

Further Challenges

1. There are several hypotheses about how dinosaurs became extinct. One of these hypotheses involves a climatic catastrophe that has some similarities to the one presented in step 1. See what you can find out about this hypothesis. Report your findings to the class.
2. Write and perform a skit that depicts organisms defending their role in a community. In it, describe the advantages, the disadvantages, and the importance to the community of a producer, a predator, and a decomposer.
3. DDT (a pesticide) was widely used for many years in the United States. This chemical does not break down in the bodies of organisms that ingest it. Therefore, it accumulates over an organism's lifetime. When DDT was being used, the populations of birds of prey, such as bald eagles, peregrine falcons, and ospreys, declined more than the populations of small songbirds. Explain why you think that was so.
4. Select a local ecosystem to investigate. Possible ecosystems include the intertidal environment at Fitzgerald Marine Reserve, wetlands found along San Francisco Bay, or the grasslands located in the San Bruno Mountain Ecological Reserve. Construct a food web that illustrates the flow of energy among the organisms that reside within the ecosystem you chose.

CHAPTER 4

Matter and Energy for Organisms

Life's splendor forever lies in wait about each one of us in all its fullness, but veiled from view, deep down, invisible, far off.

Franz Kafka

Gasoline, nuclear power, and electricity. All are obvious forms of energy. But do you also think of energy when you see sunlight or green plants? In fact, energy is found wherever matter is organized, from the molecules of rocks, to cells, to entire living organisms. Most people are not aware of the complex role that energy plays in living systems. Without energy, life would not be possible.

In chapter 4, *Matter and Energy for Organisms*, you will begin to investigate the important relationship between matter and energy. You will consider examples of where and how energy is stored and released. In chapter 3, *The Cycling of Matter and the Flow of Energy in Ecosystems* you studied the concept that food contains energy that humans and other organisms require. Eating and digestion, however, do not explain exactly how food molecules become useful as energy. How is energy in food converted to fuel the chemical reactions that keep cells, tissues, and organisms alive? What is the original source of the energy in food? Once you understand the basic connections between energy and matter, you will explore some of the specific reactions that occur in organisms to convert energy from food into a form that their bodies can use. You will also study the chemical reactions that convert energy from light into energy in food in the first place.

GOALS FOR THE CHAPTER

Throughout chapter 4, you will work to gain an understanding of the following concepts:

- ✔ Matter and energy are related.
- ✔ Energy exists in different forms, such as heat, light, and chemical.
- ✔ Photosynthesis converts light energy into chemical energy, and it converts carbon dioxide and water into energy-storage molecules.
- ✔ Cellular respiration breaks down food molecules and releases energy.
- ✔ Biological and geological processes have influenced each other throughout geologic time.

To help you develop these understandings, you will complete the following activities.

ENGAGE	Releasing Energy
EXPLORE, EXPLAIN	Energy in Matter
EXPLAIN	Keep a Body Running!
EXPLAIN, ELABORATE	Using Light Energy to Build Matter
ELABORATE	Modeling Photosynthesis and Cellular Respiration
ELABORATE, EVALUATE	Coevolution of Earth Systems and Life on Earth

Throughout chapter 4, remember to refer to the chapter organizer to help you remember what you have learned and where you are headed.

ENGAGE

Releasing Energy

Key Idea: Food contains enormous amounts of energy.

Linking Question:
What is the relationship between matter (such as food) and energy?

CHAPTER 4

Matter and Energy for Organisms

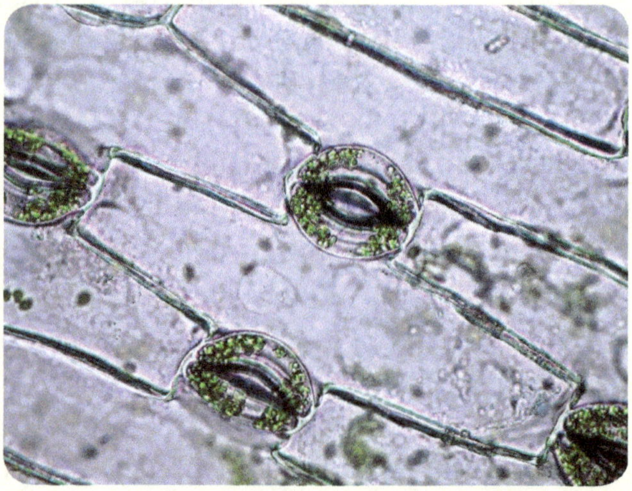

EXPLORE EXPLAIN

Energy in Matter

Key Idea: Changes in matter result in changes in energy, yet both energy and matter are conserved.

Linking Question:
How do organisms transform energy from food to build and fuel their bodies?

EXPLAIN

Keep a Body Running!

Key Idea: Food contains energy. Organisms use cellular respiration to generate small molecules that carry usable energy from food.

Linking Question:
Where does the energy stored in food come from, originally?

ELABORATE EVALUATE

Coevolution of Earth's Systems and Life on Earth

Part A: Take a Stand
Part B: The Intersection of Geology and Biology
Part C: Geological History of Oxygen
Part D: Defend Your Stand

Key Idea: Earth is a dynamic planet and geological and biological processes have influenced each other over geologic time. Scientific evidence and reasoning can be used to support a developed claim.

Linking Question:
How have the processes of photosynthesis and cellular respiration influenced geological and biological changes on Earth?

MAJOR CONCEPTS

- ✔ Matter and energy are related.
- ✔ Energy exists in different forms, such as heat, light, and chemical.
- ✔ Photosynthesis converts light energy into chemical energy, and it converts carbon dioxide and water into energy-storage molecules.
- ✔ Cellular respiration breaks down food molecules and releases energy.
- ✔ Biological and geological processes have influenced each other throughout geologic time.

ELABORATE

Modeling Photosynthesis and Cellular Respiration

Part A: Modeling Photosynthesis
Part B: Modeling Cellular Respiration in Plants and Animals
Part C: Energy and Matter for Your Critter

Key Idea: The process of photosynthesis produces sugar molecules that store energy. Plants and animals can use that energy after breaking apart the sugar molecules through the process of cellular respiration.

Linking Question:
What is the connection between photosynthesis and cellular respiration?

EXPLAIN ELABORATE

Using Light Energy to Build Matter

Part A: Seed to Tree
Part B: Measuring Photosynthesis

Key Idea: Photosynthesis transforms light energy into chemical energy in the form of sugar.

ENGAGE

Releasing Energy

A runner in a marathon is pushing hard several miles before the finish line. But then she suddenly slows to a walk, clutches her side, and sways dizzily, about to fall. This athlete is said to have "hit the wall." She did not actually slam into a brick barrier, but she may feel as if she has done so. In this case, hitting the wall means that the runner has exhausted the energy supplies necessary to keep running.

After eating food and resting, those energy supplies will be replenished. Yet how does this additional food become the energy needed for physical activity? In this Engage activity, *Releasing Energy*, you will begin to examine the relationship between matter and energy. In the scenario, *A Matter of Explosions*, the matter you will examine is the molecules from a plant people use for grain (see figure 4.1). Where is the energy?

PROCESS AND PROCEDURES

1. Read or listen to the scenario, *A Matter of Explosions,* as your teacher instructs.
2. Consider the following questions. In your science notebook, write your ideas about how to answer them.

   Remember that this is the Engage section of the chapter. You are not expected to know the answers to all of these questions now, but you are expected to write down your best ideas at this time.

   a. Where did the energy for the grain explosion come from?
   b. How can energy be stored in grain?
   c. What started the explosion that released the energy?
   d. Why don't you explode when you eat grain products such as cereal or bread?
   e. What would you do to decrease the danger of explosion and better protect a grain-storage facility and its employees?
3. Contribute your thoughts about the questions to a class discussion.

Figure 4.1 Grain-handling facilities are required by the Occupational Safety and Health Administration (OSHA) to meet certain standards that reduce the risk of safety and health hazards.

SCENARIO

ENGLISH SPANISH

A Matter of Explosions

Brad celebrated his 20th birthday by starting a new job at the town's grain-storage facility. He was eager to make a good impression. Washington County Grain Cooperative was the largest employer in town. The company offered good wages, great benefits, and opportunities for advancement. The facility consisted of three huge grain elevators (see figure 4.2) that stored wheat. Each storage elevator was shaped like a

Figure 4.2 Grain-storage elevators hold massive amounts of grain.

giant cylinder that rose 30 m (about 100 ft) aboveground and descended 6 m (20 ft) underground. A series of tunnels connected the underground portions of the cylinders. The tunnels contained machinery that dropped the grain onto concave conveyer belts and moved the grain from one storage tower to another. In the storage cylinders, giant elevating platforms lifted the grain to the top of the tower.

Brad was in a small underground storage room next to one of the towers. He was waiting for his supervisor to arrive and give him instructions. The floor was covered with several inches of dust from the grain, and he was making a mess as he paced impatiently. Noticing a metal snow shovel by the door, Brad decided to get busy and clean up the place a bit. He was scraping the shovel along the cement floor when his boss and another worker appeared at the door. In an instant, they jumped at him, one grabbing his arms while the other grabbed the shovel. Brad was dumbfounded.

"Sorry to startle you," the boss said. "I know you haven't completed your orientation yet, but you've got to learn some critical rules. First and most important, don't do anything that could cause sparks!"

"But there's no gasoline around, just wheat. What difference would a few sparks make?" Brad responded, feeling embarrassed.

"Here, read this," the boss stated grimly as he removed a worn letter that had been posted on a bulletin board. This is what Brad read:

10 August

Dear Sarah,

I am amazed to be alive and able to write this letter to you—or actually to dictate it; my hands are burned too badly to write it myself.

As you heard on the news, the worst happened: an explosion in elevator number 2. We all had been warned, but I never imagined anything like this could actually happen. The whole terrible thing took only seconds. First, there was an odd "whoosh" noise, and my right side was seared with heat. Before I could think about what was happening, there was a deafening noise. I still have ringing in my ears.

I was lucky not to be at the center of the explosion. Four guys in there didn't make it. It's hard to believe that the whole thing probably started with a tiny spark from the conveyer belt gears while we were shifting grain—we'll never know for sure. The explosion was like dynamite. It blew the top right off the number 2 elevator, and then the fire took over.

I know your facility is similar to ours. Whatever you do, don't let this happen there.

Yours truly,

Mike

Brad looked silently at the others in the room. He had no idea that the flourlike dust from simple wheat could produce such an explosion and fire. He certainly couldn't explain *why* such a thing could happen or *how*.

EXPLORE

EXPLAIN

Energy in Matter

Explosions are dramatic examples of the release of a tremendous amount of energy. But what is the source of all that energy? In the case of the grain elevator explosion, the energy was stored within the molecules that made up the grain dust and gases in the air. These are two forms of matter that few people think of as energy sources. Grain and air, however, are not exceptional forms of matter; all matter contains energy. Not all matter contains the same *amount* of energy, however. The particular organization of atoms—the building blocks of matter—in a substance determines its precise amount of energy (see figure 4.3). In this Explore-Explain activity, *Energy in Matter*, you will investigate some of the links among matter, energy, and organization and attempt to describe the close relationship among them.

Materials (per team of 2)

- 2 pairs of safety goggles
- 2 lab aprons
- Pyrex test tube
- 250-mL flask
- 2 microscope slides
- graduated cylinder
- dropping pipet
- test-tube rack
- spatula or spoon
- thermometer or thermistor
- hand lens or microscope (optional)
- 2-cm piece of magnesium ribbon
- baking soda
- vinegar
- hydrochloric acid solution
- saturated urea solution

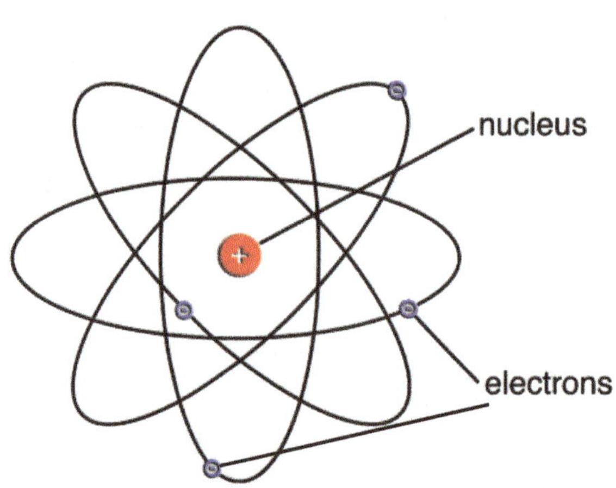

Figure 4.3 This is a graphic model of an atom. What do you already know about atoms?

PROCESS AND PROCEDURES

Matter and Energy Interactions

1. You and your classmates have examined how energy and matter flow through ecosystems. What is matter? Discuss with your partner how you would define "matter." In your science notebook, write your definition and 3 examples of matter.
2. Discuss with your partner how you would define "energy." In your science notebook, write your definition and 3 examples of energy.
3. Explain how matter and energy are related. Record your explanation in your science notebook. Include 1 or 2 examples to illustrate your ideas.

SAFETY: Put on your safety goggles and lab apron.

4. Carry out the following steps with your partner. In your science notebook, record your observations at each step of the task.
   a. Use a graduated cylinder to measure 5 mL of 0.7 M hydrochloric acid. Pour this into a test tube that is firmly seated in a test-tube rack. Do not remove the test tube from the rack.

WARNING: 0.7 M HCl is a *strong irritant*. Avoid skin and eye contact; do not ingest. If contact occurs, flush affected area with water for 15 minutes; rinse mouth with water. Call the teacher.

   b. Feel the outside of the test tube to note the relative temperature, as shown in figure 4.4.
   c. Carefully place 2 cm of magnesium ribbon into the acid in the test tube. Observe what happens as the solid reacts with the acid.
   d. Feel the test tube again to determine if the temperature has changed.
   e. Discuss the energy changes that you observed. Compare these changes with the events in the scenario, *A Matter of Explosions*.
   f. Discuss the changes in matter that you observed. Compare these changes with the events in the scenario.

Figure 4.4 This team is noting the relative temperature as magnesium reacts with hydrochloric acid.

Some chemical reactions release energy to the surrounding environment. Others absorb energy from their surroundings (see figure 4.5).

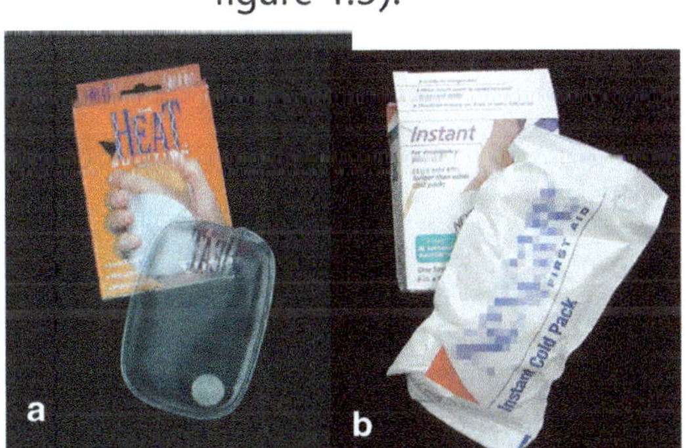

Figure 4.5 (*a*) **Some reactions give off energy.** The chemical reactions in hand warmers release energy to their surroundings. What form of energy do they give off? (*b*) **Some reactions absorb energy.** The reaction that takes place inside this cold pack absorbs energy from the surroundings. What form of energy is absorbed?

5. Carry out the following steps with your partner. In your science notebook, record your observations at each step of the task.
   a. Put approximately 40 mL of room-temperature vinegar in a 250-mL flask.
   b. Use the thermometer or thermistor to measure the starting temperature.
   c. Carefully put about 2 spatulas of baking soda into the vinegar in the flask. Swirl it gently to dissolve the powder.
   d. Record the final temperature.
   e. Discuss the energy changes that you observed.
6. Carry out the following steps with your partner. In your science notebook, record your observations at each step.
   a. Use a dropping pipet to place 1 drop of saturated urea solution on a microscope slide.

WARNING: Urea solution is an *irritant*. Avoid skin and eye contact; do not ingest. If contact occurs, flush affected area with water for 15 minutes; rinse mouth with water if it is affected. Call the teacher.

   b. Use the edge of the second slide to spread the drop in a thin layer across the surface of the first slide.
   c. Watch the layer closely for changes. This may take several minutes to develop. Use a hand lens or microscope, if available.

      Include drawings to record the changes that you see.

   d. Read the following information. Then, in your science notebook, answer the questions.

      In the urea solution that you used, urea was dissolved in the water. The urea molecules in the solution move about freely, from areas where there are many molecules to areas where there are fewer molecules. These molecules are not highly ordered, and energy is dispersed throughout the urea solution. This system can be described as being very disordered. Explain what appears to happen to the organization of urea molecules as the water in the urea solution evaporates. What do you think happened to the amount of disorder? Do you think this requires or releases energy?
7. Clean up your area. ***Wash your hands thoroughly with soap and water.***
8. Record the following questions in your science notebook, leaving enough space to write answers. Then read the essay, *Matter and Energy Are Related,* to find information that will help you answer the questions.

   Be prepared to participate in a class discussion about the relationship between energy and the organization of matter.

   Answer these questions thoroughly, because you will refer to your notes throughout this chapter.

   a. Did the reaction you carried out in step 4 release energy to or absorb energy from the surrounding environment? What about the reaction in step 5? State your evidence in each case.

b. Describe the original source of the energy for the reaction that absorbed this energy. In what form was the energy after the reaction?
c. Describe where the energy that was released in the other reaction came from and where it was transferred.
d. What do you think happened to the organization of the molecules of baking soda when they reacted with vinegar? How did that differ from what happened to urea?
e. Did any of the atoms of the chemicals involved in these reactions disappear as a result of the reaction? Explain your reasoning.

Matter and Energy Are Related

ENGLISH SPANISH

If someone asked you to describe how skin, plant roots, water vapor, and a plastic cell phone are different, could you do it? Humans are usually good at identifying differences between things in our physical world. This is because we observe many characteristic features through our senses. If we can see, touch, smell, or hear differences, then we use those characteristics to help recognize our surroundings. However, identifying similarities can be more challenging. What do skin, plant roots, water vapor, and a plastic cell phone have in common? One similarity is that all are examples of matter, meaning that they are all made of atoms.

Atoms

Atoms are the basic building blocks of all matter. They are composed of three different types of subatomic particles. Electrons are extremely lightweight, negatively charged subatomic particles (symbol, e^-). They orbit rapidly around the nucleus of an atom. The nucleus is located at the center of an atom. It is composed of two types of particles: protons, which are positively charged, and neutrons, which have no charge at all and thus are neutral. Protons and neutrons are heavier than electrons. You cannot see individual atoms with your eye or even with powerful microscopes. Atoms make up everything from the smallest virus to the stars in the largest galaxy (see figure 4.6).

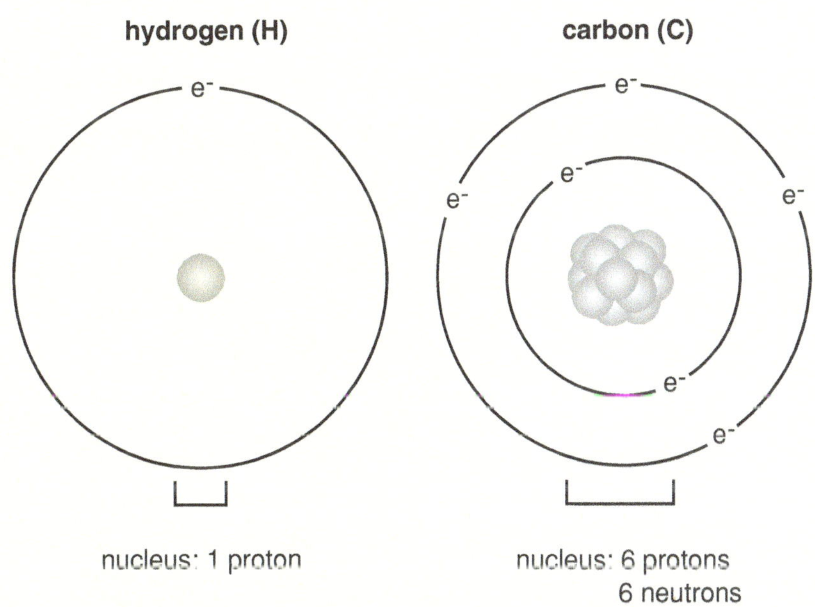

Figure 4.6 Atoms. All matter is composed of one or more elements. Elements are forms of matter that contain just one type of atom. An atom is the smallest particle of an element that still has the properties of that element. Differences among elements are due to different numbers and types of subatomic particles in their atoms. For instance, six electrons, six protons, and six neutrons are characteristic of the element carbon (symbol, C).

Elements

Different kinds of atoms are distinguished by the number of protons in their nucleus. These different kinds are called elements. Scientists have identified more than 100 different elements. While there are dozens of different elements that occur naturally on Earth, only six different kinds of elements make up 99 percent of the human body. These are carbon, hydrogen, oxygen, nitrogen, calcium, and phosphorus.

Organization of Atoms

Individual atoms are extremely small. But the matter that atoms make up is visible. The way that atoms assemble together determines the characteristics of the matter that we see. For example, skin is made up of a different combination of atoms than is the plastic in a cell phone. The arrangement of atoms in skin enables it to be flexible and to allow substances to move either in or out of the body; the arrangement of atoms in the plastic of a cell phone makes it rigid and waterproof. As figure 4.7 shows, the source of this important atomic organization is chemical bonds.

Energy Is Stored in Chemical Bonds

Chemical bonds hold atoms together in predictable ways to form molecules. Energy is stored within the structure of a molecule's bonds and atoms.

Because we cannot see atoms with the naked eye, we often overlook the importance of atomic interactions. However, these interactions are critical to understanding the relationship between matter

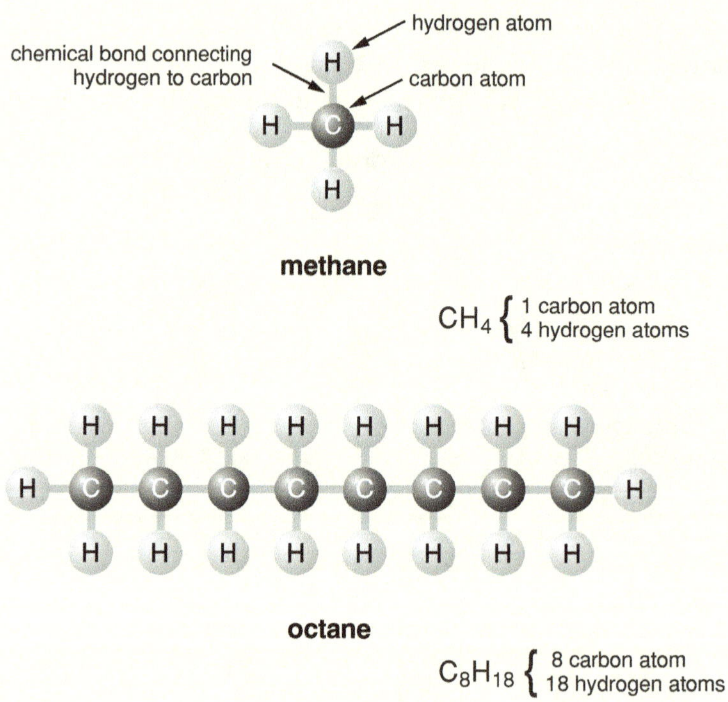

Figure 4.7 Molecules. Molecules are made of two or more atoms joined by chemical bonds. Each kind of molecule has atoms organized in a unique way. This accounts for the unique properties of materials. For instance, in one arrangement, carbon and hydrogen form a smelly gas called methane. In another arrangement, carbon and hydrogen form octane, a flammable liquid used in gasoline. Note that in both molecules, hydrogen always has one chemical bond and carbon always has four chemical bonds. The number of electrons orbiting around a particular atom determines its number of bonds.

and energy. It is difficult to describe exactly how energy exists in the structure of chemically bonded atoms. But let's try to imagine it by thinking about the following example. For this example, magnets represent atoms. Imagine holding a magnet in each of your hands. You slowly move them closer and closer together. You have to use energy to do this, right? When they are close enough, suddenly the magnets snap together (see figure 4.8). Notice a few things. First, energy was involved in rearranging the magnets. Second, energy is related to the stability of the magnets. When the magnets are stacked, they are more stable. Next, additional energy would be required to pull them apart.

What does this have to do with forming bonds among atoms? Energy must always be added to break a bond. Often (but not always), a chemical bond that has formed is fairly stable. When a chemical bond forms, energy is released.

Now consider the magnet analogy again. Imagine that you move several magnets close to each other. The forces of attraction will cause them to form a solid, cylindrical stack.

Figure 4.8 Stacked magnets. The attraction of magnets to each other in a stack is an analogy for chemical bonds. It takes energy to push the magnets close before they snap together. Then they form a stable structure. It would take energy to pull them apart.

The energy stored in this stack of small magnets is much greater than the energy stored in any of the small magnets individually. Where did this energy come from? Remember, it took energy to move the magnets closer together. The force of attraction between many magnets is somewhat similar to the energy in many chemical bonds between atoms.

Matter Is Conserved

Consider the magnet analogy one more time. When you push two magnets close enough, they snap together. But there are still two magnets. The same thing is true for atoms. Even though bonds between atoms form and break, the same number of atoms remain. In other words, matter is conserved.

Forms of Energy

Magnetic energy is one familiar type of energy. What other forms of energy can you name? Energy comes in a variety of forms. These include light, electrical, solar, nuclear, mechanical, and chemical energy. Thermal energy is another form of energy. When thermal energy is transferred out of one system into another, heat is released. Energy can be transferred between any of these forms. Each transfer always releases some heat.

Chemical Reactions That Release Energy

Chemical reactions can also release heat. In these reactions, heat is *released* when the atoms in molecules are reorganized. Let's examine what happens specifically. The molecules present at the beginning of a reaction are called **reactants**. The reactants have a certain amount of stored energy. Some energy needs to be put into the system to break the bonds of the reactant molecules. But when new bonds form, energy is released.

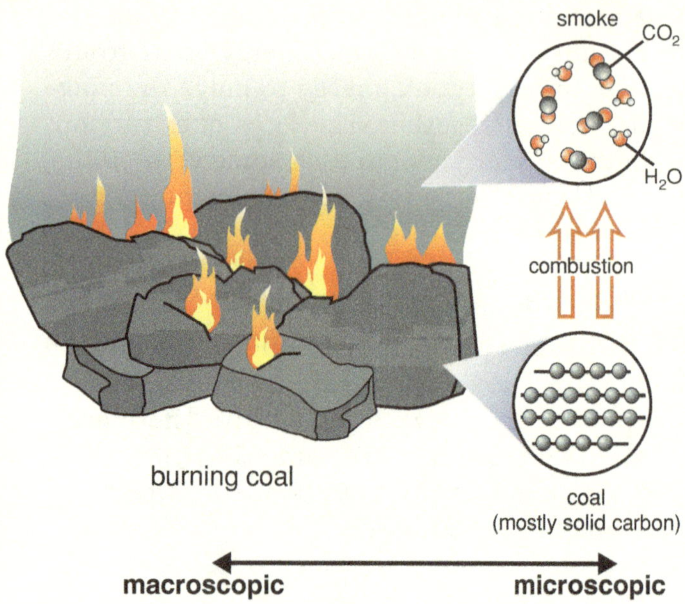

Figure 4.9 Combustion of coal. Microscopic particles of coal interact with molecules in the air to make combustion products (such as carbon dioxide, or CO_2) and heat energy. What are similarities and differences between this reaction and the combustion of grain dust?

When more energy is released by bonds forming than absorbed by bonds breaking, the reaction releases heat.

For example, the explosive combustion of grain dust is a reaction that releases heat. So is the combustion of coal (see figure 4.9). What was required for the combustion of grain dust? The molecules that make up the grain dust, oxygen (present as a gas in the air), and a spark. The spark provided the energy to break the bonds in the grain dust and the oxygen. The atoms in the grain molecules and the oxygen are quickly reorganized into new molecules. Such a reorganization of matter can cause the uncontrolled release of a tremendous amount of energy. Even though it would appear that the atoms that make up the grain molecules disappear in this reaction, they are simply transformed into other molecules of matter. In chemical reactions, matter is not created or destroyed; it only changes form.

Chemical Reactions That Absorb Energy

Some chemical reactions absorb, or take in, thermal energy from their surroundings. In these reactions, energy is again needed to break the bonds of the reactants. Energy is also released when new bonds form, but less energy is released than was required to start the reaction. Such reactions require an overall *input* of energy. For example, the oxygen atom and the hydrogen atoms in a water molecule are quite stable when bonded to one another. It is necessary to add energy, in the form of a spark or ultraviolet light, to break the bonds between the atoms. Then a molecule of oxygen (O_2) and two molecules of hydrogen (H_2) form.

Understanding the link between energy and matter is important in biology. All living systems require energy and matter for survival. In the case of humans, thousands of different chemical reactions constantly occur in our cells. Each depends primarily on chemical energy. Within living organisms, these reactions have evolved to take place in a controlled manner—without sparks and explosions.

Analysis

Read the essay, *Energy Is Converted and Conserved,* and review figure 4.10 for information that will help you complete the following tasks. Record your answers in your science notebook. Be prepared to share your responses in a class discussion.

1. If some reactions release heat, why is it inaccurate to say that some reactions release cold?
2. What is the difference between kinetic energy and potential energy? Give an example for each.
3. Explain why the energy that is stored in the chemical bonds of a molecule is not destroyed when that molecule is broken down to smaller molecules.
4. Define ATP's role as a link between matter and energy in living systems.
5. Revise your definitions of matter and energy and their relationship, which you developed in steps 1–3 of *Matter and Energy Interactions.* It may be helpful to use diagrams or other visual aids with captions to explain your definitions.

a b c

Figure 4.10 Examples of energy conversion technology that generate electricity include (*a*) wind farms, (*b*) water dams, and (*c*) solar farms.

Energy Is Converted and Conserved

ENGLISH SPANISH

When electricity flows through the metal coils on a stove burner, it makes the coils red hot. But it does more than make the teapot ready for tea. It illustrates an essential property of energy. *Energy can be converted from one form into another.*

Energy Conversions

In this case, electrical energy is converted into thermal energy. This property of energy has enabled scientists and engineers to develop techniques and tools for making energy accessible to humans. For example, power-generating dams harness the mechanical energy of water by passing water over turbines. The turbines rotate and convert the water's movement into electrical energy.

Without energy conversion, gasoline would not be much more than a smelly, toxic liquid. In fact, gasoline is an important energy source used in most societies. Gasoline stores a large amount of chemical energy in its molecular structure. Gasoline can be a dangerous substance because it is highly flammable. It is exactly that property, however, that makes it valuable as a fuel for industry and transportation. As figure 4.11 suggests, we have learned how to control the explosive property of gasoline. A gasoline engine converts the gasoline's stored chemical energy into a form of mechanical energy—a form that can be used to power machinery.

Figure 4.11 **Energy conversion.** Molecules of the chemical gasoline store energy. Combustion can convert this chemical energy into mechanical energy in an automobile. What other energy conversions can you think of?

Potential and Kinetic Energy

How is an exploding grain elevator an example of energy conversion? Such explosions illustrate that energy exists in two forms: an inactive form and an active form. The inactive form is called **potential energy**. Potential energy can be stored in the structure of matter and is available for use. When the potential energy in grain dust is released, it becomes active.

Active energy is called **kinetic energy**. A boulder sitting at the top of a hill contains a great deal of potential energy simply because of its position, as figure 4.12 shows. When the boulder is pushed off the cliff, however, its potential energy is transformed into kinetic energy, or the energy of movement. The potential energy poses no threat to you. You can sit on the boulder as it rests at the top of the hill. However, you would not want to come into contact with the boulder's kinetic energy as it rolls down the hill. Likewise, you are not afraid to pump gas, which has a tremendous amount of potential chemical energy, into a car's gas tank. But you don't want to light a match at the same time. If you do, you will see all of that potential energy converted into a dangerous amount of heat, a form of kinetic energy, very quickly.

Once potential energy has been converted into kinetic energy, one of two things can happen. The kinetic energy may be captured and made useful, or it may escape to the universe as heat. For example, what if the boulder at the top of the hill

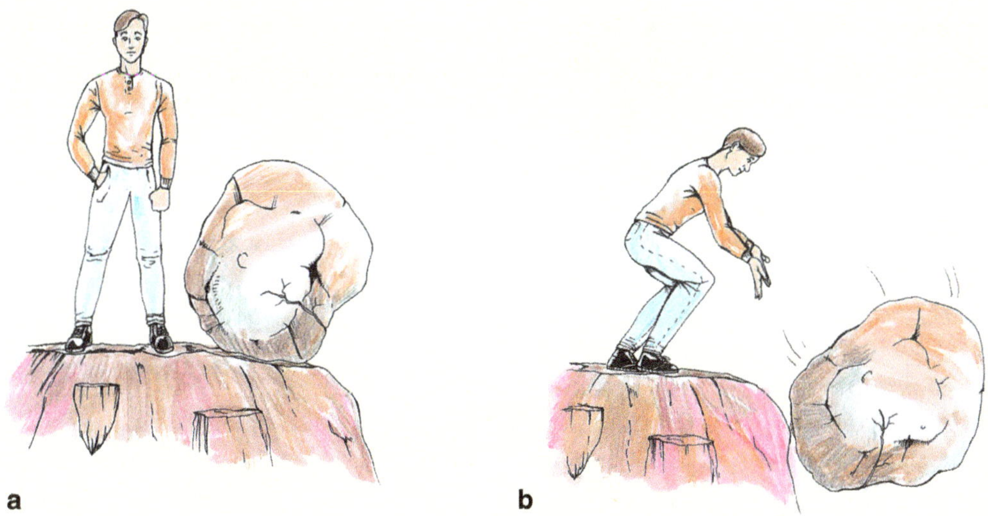

a b

Figure 4.12 Potential energy and kinetic energy. (*a*) The boulder resting at the top of the cliff has a great deal of potential energy and no kinetic energy. (*b*) The falling boulder has a great deal of kinetic energy but less and less potential energy. After the boulder comes to rest at the bottom of the hill, it has less potential energy. Did that energy disappear? Where did it go?

were attached to a rope that in turn, were attached to a wheel on a well? As the boulder rolls down the hill, it would turn the wheel and raise water from the well, spilling most of it as the bucket smashes against the wheel axle. In this example, some of the kinetic energy of the boulder would be stored as potential energy, in the raised bucket. On the other hand, if the boulder were not attached to the rope, it would simply roll until it stopped. What happens to the kinetic energy? The fact that the boulder would eventually stop indicates that yet another energy conversion occurs. Each time the boulder turns, its kinetic energy would cause molecular movements in the grass, soil, and stones. The kinetic energy would be converted to heat. The heat would be absorbed by the grass, soil, and stones as well as the air. In this case, the release of heat would be wasted energy because it would not be captured and used to do work. No energy would be created or destroyed. The energy would simply be *transferred* from the boulder to the surrounding grass, soil, stones, and air. Most of it would eventually escape into space.

Energy Is Conserved

Energy transfer also takes place in your body when your digestive system removes nutrients and energy from food and supplies them to your body's cells. As with every energy transfer, some heat is produced in this process. Even through all the intermediate steps necessary for this transfer, no energy is lost or created. This property of energy is called the **conservation of energy**. It means that even though the amount of energy at one location can change, the total amount of energy in the universe remains the same. Another way of stating this principle is that energy is neither created nor destroyed, only changed in form. The conservation of energy applies across the entire universe. Let's examine how it applies to the chemical bonds between the atoms of molecules in living organisms. Suppose a large molecule contains 8 units of energy. What happens to the energy in this molecule when it is broken down to two smaller molecules? Together, the two smaller molecules plus any small amount of heat from the breakdown will contain 8 units of energy as well.

What is it about a molecule that enables it to store energy? The arrangement of atoms within a molecule affects its energy properties. The arrangement of atoms within a molecule is determined by the chemical bonds holding it together. There are several types of chemical bonds, or forces of attraction, between atoms. The forces vary in strength. The stronger the bond, the greater the energy needed to break it.

The conservation of energy and the ability to transfer energy both play key roles in maintaining the energy balance in organisms. The energy needs of organisms vary from time to time. The ability to convert energy from one form to another means that organisms can use the form that best matches their current energy needs. As figure 4.13 illustrates, when energy needs are low (for example, during sleep) organisms store potential energy in large storage molecules, such as glycogen or starch. Glycogen and starch are types of carbohydrates made up of many sugar molecules bonded together. These molecules contain many bonds and atoms, and thus, much energy. When energy needs are high (for example, during exercise), organisms convert some of this stored potential energy into other forms of energy, such as the mechanical energy used for the contraction of muscles.

Energy for Organisms

Nearly all biological reactions require the input of some energy before they can proceed. Enzymes are large molecules found in the cells of all organisms. They reduce the energy needed to start chemical reactions. Could a cell use a molecule such as glucose directly? Breaking down glucose releases a great deal of potential energy—so much that a cell would lose control.

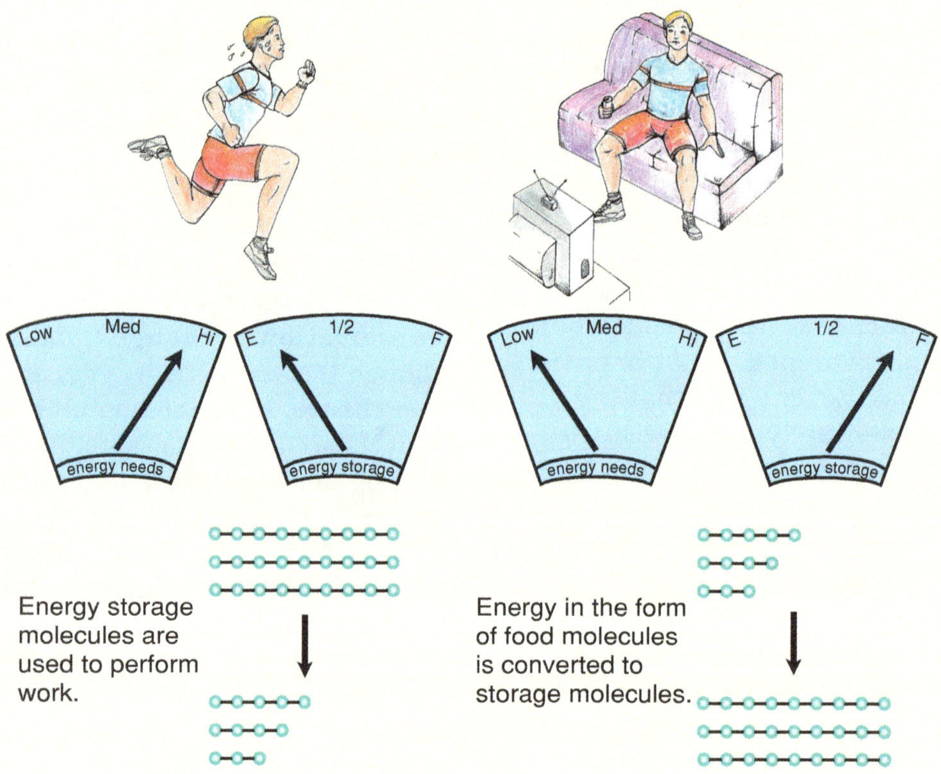

Figure 4.13 Living systems regulate energy storage. When the energy needs of humans are high, our bodies convert energy storage molecules such as glycogen into a form of chemical energy. This energy can be used to perform work. When the energy needs are low, our bodies convert food molecules into energy storage molecules. These molecules can be used to meet future energy needs. Are your energy needs high or low right now? Is your body trying to store energy? Or is your body using previously stored energy to perform work?

To make the energy stored in molecules such as glucose useful, the energy must be converted into a form that cells can use *directly*. This form is often a special type of molecule called adenosine triphosphate, or ATP. The chemical structure of ATP is shown in figure 4.14a. ATP connects many energy-conversion reactions and is called the "energy currency" of living cells (figure 4.14a). When ATP breaks down, it releases a smaller, but more easily used, amount of energy compared with glucose.

Imagine foreign tourists who arrive in the United States with no American dollars, only money from their home. The tourists must pay a fee to change their money into dollars, which they can use to pay for their activities. Similarly, the cells in organisms transform or change the chemical energy in various molecules into the chemical energy of ATP. Then ATP "pays" for activities that require energy in cells. The "fee" is the energy released as heat during the conversion.

ATP is constantly made and broken down in cells, forming a cycle. When a molecule of ATP (adenosine *tri*phosphate) gives up one phosphate group, it becomes ADP (adenosine *di*phosphate) and releases a usable amount of energy (see figure 4.14b). The energy is available for cellular work. When ADP combines with a phosphate group, it forms ATP. This step requires energy. Thus, ATP acts as an energy carrier. It can be used to release energy (ATP breaks down) and to store energy (ATP forms).

Cells need to use chemical energy for many things. They need it to make new molecules the organism needs and to remove wastes. They need it to take in nutrients and to move molecules in and out of cells. Cells also need energy to move from place to place, to move their internal structures, as well as for many other processes. ATP supplies much of the energy for this work.

The chemical energy in the energy-yielding nutrients in food is converted to chemical energy in the form of ATP. There are other molecules that exist that have similar energy-transferring properties as ATP, but ATP is the most common. All known living cells use ATP as an energy carrier. This includes the cells of fungi, plants, humans, bacteria, and even archaea. The first organisms on Earth most likely used ATP as an energy source.

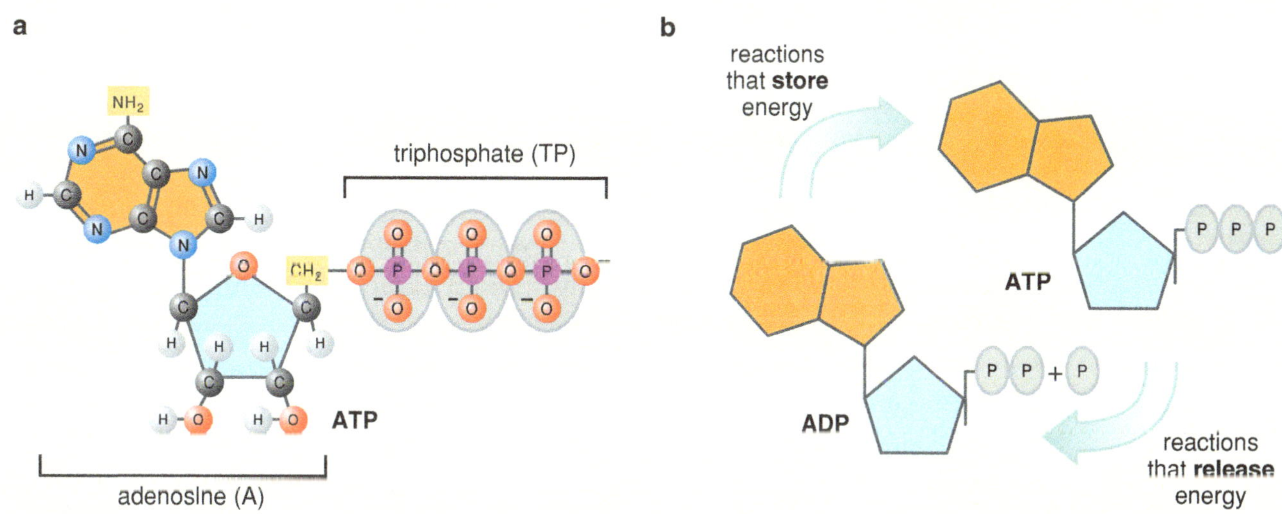

Figure 4.14 ATP cycle. Adenosine triphosphate (ATP) stores potential energy in its chemical bonds. Notice that ATP has three phosphate groups. A phosphate group has a phosphorus atom and four oxygen atoms. It is represented by the letter *P* in a circle. When the bond between the second and third phosphate groups breaks, ATP becomes adenosine diphosphate (ADP). ADP has two phosphate groups. The energy released from breaking bonds in ATP and forming new bonds in ADP becomes available for cells to use.

EXPLAIN

Keep a Body Running!

One of your classmates was explaining what she learned about the connection between food and energy to her older sister, Lily. Lily is a senior on the varsity track team and she knows first-hand that she and her friends need more food when they are training so they have enough energy to race well. Lily listened carefully to your classmate and said "I understand that the chemical bonds in food molecules store energy. And I understand that the body converts the energy stored in food into ATP molecules. But *how* does the body do that? How can I explain this to friends on the track team?" Lily asked her sister to make a poster or slide show that she could use to explain how this happens to her friends.

This Explain activity, *Keep a Body Running!*, provides the tools for you and your teammates to answer Lily's question. To help her explain the connection between food molecules and ATP molecules to her friends, you will develop a poster or set of slides to answer the focus question: "How does the body harness the potential energy in food?"

Materials

Cellular Respiration on the online resource (optional)
materials to design your poster or slide show

PROCESS AND PROCEDURES

Explaining Our Body's Energy

Running races and other track events require the transformation of an enormous amount of energy. How does your body transfer the stored energy in food into chemical energy to use for running, thinking, and everything else you do? How does your body release this energy without starting an actual fire in your body? (Or an explosion—remember the grain elevator in the Engage activity, *Releasing Energy*?)

1. Read the essay, *Cellular Respiration: Controlling the Release of Energy*. As you read, write the terms "cellular respiration" and "ATP" into your science notebook and answer the following questions.
   a. Describe how the overall process of cellular respiration releases energy in a controlled way.
   b. Examine the caption for figure 4.16. Why do plants, animals, and fungi each require cellular respiration to live?
   c. Compare the 3 stages of cellular respiration in terms of how much ATP is produced in each stage.

d. When energy breakdown occurs anaerobically, the first stage of glycolysis is completed, but molecules do not enter the Krebs cycle or the electron transport chain. Explain why it is more efficient for organisms to use aerobic cellular respiration to release energy from food.

Cellular Respiration: Controlling the Release of Energy

ENGLISH SPANISH

In the example of the grain-storage explosion, energy was released from matter suddenly and dramatically. When you eat bread or cookies, however, you do not explode. Clearly, your body must release the energy stored in grain in a more controlled way. When you eat bread made from grain, the cells in your body release the energy one small step at a time. Enzymes help start a series of specific chemical reactions. Cellular respiration is the name of this series of reactions. There are two kinds of cellular respiration: aerobic, which requires oxygen, and anaerobic, which doesn't require oxygen. We will focus on aerobic cellular respiration, which we will simply refer to as cellular respiration.

Converting Energy of Food Molecules into Energy of ATP

During cellular respiration, the potential energy in food molecules is transformed into molecules of ATP and heat (see figure 4.15). The ATP molecules store energy in smaller

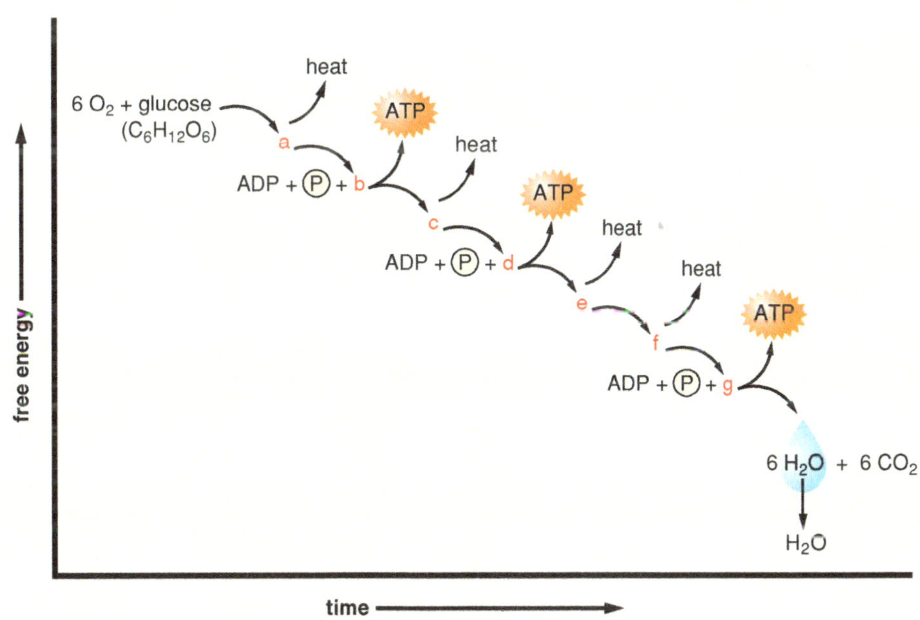

Figure 4.15 Energy transfer in cellular respiration. A series of reactions transfers the energy of glucose to ATP. The labels *a* through *g* represent intermediate compounds in the decomposition of glucose to carbon dioxide and water. Notice what happens to the relative amount of free energy.

amounts, appropriate for fueling chemical reactions in the body. Molecules such as starch, glycogen, or glucose (a type of sugar) may be used in this process. Each type of molecule contains atoms of carbon, hydrogen, and oxygen. The molecules differ only in how those atoms are organized and how many of each atom are present. Other types of molecules from food can also be used in the process. Just like the grain explosion, aerobic cellular respiration requires oxygen. You need oxygen because it allows your cells to carry out cellular respiration. To explore this process, we will focus on the breakdown of the sugar glucose.

The First Stage: Glycolysis

Three main stages divide the steps of cellular respiration, as illustrated in figure 4.16. The stages are known as **glycolysis**, the **Krebs cycle**, and the **electron transport system**. The first stage, glycolysis ("glyco-" = "sugar," "lysis" = "to split"), occurs in nearly every living cell. Glycolysis is a series of reactions that splits a glucose molecule into two smaller molecules. This produces a small amount of ATP. These two molecules still contain much stored energy, however.

The Second Stage: Krebs Cycle

In the Krebs cycle, a series of reactions breaks down the two small molecules to carbon dioxide. This produces several more ATP

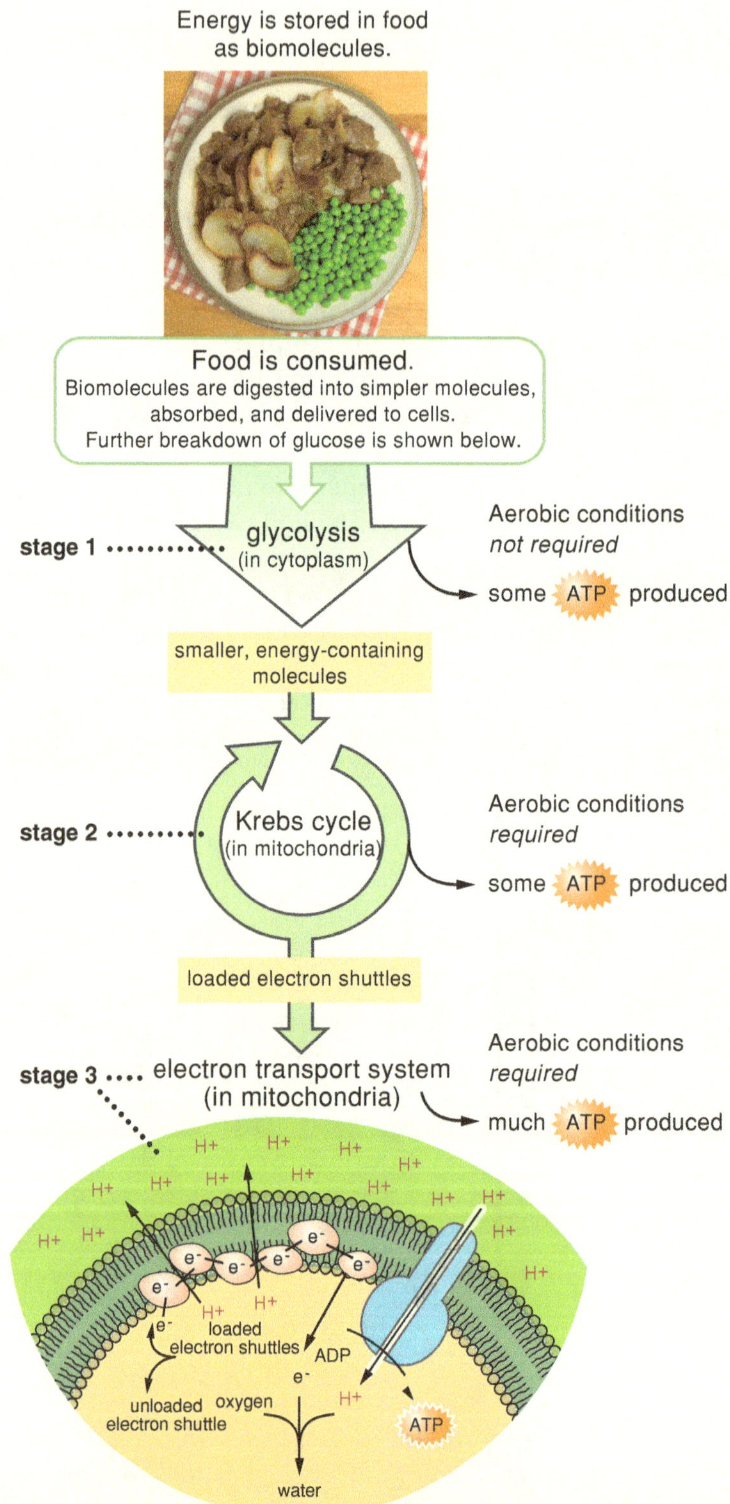

Figure 4.16 Stages of cellular respiration in plants, animals, and fungi. Glycolysis breaks down glucose to smaller molecules. They are further broken down to carbon dioxide in the Krebs cycle. In the electron transport system, the electron shuttle molecules made during glycolysis and the Krebs cycle are used to form many molecules of ATP.

molecules. When you breathe out, most of the carbon dioxide comes from the chemical reactions that occur in the Krebs cycle. In fact, when you lose weight, most of your "lost" weight leaves your body as you exhale carbon dioxide. After the Krebs cycle, some of the glucose molecule's energy has been captured in the form of ATP. Energy was also transferred to other molecules known as loaded electron shuttles. These molecules received high-energy electrons during the Krebs cycle.

The Third Stage: Electron Transport System

The loaded electron shuttle molecules from the Krebs cycle move to the last stage of cellular respiration, the electron transport system. In this stage, the energy from the electrons carried by the electron shuttles is ultimately transformed into ATP. In fact, most of the ATP made from glucose is generated during this last stage of cellular respiration. At the end of the electron transport system, the electrons from the electron shuttles combine with oxygen (O_2) and hydrogen ions (H^+) to form water (H_2O). Finally, a significant amount of energy from glucose is released as heat.

End Products of Cellular Respiration

Overall, the reactions of aerobic cellular respiration can be summarized as follows:

$$\underset{\text{glucose}}{C_6H_{12}O_6} + \underset{\text{oxygen}}{6O_2} \xrightarrow{\text{enzymes}} \underset{\text{carbon dioxide}}{6CO_2} + \underset{\text{water}}{6H_2O} + \underset{\text{energy}}{ATP}$$

There are four end products of cellular respiration. First, ATP molecules are made in each stage. Second, carbon dioxide is a by-product of the Krebs cycle. Third, heat is produced. Finally, water is formed at the end of the electron transport system. Cells can use ATP directly to fuel cellular work, such as muscle contractions.

2. Begin developing your poster or slide show for Lily to use when she talks with her running club by drafting a paragraph that explains how the body's cells convert glucose into energy that the body can use to move muscles.

   Consider this a draft paragraph that will help you organize your thinking and help you commit information to memory. Draw on the information and diagrams in the essay *Cellular Respiration: Controlling the Release of Energy*. You will have a chance to revise and add to this paragraph in the *Analysis* of this activity.
3. To deepen your understanding about the 3 stages of cellular respiration, your teacher may have you watch the animation, *Cellular Respiration,* on the online resource. Copy into your science notebook each of the 3 review diagrams in the animation.
4. To solidify your learning about the 3 stages of cellular respiration, make a table with 4 columns in your science notebook. Label the columns as follows: "stage of cellular respiration," "energy going into the stage," "what is happening to the energy," and "energy going out of the stage." Fill in the energy transformations based on the essay, *Cellular Respiration: Controlling the Release of Energy* and your team and class discussions.
5. If you would like to know more about the process of cellular respiration, read the optional Sidebar, *A Closer Look at Cellular Respiration.*

SEP

http://www
Online Resource

Analysis

Using the information from the essay and animation, as well as your table from step 4, complete the following task.

1. Complete your poster or slide show for Lily's track team. Your final product should explain how the potential energy in food is transformed into energy that can be used by a runner's muscles. In addition, it should be an attractive and engaging presentation. Your presentation should include the following three sections:
   a. **Energy from Food.** In this section, describe the type of energy in food and in ATP. Explain why the energy from food cannot be released all at once in the body.
   b. **Cellular Respiration: Making ATP for the Body.** In this section, discuss how the body releases energy from food in a controlled way. Describe generally what happens in each of the stages of cellular respiration. Finally, list the overall products of cellular respiration (be sure to include the by-products).
   c. **Intriguing Questions and Answers.** In this section, answer the following questions for Lily's track team:
      i. Why do we need oxygen?
      ii. Matter is conserved. What happens to the matter in our bodies when we lose weight?
      iii. Energy is also conserved. Some of the energy in food molecules is transferred to ATP. What happens to the rest of the energy?

   Your poster or slide show should build on the paragraph you developed for the activity, *Explaining Our Body's Energy*, step 2, as well as the table you created in step 4.

SIDEBAR

ENGLISH SPANISH

A Closer Look at Cellular Respiration

Let's take a deeper look at the process of cellular respiration. A unifying principle of biology is that all living systems require energy. What is the energy needed for? It is needed to move things, such as when muscles contract. It is used when active transport carries molecules across membranes. Organisms also need energy for synthesis, that is, to build things. Tissue repair and growth require all of these processes. Where does the energy come from? The potential energy in food molecules supplies the energy needs of organisms. However, the structure of these molecules does not store energy in a form that can be used *directly* for cellular work. Food energy must be converted into a more usable form.

Through natural selection, such processes evolved in the first organisms. The simplest processes for converting food to usable energy are remarkably similar in all living organisms. Organisms that are as different as bacterial cells and humans have nearly identical processes for producing ATP. How do organisms convert food energy into energy that cells use easily? Let's follow the path of a glucose molecule to see how and where a cell harvests its potential energy. Remember, specific enzymes control every chemical reaction in cellular respiration.

The process begins in the cytoplasm. Glucose in the cytoplasm can go through glycolysis. This process does not require oxygen (see figure 4.17). A glucose molecule has six carbon atoms. It is quite stable. That is, the bonds holding its atoms together are not easily broken. Because of this stability, the cell uses a small amount of energy to begin the glucose-splitting reactions (figure 4.17a). It takes energy from two ATP molecules. The reaction splits each ATP molecule into one stable ADP (adenosine diphosphate) molecule and one phosphate (P) molecule. Each phosphate molecule from the two ATP molecules is attached to one end of the glucose molecule.

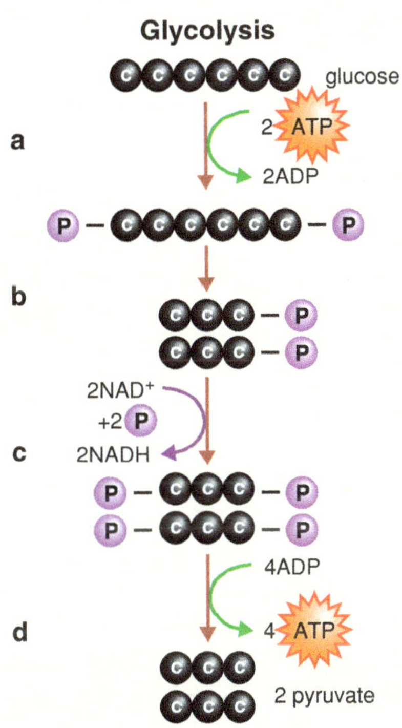

Figure 4.17 A molecular view of glycolysis. In glycolysis, glucose is broken down to two molecules of pyruvate. This takes place in many enzyme-catalyzed steps, illustrated by each red arrow. (*a*) The energy from some ATP is needed to begin the glycolysis reactions. Two ATP molecules each transfer a phosphate to glucose. (*b*) The glucose molecule is then broken down into two molecules that contain three carbons each. (*c*) Next, one more phosphate is transferred to each 3-carbon molecule, using energy from the reaction that converts NAD+ to the electron shuttle molecule NADH. (*d*) In the final steps, carbon atoms are rearranged to form pyruvate and the phosphates are transferred to ADP to make ATP.

Glycolysis breaks glucose down to two molecules that have three carbon atoms each (figure 4.17b). Enzymes rearrange the atoms in these molecules to form two molecules of pyruvate (figure 4.17c). To do this, the electron shuttle NADH is first made from NAD+. The energy from that reaction is used to transfer a phosphate molecule from the cytoplasm of the cell to the other ends of the 3-carbon molecules. Next, the four phosphate molecules (two from each of the two 3-carbon molecules) are transferred to ADP to make four molecules of ATP. This leaves two molecules of pyruvate.

Glycolysis produces a small amount of ATP. Most of the energy of the original glucose molecule, however, remains in the two pyruvate molecules. Glycolysis may be followed by a pathway that does not require oxygen (figure 4.18). This is the only way that some bacteria and archaea access the energy they need. But many organisms proceed to the next two stages of aerobic cellular respiration.

In addition to the ATP, glycolysis produces two molecules of NADH (nicotinamide adenine dinucleotide with a hydrogen atom). NADH is an example of a loaded electron shuttle. Some of the energy from the glucose molecule transfers to these electron shuttles. Cells sometimes use energy directly from loaded electron shuttles to do work.

The pyruvate molecules now enter the second stage of cellular respiration, the Krebs cycle. The Krebs cycle is a stage of aerobic

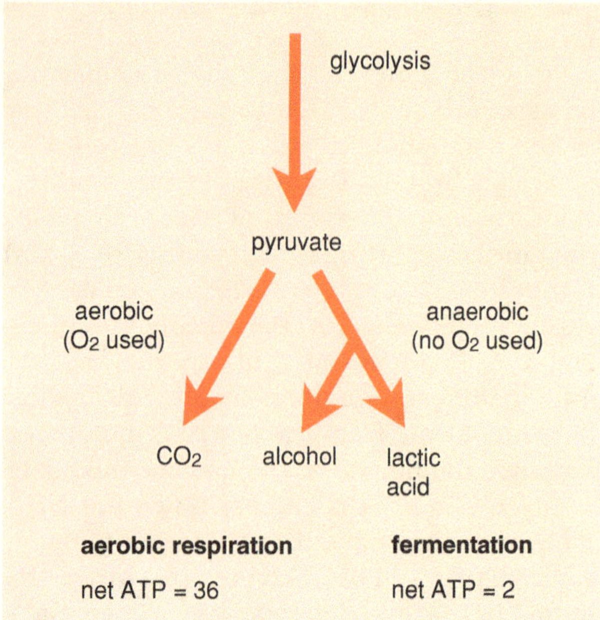

Figure 4.18 Aerobic and anaerobic energy release. When oxygen is present, pyruvate enters the Krebs cycle. There, the carbon in pyruvate is converted into carbon dioxide. Next, electron shuttle molecules transport electrons to the electron transport system. This system produces much ATP. In the absence of oxygen, pyruvate can be converted into alcohol or lactic acid through fermentation. What are the advantages of each process?

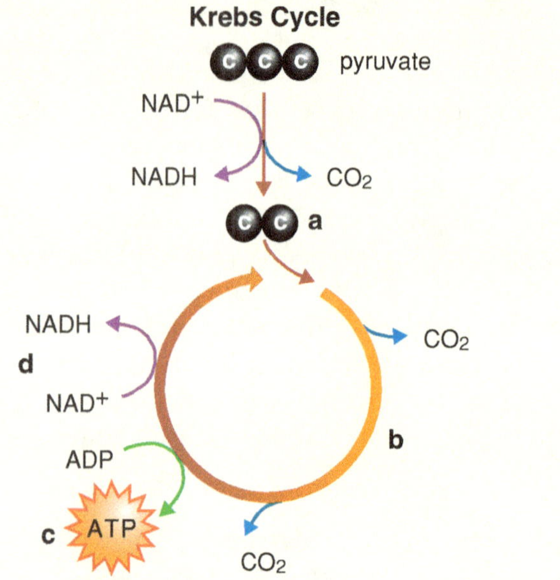

Figure 4.19 The Krebs cycle. (*a*) In preparation for the Krebs cycle, each pyruvate molecule produced by glycolysis is converted into a 2-carbon molecule and one molecule of carbon dioxide. Many enzyme-catalyzed reactions occur that release (*b*) two molecules of carbon dioxide and (*c*) one molecule of ATP. (*d*) In addition, loaded electron shuttles (NADH) are produced at multiple points in the cycle.

respiration that involves many enzymes and molecular rearrangements. These reactions release most of the remaining energy in pyruvate. Because the Krebs cycle only operates when oxygen is available, it is a part of *aerobic respiration*.

In preparation for the Krebs cycle, enzymes convert pyruvate into a 2-carbon molecule. This process releases a molecule of carbon dioxide. Next, the 2-carbon molecule combines with a 4-carbon molecule and is carried through a cycle of reactions (figure 4.19). At the end of the cycle, carbon dioxide, a small amount of ATP, and several more loaded electron shuttles are formed. What do you think is released from the body as a result of these reactions? As with all transformations of energy, heat is produced. In humans, the circulatory system transports the carbon dioxide that was produced to the lungs. You exhale the gas as a waste product.

By the end of the first two stages of cellular respiration, the energy from glucose has been converted into energy in a few ATP molecules, loaded electron shuttles, and heat. Next, the energy in the loaded electron shuttles is converted into many more ATP molecules. This takes place in the third stage of cellular respiration, the electron transport system. Figure 4.20 outlines the electron transport system.

The electron transport system consists of a series of electron carrier molecules (different from the electron shuttles). These carrier molecules are embedded in the inner membrane of a mitochondrion. The electron shuttles deliver their high-energy electrons to the system. The electrons are passed to the chain of electron carrier molecules. As the electrons move from one carrier to the next, they release energy. Some of the energy pumps protons (H^+) across the

Electron Transport System

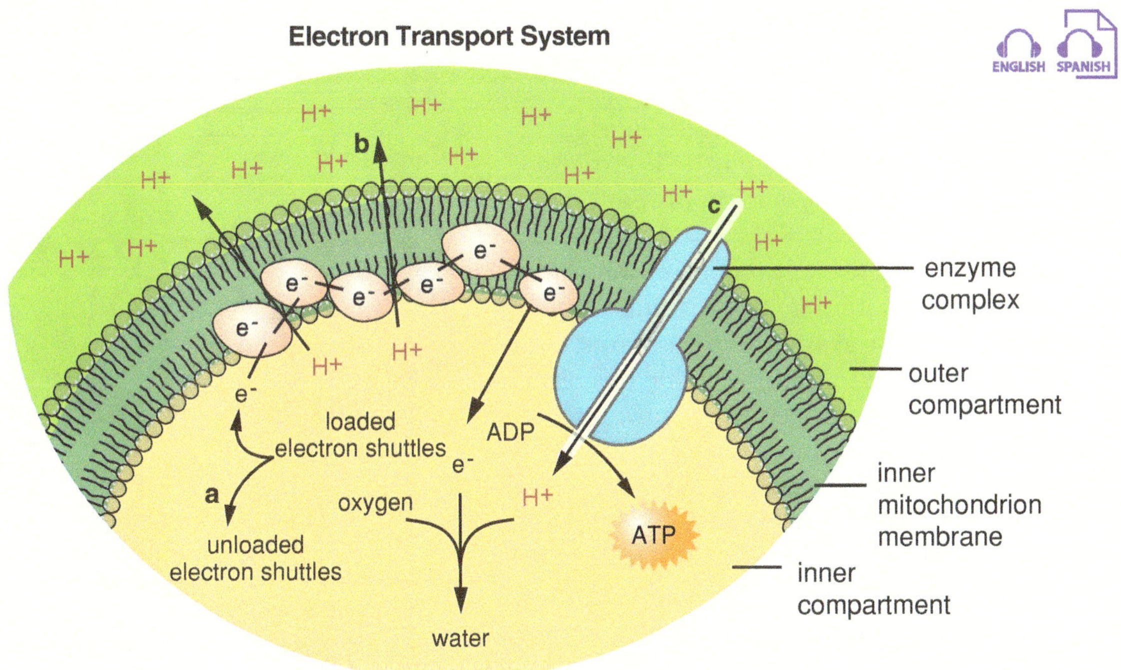

Figure 4.20 Electron transport. The final step in aerobic respiration is the transfer of electrons. (*a*) Electrons transfer from loaded electron shuttles to a chain of electron carrier molecules embedded in the inner membrane of the mitochondrion. The electrons pass from one electron carrier to the next. (*b*) As this happens, energy is released and hydrogen ions are pumped into the outer compartment. The resulting proton concentration gradient drives the production of ATP. (*c*) Protons flow back into the inner compartment of the mitochondrion, through an ATP-producing enzyme complex. At the end of the electron transport chain, the electrons join with oxygen and protons to form water.

inner membrane of the mitochondrion. The protons accumulate in the outer compartment of the mitochondrion. This results in a higher concentration of protons in the outer compartment than on the inside (a concentration gradient).

A concentration gradient is a source of potential energy. The protons diffuse down the concentration gradient into the inner compartment. To diffuse into the inner compartment, the protons can only pass through an enzyme complex located in the membrane. The enzyme complex works much like a waterwheel (figure 4.21). Waterwheels are sometimes used to transfer the energy of a flowing stream to a wheel used to grind wheat. In the electron transport system, the energy from the flow of protons through the enzyme complex is used to produce ATP. In this way, the energy carried into the electron transport

Figure 4.21 Waterwheel. Waterwheels transfer energy from a flowing stream to a wheel used to grind grain. In the electron transport system, an enzyme complex in the membrane transfers the energy from the flow of protons to make ATP from ADP.

system by electron shuttles is transferred to ATP. ATP is the ultimate product of cellular respiration. Carbon dioxide, water, and heat are by-products that we release when we exhale (figures 4.22 and 4.23). ATP can serve as the energetic *push* that starts many chemical reactions in the cell.

But where do the electrons from the electron transport system go? At the end of the electron transport system, the electrons are accepted by oxygen molecules. The oxygen then reacts with hydrogen to form water. Imagine what would happen if there were no oxygen present for use in the electron transport system. What would build up, and what would happen to the body? You breathe because you need oxygen to keep the electron transport system and all its transformations of energy going. Humans, as well as other animals, plants, fungi, protists, and many prokaryotes, require oxygen for their survival. Transformations of energy are necessary for life.

Figure 4.22 How is perspiration related to electron transport?

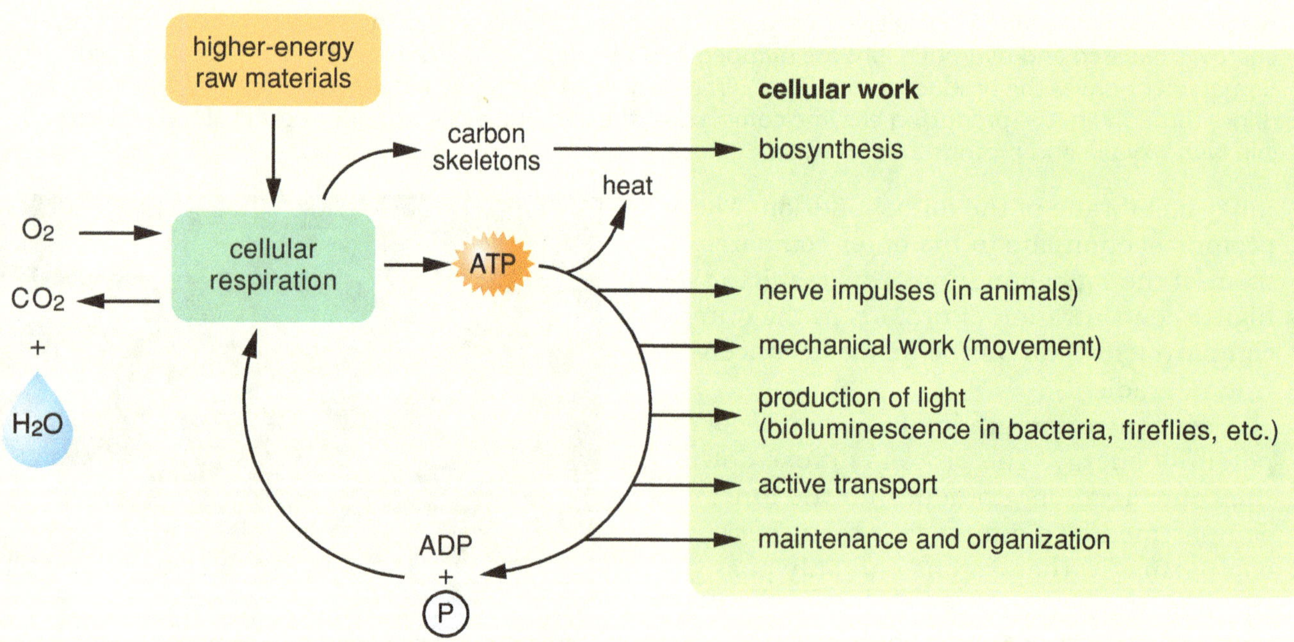

Figure 4.23 **Products of cellular respiration.** Cellular respiration breaks down sugars and other compounds. Heat and water are also produced. What kind of energy transformations occur in cellular respiration?

Using Light Energy to Build Matter

EXPLAIN

ELABORATE

In the last activity, you studied cellular respiration. Cellular respiration is a method for obtaining energy from matter in a form that is useful for cell activity. The process depends on glucose molecules, which can be generated by breaking down glycogen, starch, and other carbohydrates. In humans, the glucose, and indeed all of the carbon in humans' organic molecules, originally came from molecules produced by other organisms. Organic molecules have carbon atoms as the base of their structures. Many of these molecules come from the plants we eat. The rest come from animals or other organisms that depend on plants.

Humans, and all organisms that depend on previously assembled molecules for their carbon and energy, are called **heterotrophs** ("hetero-" = "other," "troph" = "to feed," figure 4.24). The dependence on other sources raises interesting questions about the *original* source of energy and matter. For example, how did energy first come to be stored in these molecules that are essential for life? The laws of nature tell us that energy cannot be created or destroyed, it can only be transferred. Since energy cannot be created, what is the original source of the energy that became stored in the macromolecules that build the bodies of heterotrophs?

The answer is that the energy came from green plants, which are quite different from heterotrophs. They are called **autotrophs** ("auto-" = "self," "troph" = "to feed"). This means they do not have to rely on other organisms for the complex molecules necessary for life. Autotrophs can make all of their own organic molecules. In other words, plants can make their own food. They do this through a process called photosynthesis ("photo-" = "light," "synthesis" = "to make").

There are two focus questions for this Explain-Elaborate activity: "Where does the mass of a plant come from?" and "What factors affect how quickly a plant can carry out photosynthesis?"

Figure 4.24 This bee obtains matter from the nectar it drinks. Which is the heterotroph, the bee or the plant with the nectar-bearing flower?

Materials

Part A

Seed (one per student)
Mature tree on campus

Part B (per team of 4)

250-mL beaker
2 250-mL flasks
250-mL graduated cylinder
thermistor or thermometer
1 pH probe (optional; if using probeware) or pH strips sensitive to pH 1–2 and narrow-range pH strips (pH 4–11)
1 dissolved-oxygen probe (optional; if using probe software)
bromothymol blue
distilled water
4–6 sprigs of young, healthy *Anacharis* (*Egeria densa* or *Elodea spp.*)
jar of tap water for storing pH probes (optional; if using probeware)
drinking straw with hole punched in it
fluorescent lamp
red, blue, and green cellophane (optional)
aluminum foil
tape
ruler

PROCESS AND PROCEDURES

Part A Seed to Tree

1. You and your classmates have been given a seed and are visiting a mature tree on your school's campus. Look at the seed and then shift your gaze to the mature tree. It's huge! Think about the first focus question of this activity and write it in your science notebook: "Where does the mass of a plant come from?"
2. Take a moment to think on your own and journal in your science notebook about where you think the mass of the tree you are visiting came from.
3. Partner with a classmate and use the Pair-Share Sentence Stems to share your thoughts on where the mass of the tree comes from.

   HOW TO

   Pair-Share Sentence Stems can be found in appendix B12. Use these to clarify, reinforce, or expand on something that you heard.
4. Contribute to the larger class discussion by contributing something that you and your partner discussed about the mass of the tree. Add any new ideas to your science notebook.
5. You and your classmates have shared many opinions today. Read the question and student opinions below. Then decide which of the

opinions you agree with most and WHY. Be sure to record your response in your science notebook.

Question: The tree you visited started as a little seed. What provided most of the mass that made the tree grow so large?

Rosanna:	I think it came from the water taken up by the tree's roots.
Richard:	I think that most of it came from the Sun's energy.
Freja:	I think that it came from molecules in the air that came in through holes in the tree's leaves.
Denise:	I think that most of it came from the nutrients in the soil taken up by the tree's roots.

Part B Measuring Photosynthesis

1. Write the second focus question of the activity into your science notebook: "What factors affect how quickly a plant can carry out photosynthesis?" As you consider this question, read and annotate the essay, *Photosynthesis Brings Energy and Matter into Living Systems,* using the Mark It Up! annotation strategy. In addition to using the annotation strategy, record the "big ideas" that you learn about.
2. With your teammates, develop a list of at least 3 variables that could affect the rate of photosynthesis.

   Draw on the essay, *Photosynthesis Brings Energy and Matter into Living Systems,* and figure 4.25 to help you generate your list.
3. Compare your list of variables with another team's list. Add any new ideas to your list.

Figure 4.25 To generate a list of factors that may affect the rate of photosynthesis, consider what the environmental conditions are like when plants outside are thriving.

Photosynthesis Brings Energy and Matter into Living Systems

ENGLISH SPANISH

There is a saying that you must have money to make money. A similar loop exists in biological processes: you need energy to get energy. Food supplies the energy and matter requirements of many organisms. But the food itself usually comes from other organisms. How, then, do energy and matter get into organisms in the first place? The main way is through photosynthesis. *Photosynthesis* is a series of reactions that occur in autotrophs. These reactions use light energy from the Sun to synthesize large, energy-rich molecules from smaller molecules. Plants, algae, and some eubacteria carry out photosynthesis.

Converting Solar Energy into Chemical Energy

You might think of a plant as a solar-powered factory. The factory converts the energy of sunlight (solar energy) into stored chemical energy. The energy is stored in molecules of sugar and starch. Photosynthesis uses only certain wavelengths of visible light (figure 4.26). This fact is shown by the green color of most plants. Plant pigments absorb and use energy from most colors of light. But they reflect green light, so leaves appear green. Photosynthesis also depends on the intensity of the light. The ideal intensity of light varies for different plants. Many other factors, such as the availability of water and nutrients in the soil, also affect photosynthesis.

Photosynthesis Has Two Phases

The reactions of photosynthesis take place in **chloroplasts**, small compartments inside certain plant cells. Figure 4.27 shows a plant cell and a chloroplast. Photosynthesis occurs in two phases. Two major events occur in the first phase, and one major event occurs in the second. Figure 4.28 summarizes the two phases of photosynthesis.

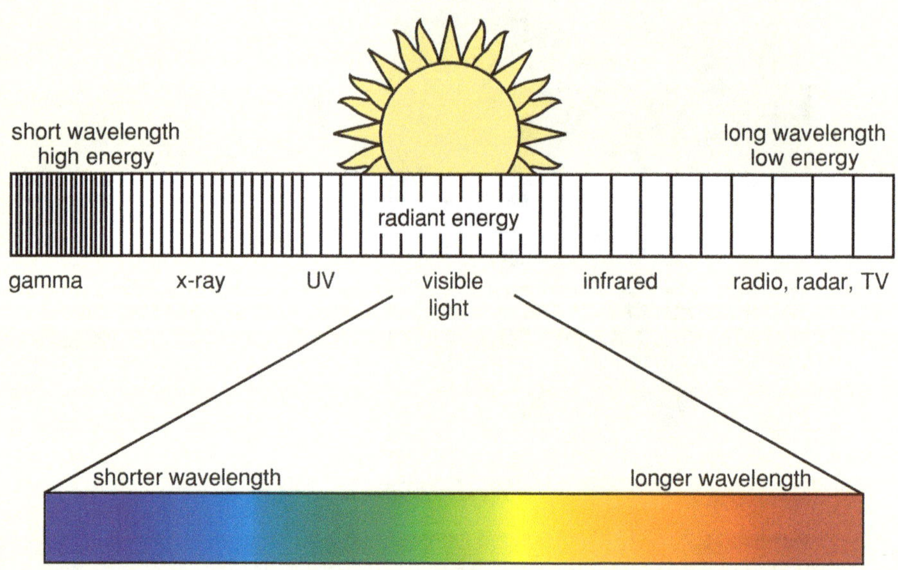

Figure 4.26 Spectrum of light energy. The Sun is the source of different types of radiant energy. The types include damaging ultraviolet (UV) light, light that can be detected by the human eye (visible light), and warming infrared light. Photosynthesis uses only a small portion of this spectrum of light energy.

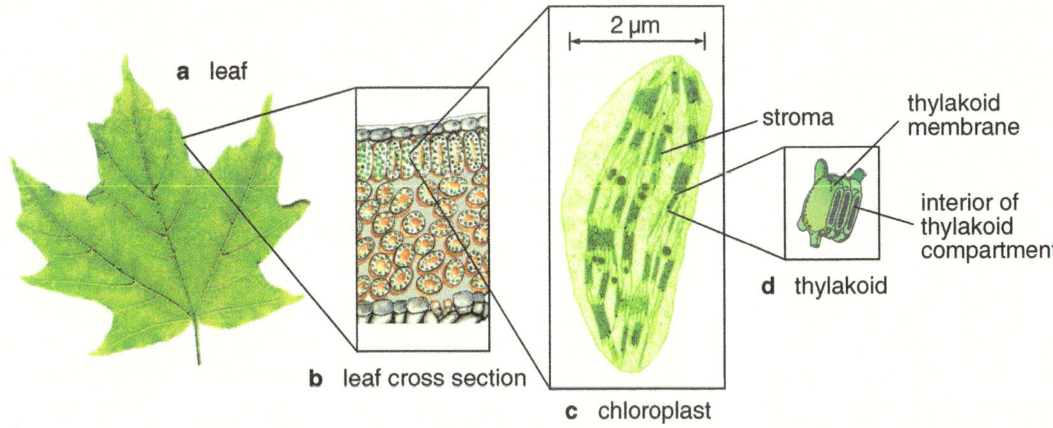

Figure 4.27 The site of photosynthesis. Where does photosynthesis take place? (*a*) This image shows a typical leaf. (*b*) The green disks in the cells in this diagram of a leaf cross section are chloroplasts. (*c*) This electron micrograph of a chloroplast shows the layers and stacks of the thylakoids. (*d*) The thylakoid membranes contain chlorophyll and other pigments. The symbol µm stands for micrometers (also called microns). One micron equals one-millionth of a meter.

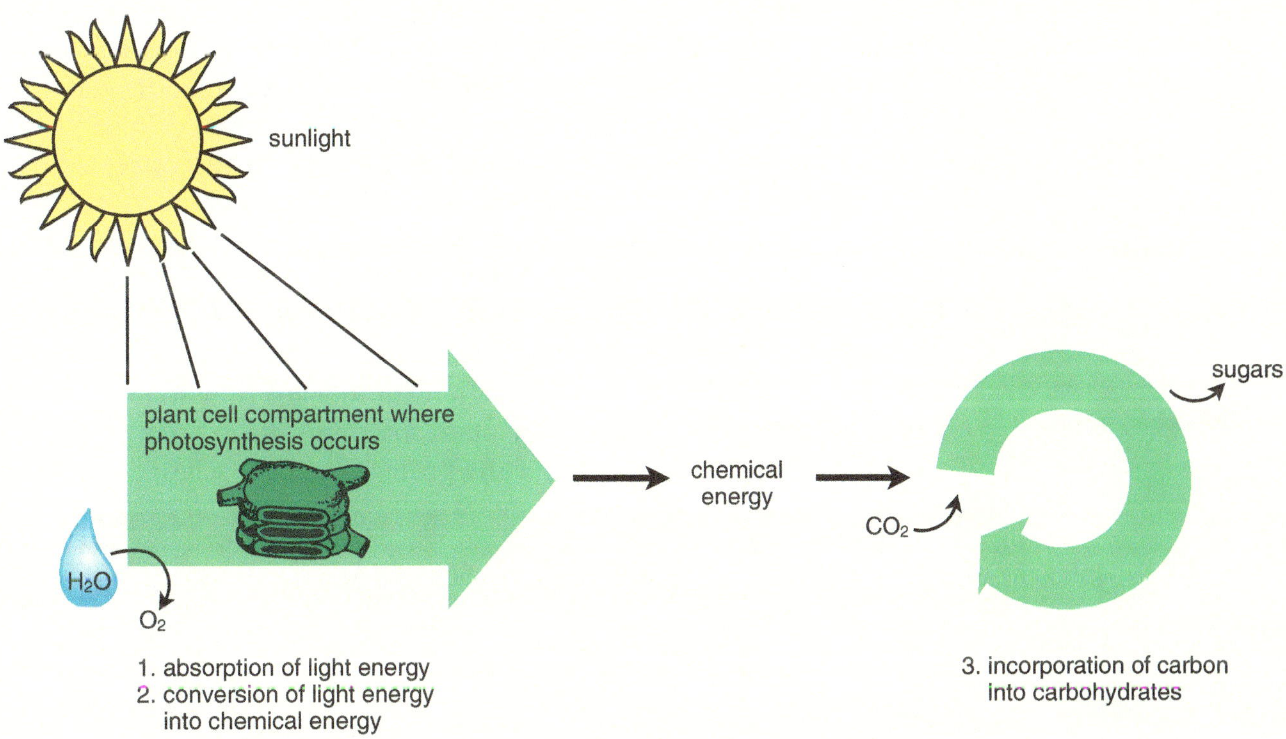

Figure 4.28 Summary of the reactions of photosynthesis. During photosynthesis, light energy is absorbed and converted into chemical energy (ATP and loaded electron shuttles). This step traps solar energy in matter. The chemical energy in these short-term storage molecules then powers the incorporation of carbon from carbon dioxide into carbohydrates. Carbohydrates are long-term energy-storage molecules. How can other organisms benefit from these long-term storage molecules?

The first phase of photosynthesis is, the "photo" part. In this phase, light energy is absorbed by a molecule called chlorophyll. Next, enzymes and an electron transport system like the one in cellular respiration help convert the light energy into chemical energy. The chemical energy is stored in ATP and loaded electron shuttles similar to ones used in cellular respiration. NADPH is very similar to the NADH used in cellular respiration. Oxygen is released as a by-product of this phase of photosynthesis.

Sugar molecules are made in the second phase of photosynthesis. This requires the chemical, energy from the ATP and loaded electron shuttles made in the first phase. It also requires carbon. The source of carbon to build carbohydrates is carbon dioxide from the air. Plants convert some of the sugars they make into starch for long-term energy storage. (Starch is made of many sugar molecules bonded together.) This is similar to the way humans use glycogen. Plants also use cellular respiration with these sugars to make ATP and other molecules they need. Of course, we and other heterotrophs also use plant sugars directly or indirectly—every time we eat.

SEP

4. With your team, read the following paragraph about the model system you will investigate.

   There are several ways to measure the rate of photosynthesis. You will use one model system of photosynthesis in this activity. Aquatic plants such as green seaweed, shown in figure 4.29, exchange gases with the environment just as land plants do. Gases such as carbon dioxide (CO_2) and oxygen (O_2) dissolve in water. When carbon dioxide dissolves in water, it combines with water (H_2O) to become carbonic acid (H_2CO_3). You can indirectly measure CO_2 concentration in an aquatic environment by measuring the pH. (pH is a measure of how acidic or basic a solution is.) When carbonic acid is added to water, the pH decreases. You can measure oxygen dissolved in water by using a dissolved-oxygen probe.

PROTOCOL

5. Read *The Photosynthesis Protocol* as a team to learn how to measure the gases exchanged by the aquatic plant *Anacharis*. Keep track of questions your team has about the protocol in your science notebook.

Figure 4.29
Green seaweed in an aquarium. Like the *Anacharis* you will use in your investigation, green seaweed grows under water. What gas do you think is present in the gas bubbles?

PROTOCOL

The Photosynthesis Protocol

1. If probeware is available, set it up according to your teacher's instructions.
   a. Calibrate the pH probe to a pH 4–11 range.
   b. Calibrate the dissolved-oxygen probe.
   c. Set up the thermistor.
   d. Choose appropriate settings to monitor the 3 probes (pH, oxygen, and temperature), save your data, and graph your results.

   Consult your teacher for help selecting the settings.

2. If you will not be using probeware, follow your teacher's instructions for locating and using the following materials:
   - pH test strips
   - thermometer
3. Put 125 mL of distilled water in a 250-mL flask.
4. Put a straw in the flask and blow gently through the straw into the water for 3 minutes.

   Blow into the end of the straw with the additional small hole in it. This helps prevent you from sucking any liquid into your mouth. Discard the straw after use. What are you adding to the water?

CAUTION: Avoid getting bromothymol blue in mouth or eyes, or on skin.

5. Add 2.5 mL of bromothymol blue to the water in the flask that you have been blowing into. Swirl the flask to mix. Note the color.
6. In the second 250-mL flask, add 2.5 mL of bromothymol blue and 125 mL of distilled water.
7. Use a pH probe or pH strips to test the pH of the water in both flasks.
8. Place 4–6 sprigs of *Anacharis*, cut end up, into a 250-mL beaker.
9. Fill the beaker three-quarters full with the mixture from the second flask (the flask you *did not* blow into).
10. Place the thermistor or thermometer in the beaker. If using, place the pH probe and the oxygen probe in the beaker.
11. Position the beaker within 25 cm of a light source, as shown in figure 4.30.

    If you use colored cellophane to modify the light source, position the light only 10 cm away from the *Anacharis* to adjust for the drop in light intensity that the cellophane causes. Do *not* attach the cellophane directly to the light source. The heat from the light source may cause the cellophane to melt.
12. Take pH, oxygen, and temperature readings for at least 30 minutes. Record values every 5 minutes.

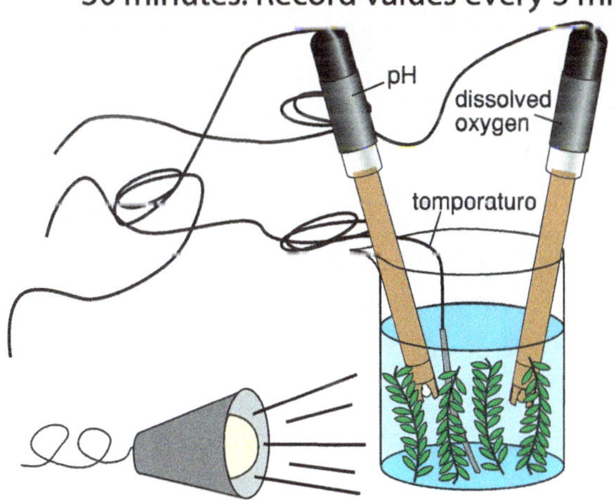

Figure 4.30 Follow this setup for measuring the rates of photosynthesis with probes.

6. Participate in a class discussion about *The Photosynthesis Protocol* and how it could be modified to test the effect of different variables on the rate of photosynthesis.
7. Use *The Photosynthesis Protocol* and the materials available to you to outline an experiment that tests the influence of 1 of the variables you discussed in step 6 on the rate of photosynthesis. In your outline, include the following:
   a. A testable question
   b. The factor you will vary
   c. How you will use and modify *The Photosynthesis Protocol* to test your question
   d. A control

   HOW TO

   For more information, refer to appendix C2, *How to Plan and Carry Out an Investigation.*
8. Predict how the following will change during your investigation. Write your predictions as hypotheses in a table like the one shown in Figure 4.31 in your science notebook.
   a. pH level
   b. Dissolved-oxygen level
   c. Temperature

   For example, if you were measuring the dissolved-carbon-dioxide level, your prediction might be "the closer the *Anacharis* is to the light, the faster the decrease in the level of carbon dioxide." Your hypothesis, written as an if-then statement, would be "If greater light intensity increases the rate of photosynthesis in *Anacharis*, then the rate of decrease in dissolved carbon dioxide will be faster when *Anacharis* is placed closer to the light source."

Factor	Predictions written as hypotheses. Use "if-then."
pH level	
Dissolved-oxygen level	
Temperature	

Figure 4.31 Predictions for each factor that could affect photosynthesis. Draw a table like this one in your science notebook. Record your predictions, written as hypotheses, for each factor.

9. Collaborate with your teacher and classmates to create a data table to record your data, including qualitative observations such as bubbling and color changes. Remember to make a column to record the time at which each set of observations is taken.

   HOW TO

   Review your design and procedure to help you decide what your data table needs to include. For more information, refer to appendix B4, *How to Use and Create Organizing Tables.*

10. Get your teacher's approval of your experimental plan and data table, arrange your setup to test your question and carry out your experiment.

    Remember to record all data and observations in your data table.

11. Remove the probes and rinse them with distilled water. Follow your teacher's instructions about where to return the *Anacharis* plants. ***Then wash your hands thoroughly with soap and water.***
12. Turn to a new page in your science notebook. You will be making 6 graphs on this notebook page. Using your control data, generate 3 separate line graphs down the left side of the page showing pH, oxygen, and temperature (on the y-axis) against time (on the x-axis).
13. Using data from your experimental setup (where you varied a factor), generate another 3 separate line graphs on the right side of the page showing pH, oxygen, and temperature (on the y-axis) against time (on the x-axis). Make sure the two graphs showing the same information, like pH, are next to one another.
14. Title each of your 6 graphs to indicate the design of your investigation. Then add highlight comments to each set of graphs. In addition to placing comments on each graph, write 1 or 2 highlight comments that *compare* what you see between your control and experimental graphs.

    An example title is "pH of water when *Anacharis* is placed 10 cm from light." For more information about highlight comments, refer to appendix B6, *How to Write Highlight Comments and Captions.*

    HOW TO

15. Compare your results with your predictions from step 8. Discuss your results with your team and be prepared to share your results with the class.
16. Participate in a class discussion about the results of the different teams and the effect of light on gas exchange in *Anacharis*.
    a. Present your team's results, including your graphs, to the class.
    b. How did pH, oxygen, and temperature change in the different investigations?
    c. How did the color of bromothymol blue change over the course of the lab? What does this chemical indicate?
17. Your teacher has a test tube containing *Anacharis* and bromothymol blue solution that has been in the dark for at least 24 hours. The solution in it was blue initially. Predict the pH and color of the solution now. Think about the evidence you have from the investigation to inform your prediction.
18. Study the test tube that your teacher will show you. Answer the following questions in your science notebook.
    a. What color is the solution? What pH is the solution?
    b. Which teams observed similar changes in their solutions? Propose reasons why they observed similar changes.

Analysis

Using the data from your investigation of photosynthesis and the information from the text, complete the following tasks and questions.

1. In part A of the activity, you visited a tree on your school's campus that started as a little seed and were asked, *"What provided most of the mass that made the tree grow so large?"* Review the opinion you agreed with and your reasoning. How has your thinking changed about what provided most of the mass that made the tree grow so large? Use evidence from the investigation and reading to support your answer.
2. Create a concept map of the process of photosynthesis by using the following words and any other words you would like to add. Remember to use verbs to connect the concepts to show relationships. Arrange the concept map so that the most universal concept is in the middle.
   a. Light energy
   b. Photosynthesis
   c. Chemical energy
   d. Chloroplast
   e. Water
   f. Carbon dioxide
   g. Oxygen

HOW TO

For more information about concept maps, refer to appendix B7, How to Construct a Concept Map.

Further Challenges

1. Begin to understand the intricacies of photosynthesis such as the flow of electrons and carbon fixation by reading sidebar, *Photosynthesis: A Closer Look.*
2. Explore the diversity of photosynthesizing organisms by reading the sidebar, *Photosynthesis Isn't Just for Plants.* Then use the Web to research the scientist Lynn Margulis and her theory of endosymbiosis. Your research will help you deepen your understanding of the evolution of plants.

SIDEBAR

ENGLISH SPANISH

Photosynthesis: A Closer Look

Plants, algae, and some eubacteria achieve something that humans and other animals can only ever dream about: synthesizing food from sunlight and air. Wouldn't it be wonderful to have an organ sticking out of your head that carried out photosynthesis? You would never suffer hunger pains sitting in class, just before lunch. Lunch would be "served" anytime you were exposed to enough sunlight. Let's look in detail at what goes on inside the cells of photosynthesizers such as *Anacharis*.

Bringing Solar Energy into Living Systems

The reactions of photosynthesis take place in chloroplasts, small compartments inside certain plant cells. Figure 4.27 shows a plant leaf and its chloroplasts. A system of membrane-bound compartments called the **thylakoids** is found inside the chloroplast. The thylakoids provide a lot of membrane surface area. Surrounding the thylakoids is a colorless fluid known as the **stroma**.

Embedded in the thylakoid membranes are pigment molecules such as **chlorophyll**. Chlorophylls and other pigments give plants their color. These pigments absorb light energy in the visible wavelengths. The absorbed energy causes a flow of electrons (e^-), as shown in figure 4.32. Electrons are removed from water molecules and passed to chlorophyll molecules, then passed to an electron transport system in the same membrane. Some of the energy from the electrons is used to pump hydrogen ions (H^+) across the thylakoid membranes. The protons accumulate inside the thylakoids and form a concentration gradient. Just as in aerobic cellular respiration, the protons then diffuse down the concentration gradient through an enzyme complex in the membranes. The enzymes synthesize ATP as the protons pass through the complex. In addition to producing ATP, the flow of electrons produces loaded electron shuttles, similar to those formed during cellular respiration.

You might have noticed that chlorophyll loses electrons during photosynthesis. How are they replaced? In fact, new electrons are "stolen" from water. Photosynthesis cannot take place without water. Enzymes in the membrane remove electrons from water molecules. However, when the electrons are removed, water splits into oxygen (O) and hydrogen ions (H^+). The plant releases the oxygen from the water as oxygen gas (O_2). Oxygen is a by-product of photosynthesis. But it is no ordinary waste product. It is the same oxygen upon which nearly all organisms depend for aerobic cellular respiration.

The first phase of photosynthesis results in three products. They are oxygen gas, ATP, and loaded electron shuttles (in the form of NADPH, a molecule very similar to NADH). The ATP and electron shuttles are now available to provide energy for the second phase in photosynthesis. In the second phase, the reactions form carbohydrates. This important process converts the energy trapped in ATP and electron shuttles into long-term energy storage (sugars and starch).

Bringing Carbon into Living Systems

The final reactions of photosynthesis use ATP and loaded electron shuttles from

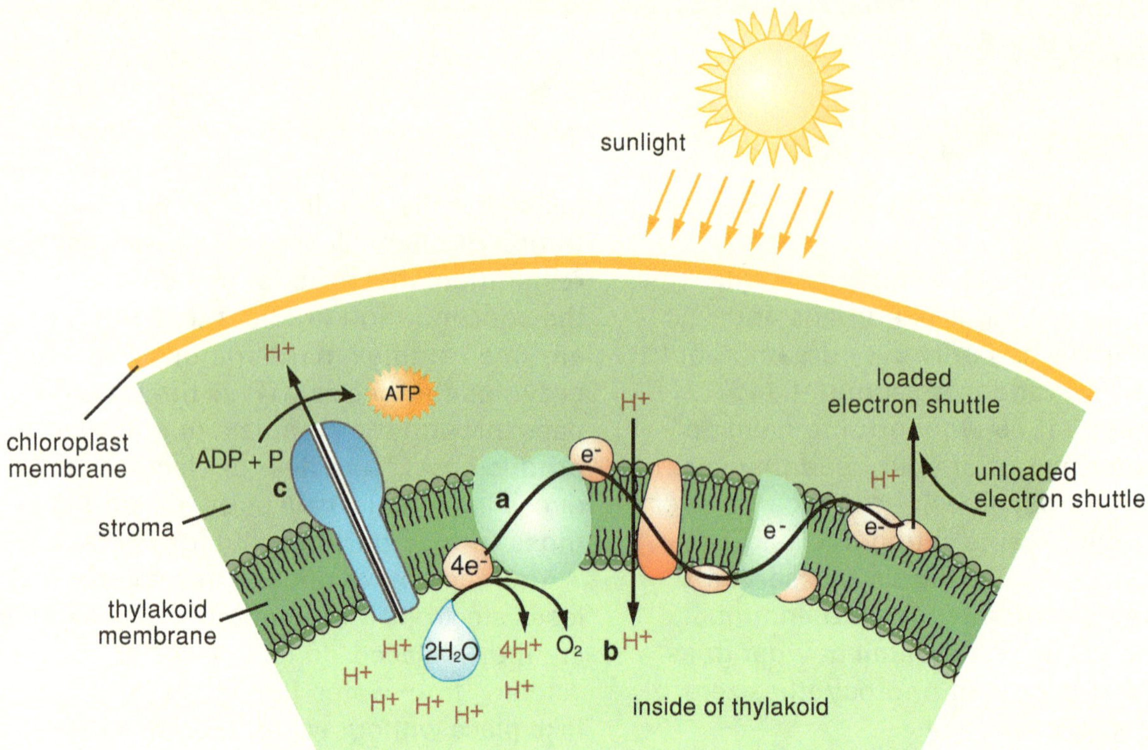

Figure 4.32 **ATP production in the chloroplast during photosynthesis.** (*a*) Absorption of light energy sets up a flow of electrons from water through pigments like chlorophyll and other molecules in the thylakoid membranes. (*b*) Protons accumulate on the inside of the thylakoids. The resulting proton gradient supplies potential energy. (*c*) This energy enables a membrane-spanning enzyme to synthesize ATP from ADP and phosphate.

the first phase of photosynthesis. The final phase rearranges atoms from carbon dioxide to form carbohydrates. Think about this fact for a minute. Carbon dioxide, an invisible gas in the air around you, is made into carbohydrates. These carbohydrates can be used to form the other molecules of life that become plant leaves, stems, and roots. Photosynthesis makes the invisible visible.

The process of converting carbon dioxide into carbohydrate molecules is called **carbon fixation**. The reactions incorporate, or *fix*, carbon into carbohydrates. Figure 4.33 is a simplified diagram of those carbohydrate-producing reactions. These reactions take place in the stroma of the chloroplast. First, carbon dioxide from the air enters the plant through the stomata. The carbon dioxide is added to an existing 5-carbon sugar, which creates a 6-carbon sugar. The 6-carbon sugar quickly splits into two 3-carbon sugars. These 3-carbon sugars have several possible fates.

In a cycle of reactions, these 3-carbon sugars are rearranged into a variety of other sugars. Some become the 5-carbon sugar that first combines with carbon dioxide. Others are exported from the chloroplast and used to form sucrose and starch. The carbon also can be used to form lipids, proteins, chlorophyll, and

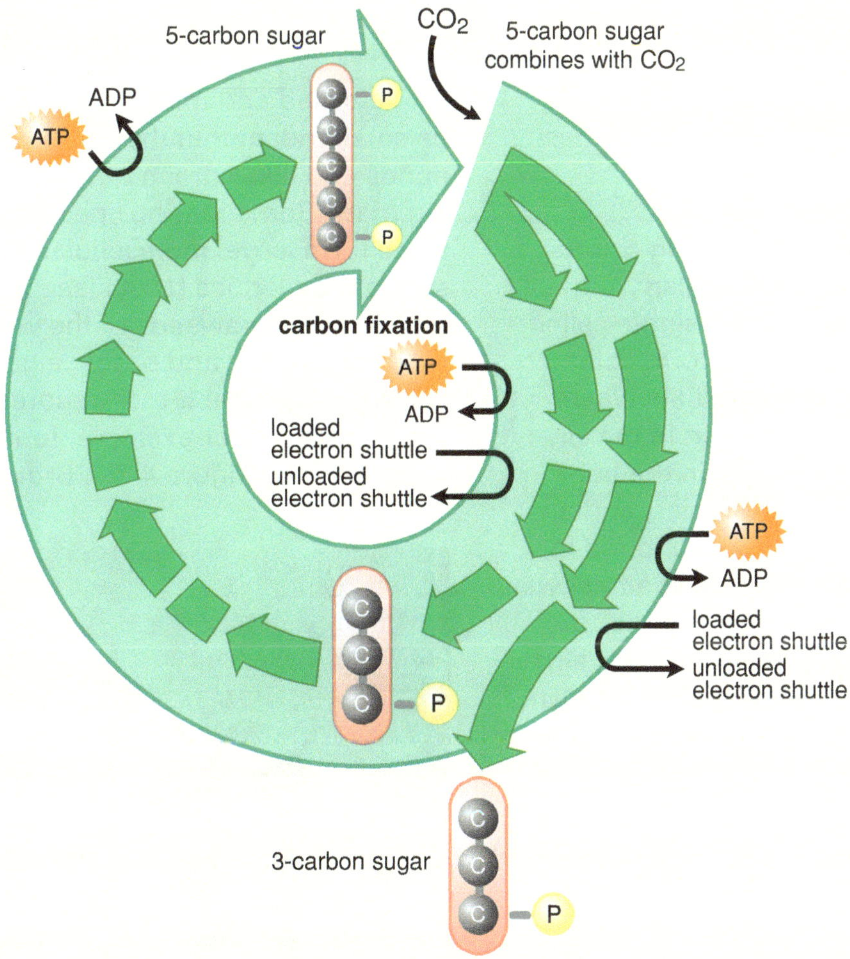

Figure 4.33 Carbon fixation. Carbon fixation takes place in the stroma of chloroplasts. The process uses carbon from carbon dioxide to build carbohydrates. The carbon fixation reactions organize matter and increase the total amount of carbon in the plant. ATP and NADPH provide energy for these reactions. The 3-carbon sugars that are produced can be exported from the chloroplast. Or they can proceed through many steps that regenerate the 5-carbon sugar that combines with carbon dioxide.

other molecules the plant cell needs. Sucrose can be transported to provide fuel to distant tissues of the plant such as the roots. Plant cells use sucrose as a source of matter (carbon) for producing new tissues and for energy. In that way, plant cells that cannot perform photosynthesis can obtain energy to live. All plant cells use the process of cellular respiration to break down sucrose and produce ATP. For that reason, plants are said to be producers. They produce their own "food" as well as food for other organisms.

SIDEBAR

ENGLISH SPANISH

Photosynthesis Isn't Just for Plants

Like plants, most algae contain chlorophyll and other pigments and carry out photosynthesis. They may be single-celled or multicellular, or they may form colonies. Algae can be found just about anywhere you find tiny droplets of water: in the air, on tree trunks and branches, in the bottoms of streams, on soil particles, and on rocks at the seashore. Algae are found in diverse environments that include freshwater habitats, the ocean, desert sands, and hot springs. In aquatic environments, algae are the dominant photosynthetic organism. There are many different groups of algae. Although all algae are photosynthetic, not all of them appear green. Examples of algae include green algae, red algae, brown algae, diatoms, and dinoflagellates (figure 4.34).

Green algae are the most widely studied and the most similar to plants. You can find them in your aquarium at home or on the shady side of trees. Many green algae are single-celled and microscopic, but they may be so abundant that they color the water of ponds and lakes green or form huge floating mats. Some marine species of green algae form large, multicellular seaweeds, such as sea lettuce (*Ulva*) (see figure 4.35).

The major seaweeds of the world are multicellular red and brown algae. The brown algae inhabit rocky shores in cooler regions. Kelps, for example, form extensive offshore beds (figure 4.36). Brown algae

Figure 4.35 Sea lettuce. Sea lettuce (*Ulva lactuca*) is a type of multicellular algae found on rocky shores.

Figure 4.34 Mixed algae. How many types of algae can you see?

Figure 4.36 Kelp. Kelp forests provide shelter and protection for many marine organisms.

appear greenish-brown because they contain gold and brown pigments as well as chlorophyll. Most seaweeds in warm oceans are red algae. They generally grow attached to rocks or other algae. Because red algae absorb blue light, which penetrates farther into the ocean water than other wavelengths of light, they can grow at greater depths than any other algae.

Diatoms are abundant in both freshwaters and marine waters. They are the primary photosynthetic organism in oceans. Diatoms have glasslike cell walls made of silica. They are some of the most intricately patterned creatures on Earth. Electron microscope studies have shown that the patterns on their cell walls are created by pores that connect the inside of the cell with the outside environment. As shown in figure 4.37, the silica cell walls of diatoms remain intact for many years after the organism has died. Great masses of shells that have slowly accumulated over many years are now mined as diatomaceous earth. That material is used in many ways, for example, as a filter (for swimming pools), as an abrasive in silver polish, and as a brightener for the paint used to mark highways.

Dinoflagellates are found primarily in marine waters. About half of the dinoflagellates are photosynthetic. In mid- to late summer, a surplus of nutrients can cause blooms of photosynthetic dinoflagellates. When dinoflagellates occur in large numbers, the water may appear golden or red, producing a *red tide*. When this happens, marine species may suffer because some species of dinoflagellates produce a neurotoxin, a poison that damages the nervous system. Humans are sometimes affected by red tides after eating fish or shellfish containing neurotoxins.

Some eubacteria are photosynthetic. However, most photosynthetic eubacteria do not produce oxygen as a by-product of photosynthesis. Some examples are purple bacteria, green sulfur bacteria, green nonsulfur bacteria, and heliobacteria. Cyanobacteria are the only group of eubacteria that produce oxygen through photosynthesis.

Purple bacteria are widespread, especially in anaerobic (without oxygen) environments such as sewage treatment ponds (figure 4.38). Green sulfur bacteria are found in the anaerobic zone at the bottoms of lakes.

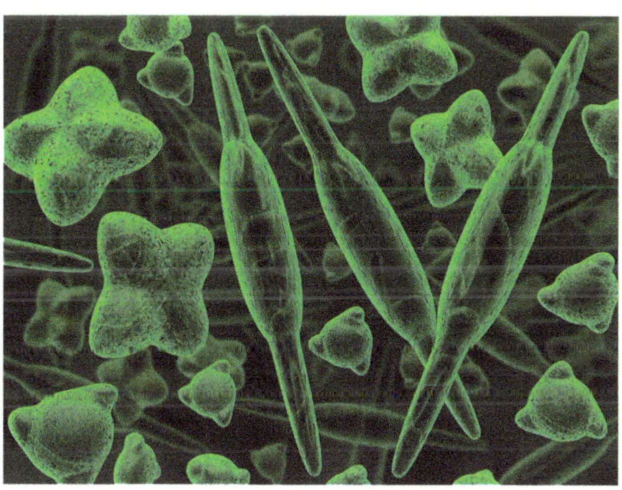

Figure 4.37 Diatoms. Each species of diatom has a characteristic shape. Notice the striking geometric patterns.

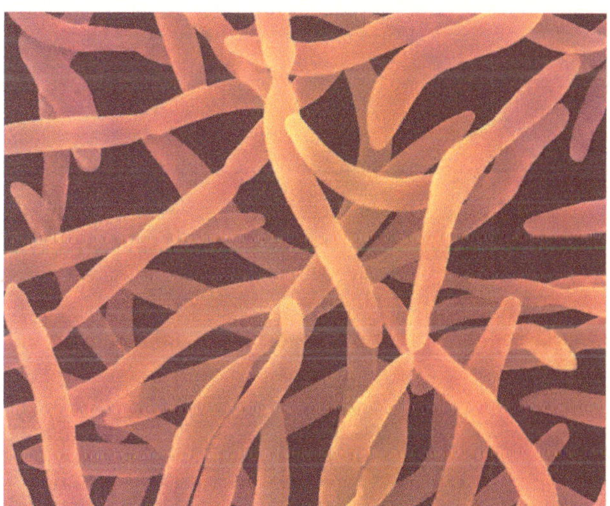

Figure 4.38 Purple bacteria. Most species of purple bacteria are anaerobic. Unlike cyanobacteria, purple bacteria do not produce oxygen as a by-product of photosynthesis.

Green nonsulfur bacteria are often found in microbial mats with cyanobacteria. Heliobacteria live in soil environments. Cyanobacteria are hardy and are found in almost every environment where light is available. They live in freshwater, marine, and terrestrial environments. They also live in extreme environments such as hot springs and rocks found in Antarctica, as well as scorching deserts.

ELABORATE

Modeling Photosynthesis and Cellular Respiration*

Although you have learned about cellular respiration and photosynthesis, it can be difficult to visualize these cellular processes. Imagine being small enough to fit inside of a plant or animal cell. What would you see? In this activity, you and your classmates will model the processes of photosynthesis and cellular respiration to help you better understand the relationship between the two. Follow the link provided below to complete this activity: https://www.calacademy.org/educators/lesson-plans/modelling-photosynthesis-and-cellular-respiration

*Pages 169–172 have been removed.

Coevolution of Earth Systems and Life on Earth

ELABORATE

EVALUATE

Throughout this unit you have investigated how energy and matter move through biotic and abiotic systems. In this section, you will consider how geological and biological factors on Earth influence each other. Earth's early atmosphere was very different than that of today—there was no oxygen! Over Earth's 4.5 billion year history, the atmosphere has changed radically. Today, the composition of the atmosphere is 78% nitrogen, 21% oxygen, 0.9% argon and 0.1% trace gases. In this activity, adapted from HHMI BioInteractive, you and your team will investigate how both atmospheric concentrations of oxygen as well as how life and geological processes on Earth have changed in response to each other. You'll collect evidence throughout your investigation to determine which has a more powerful influence on the other—Earth on life or life on Earth?

Materials

Part A

Statement Posters

Part B

Click and Learn: *Deep History of Life on Earth*, HHMI BioInteractive

Post-Its (per student)

Poster: Earth Evolution: *The Intersection of Geology and Biology*, HHMI BioInteractive

Part C (per team)

Team directions and tasks document

Click and Learn: *Geological History of Oxygen*, HHMI BioInteractive

Computer and Internet Access

Poster: *Earth Evolution: The Intersection of Geology and Biology*, HHMI BioInteractive

Poster: *The Making of Mass Extinctions*

Part D (per student)

Appendix B11: *How to Use the Scientific Argument Writing CER Strategy*

Access to chapter resources and science notebook

CER Graphic Organizer (optional)

PROCESS AND PROCEDURES

Part A: Take a Stand

1. Take a moment to think on your own about two questions: What would happen if there were no photosynthesis? What would the consequences be? Respond to the questions by journaling in your science notebook.

2. Partner with a classmate and use Pair Share Sentence Stems in appendix B11 to share your thoughts about the two questions and to clarify, reinforce, or expand on something that you heard.
3. Contribute to the larger class share-out by contributing something that you and your partner discussed about the consequences of photosynthesis ceasing to occur. Add any new ideas in your science notebook.
4. You and your classmates have shared many opinions today. Read this question and be prepared to take a stand by physically standing near the statement in your classroom that you agree with most. **Question: Which has a more powerful influence on the other—Earth on life or life on Earth?**
5. Turn to 1-2 classmates that are gathered with you and provide the reasoning behind your choice.
6. Take a moment to journal in your science notebook about the stance you took as well as the thinking and reasoning that was shared by classmates.

Part B: The Intersection of Geology and Biology

1. You've just taken a stance. Now it's time to investigate which has a more powerful influence on the other—Earth on life or life on Earth. Your teacher will be using a HHMI BioInteractive Click and Learn, *Deep History of Life on Earth*, to highlight and walk you through events during the 4.5 billion year history of Earth. The highlighted events or evidence range from chemical evidence for life found in rock samples dating back to 3.8 billion years ago to the diversification of land plants 360 million years ago.
2. Your job is to collect evidence for Earth or life having a more powerful influence on the other. Record that evidence as it is highlighted on Post-It notes with the date (how many billion or million years ago).
3. After walking through the Deep History of Life on Earth, look at the evidence you have recorded. With your team, combine like evidence into bundles. Then, determine if each bundle of evidence supports that Earth had a greater influence on life or if life had a greater influence on Earth.
4. Participate in a class share and discussion of evidence. Be prepared to share:
   a. Evidence that supports that Earth had an influence on life and the source for that evidence
   b. Evidence that supports that life had an influence on Earth and the source for that evidence
5. Look at the poster, *Earth Evolution: The Intersection of Geology and Biology*, that has been distributed to your team. Look at the left side of the poster and read the blue notes. Then, look at the right side and

read the green notes. Which side would you title "Earth" and which side would you title "Life?" Use a Post-It to title each side at the top of each timeline.

6. When were the events that you documented during the *Deep History of Life on Earth* Click and Learn? Were they caused by the Earth changing (Geology) or by life changing (Biology)? Place the Post-Its on the side of the poster at the appropriate place on the timeline.
7. Take a picture of your poster before moving on to Part C.

Part C: Geological History of Oxygen

1. Your team has taken a peek at the geological and biological history of Earth. One thing you may have noticed during the initial collection of evidence is that the concentration of oxygen in the atmosphere has changed over Earth's history. You and your team will now learn about the biological and geological factors that have influenced these changes.
2. Look at the Team Directions your teacher has provided you with. You will have approximately 10 minutes to complete the tasks at each of 6 stages illustrated in figure 4.40. Your job is to make sure you continue collecting evidence at each stage for Earth or life having a more powerful influence on the other.

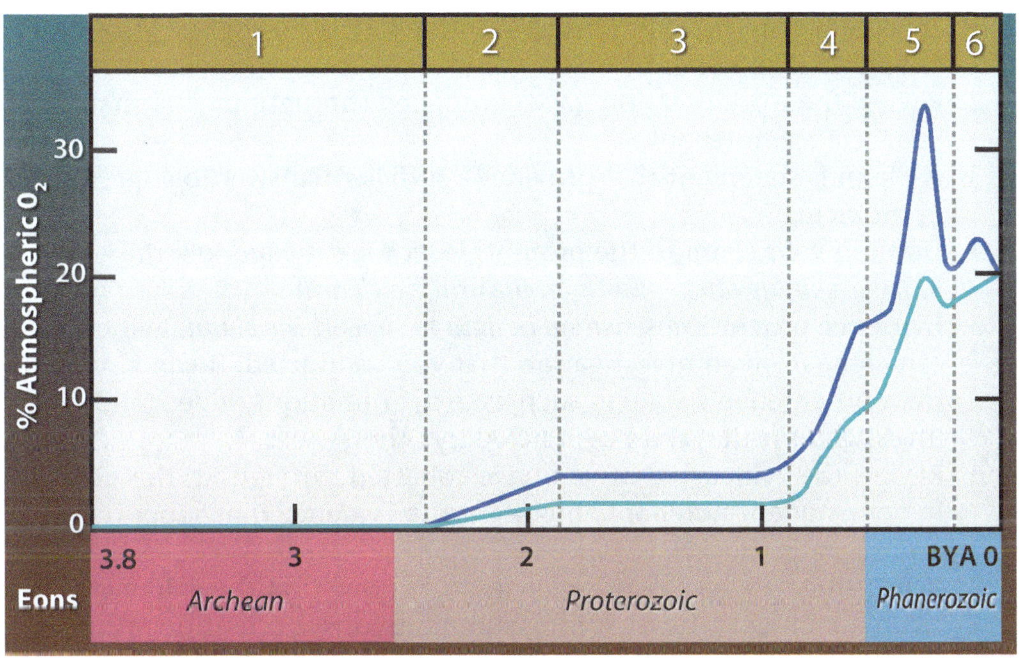

Image courtesy of Howard Hughes Medical Institute–HHMI BioInteractive, http://www.hhmi.org/biointeractive, as reproduced with permission from The Royal Society.

Figure 4.40 Atmospheric concentration of oxygen throughout Earth's history What does this graph tell you about how the concentration of oxygen in the atmosphere has varied in the history of Earth?

3. Collect evidence that will help you support your position about which has a more powerful influence on the other, Earth on life or life on Earth? Note the source of the evidence in your science notebook. Use a table like the in figure 4.41. You can also include the evidence from Part B: *The Intersection of Geology and Biology*.
4. Complete each stage with your team. Be prepared to review evidence that you collected with your classmates and teacher during class check-ins.

Evidence that Earth influences life	Source	Evidence that life influences Earth	Source

Figure 4.41 Evidence that Earth influences life and that life influences Earth. In your science notebook, record your evidence and the source of the evidence in a table like this one.

Part D: Defend Your Stand

1. In this Evaluate part of the activity, you will defend your individual stand on which has a more powerful influence on the other—Earth on life or life on Earth? You are expected to provide multiple sources of evidence and reasoning to support your claim. Note: After collecting evidence, your original stand may have changed from that of Part A: Take a Stand.
2. Review the work that you have completed throughout this chapter in your science notebook.
3. A Claim is statement that provides a sufficient answer to a question or problem.
4. Develop your claim to the prompt: *Which has a more powerful influence on the other—Earth on life or life on Earth?*
5. **Evidence** provides *observation or data to support the claim.* Evidence can come from an investigation that you conducted. It can also come from other reliable sources such as text, animations, video, and investigations that were conducted by others.
6. Review the evidence that you have collected throughout this chapter. In your student notebook, highlight the evidence that supports your claim.
7. **Reasoning** provides *justification* (why or how) that the evidence and claim are linked. The reasoning should involve a scientific principle or idea to describe why or how the evidence you collected supports your claim.

   Use the *CER Graphic Organizer* (optional) to organize your thoughts and generate a written response to the prompt in step 4. Remember, you are expected to provide multiple sources of evidence and reasoning to support your claim.

Analysis

Discuss the following questions with your team. Then, answer them in your science notebook.

1. Earth's early atmosphere had no oxygen. Use your understanding of plant photosynthesis to explain how photosynthetic organisms made the evolution of organisms that use aerobic cellular respiration possible.
2. Describe one way that life influenced Earth.
3. Describe one way that Earth influenced life
4. Which had a larger influence on the other, Earth on life or life on Earth? What evidence do you have to support your claim?

Earth is teeming with life. You are probably familiar with many common types of living things such as large animals and plants. However, there are a large number of living species that you have probably never seen. Microbes like the *Streptococcus*, or strep, bacteria in the unit opener collage are by far the most numerous organisms on Earth. Other organisms like the styptic fungus are found growing at the base of trees. This fungus is bioluminescent, emitting an eerie green glow. The tiny poison dart frog is no larger than two inches but carries a poison that can be fatal to many animals, including humans. The giant padma plant has the largest single flower in the world, which can be up to three feet across. It is sometimes called the corpse flower because of the smell given off by the flower when it blooms. During that time, the flower smells like decomposing flesh. Biologists use the term *diversity* to refer to the wide range of different types of organisms.

How did Earth come to have such a great diversity of organisms? On the other hand, what do the organisms living on Earth have in common with one another? Can we learn anything about humans from our knowledge of these diverse organisms? What can we learn from evidence about ancient humans?

These questions do not have simple answers. Scientists and philosophers have been asking these questions for thousands of years. Biologists study the diversity found in living systems by looking for patterns of evidence. For example, they study the patterns in the information, sequences of events, and chemical and physical structures of living things. Biologists also try to understand the processes that lead to the formation of diversity. Understanding these processes helps us address questions about the diversity of life and how it came to be. Scientists learn about these patterns and processes by asking questions, gathering information, and proposing explanations. Through the process of science, scientists have developed evolutionary theory, which provides an explanation for why life is so diverse yet shares a set of fundamental characteristics. These analyses also help us understand some things about humans.

UNIT 3

EVOLUTION: Change in Living Systems

In unit 3, you will explore change in living systems from a scientific viewpoint. The following goals will help you learn the big ideas in this unit. By the end of unit 3, you should be able to

- ✔ demonstrate your understanding of how evolution explains the unity and diversity of life and
- ✔ use evidence to develop explanations about the living world.

UNIT CONTENTS

CHAPTER 5

Evolution: Change Across Time

They were about to return to camp when Johanson noticed part of a fossilized arm bone lying on the ground partway up the slope of a gully.

Scenario: Digging Up the Past

Look at the fossilized bones in the photograph. What animal do you think of when you see these bones? It turns out that these bones are from a primate that lived about 3.2 million years ago. Fossil skeletons give scientists a tremendous opportunity to study the physical characteristics of organisms that lived long ago. As you look further and further into the past, most of the skeletons of ancient organisms look less and less like organisms that are alive today. Why? It is because populations of living organisms change across time, an idea you will explore in this chapter.

At the beginning of chapter 5, you will think about the fossil skeleton in the photograph. You will imagine how she may have looked and behaved at the time she lived and how the line of ancestors that led to her may have changed over time. Next, you will explore reproduction in populations and examine how individuals within a population differ. These observations will help you understand a model of how a population changes over time. Next, you will consider two different views on how the process of change in populations works. You will use your growing understanding to evaluate evidence that evolution has occurred. You will also make a timeline of Earth's history to help you appreciate the large time spans over which life has evolved. Finally, you will evaluate your understanding of evolution by analyzing an example of evolution in action.

GOALS FOR THE CHAPTER

Throughout chapter 5, you will work to gain an understanding of the following concepts:

- ✔ Many different sources of evidence support biological evolution.
- ✔ Natural selection is a major mechanism of evolution that can result in adaptations.
- ✔ Life has been shaped by evolution, extinction, and geologic events over vast periods of time.
- ✔ Scientific studies often develop models of natural phenomena to assist in making explanations and predictions.

To help you develop these understandings, you will complete the following activities.

- **ENGAGE** Lucy
- **EXPLORE** Exploring Change
- **EXPLAIN** Explaining Adaptation
- **ELABORATE** Evidence for Evolution
- **ELABORATE** Modeling Earth's History
- **EVALUATE** Evolution in Action

CHAPTER 5

Evolution: Change Across Time

ENGAGE

Lucy

Key Idea: Scientists use evidence and logical inference to understand how organisms have changed through time.

Linking Question:
What characteristics of populations could lead to changes over time?

EXPLORE

Exploring Change

Part A: Seeds of Change
Part B: Celebrate Diversity
Part C: Modeling Change

Key Idea: More offspring are born than can survive, and individuals in populations have some differences. This situation leads to a process of change over time.

Linking Question:
What process of evolution leads to adaptations?

EXPLAIN

Explaining Adaptation

Key Idea: Adaptations are a product of natural selection.

Linking Question:
What is the evidence that life has changed over time?

EVALUATE

Evolution in Action

Key Idea: Evolution by natural selection can result in changes in populations that affect humans.

Linking Question:
How do natural selection and evolution work in today's world?

MAJOR CONCEPTS

- ✔ Many different sources of evidence support biological evolution.
- ✔ Natural selection is a major mechanism of evolution that can result in adaptations.
- ✔ Life has been shaped by evolution, extinction, and geologic events over vast periods of time.
- ✔ Scientific studies often develop models of natural phenomena to assist in making explanations and predictions.

ELABORATE

Modeling Earth's History

Key Idea: Events in Earth's history took place over vast periods of time.

Linking Question:
On what scale did the changes that support evolution take place?

ELABORATE

Evidence for Evolution

Key Idea: Multiple lines of evidence support the theory of evolution.

ENGAGE

Lucy

Populations of living things on Earth change over time. Scientists use evidence and logical inference to reconstruct these past events, the same way you did when you studied the Iceman. For example, since the mid-1800s, scientists have been piecing together the long line of species and populations from which modern humans emerged. This long line of ancestors is called a **lineage**. Some of the most important pieces of this complex puzzle are the fossilized skeletal remains of individual hominids that lived millions of years ago. **Hominids** are erect-walking primates that include modern humans and extinct, humanlike species. Dated at about 3.2 million years old, the skeleton shown in the opening photo is one of the oldest and most complete hominid fossils. Scientists named the skeleton Lucy, after a song that was popular when it was recovered.

But how do populations of living things change? Writing your ideas about this question will be your main task in this Engage activity, *Lucy*. Both evidence and inference will guide your thinking on the topic.

PROCESS AND PROCEDURES

1. Read the scenario, *Digging Up the Past,* to learn about the discovery of Lucy. As you read, write a statement in your science notebook about the aspects of inquiry in which the scientists were participating.

   Use the How Science Works diagram (figure 1.3) to review all the practices used by scientists.

2. Describe in your science notebook how Lucy may have looked. Use words and sketches to convey your ideas.

   Use the photograph of the fossil skeleton in the opening photo to this chapter and your notes from chapter 1, *The Human Animal*, to help you develop your description. In particular, consider how Lucy's hands, feet, and posture might have looked and how she might have moved.

3. Describe 3 things Lucy might have done during a typical day and how she might have gone about doing them. What ideas did you use to decide how she might have done each activity?
4. Make a T-table in your science notebook with the 2 columns labeled "evidence" and "inference."

   HOW TO

   The resource *How to Use and Create Organizing Tables*, in appendix B4, can help you with this task.

SCENARIO

Digging Up the Past

No one knows how Lucy died. She apparently died quietly. If a lion or a leopard had killed her, her bones probably would have been splintered and crushed. Instead, she died by the edge of an ancient lake, where mud and sand covered her.

Lucy lived in what is today the Afar desert, a remote region of northeastern Ethiopia in East Africa (see the map in figure 5.1). The large lake that once existed there has long since dried up. The area is now hot and sandy. Even though she was fully grown and probably in her 20s when she died, Lucy stood only 1 m tall. Her head was a bit larger than a softball, and her brain could not have been much larger than that of a modern chimpanzee. The shapes of her knee joint and pelvis bones indicate that she walked upright on two legs. For more than 3 million years, Lucy remained buried in the ancient lake bed.

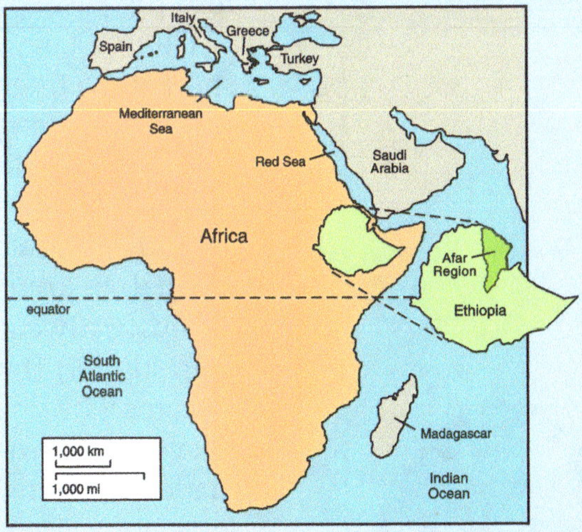

Figure 5.1 Map of the Afar region in Ethiopia.

Finding Lucy

In 1973, Donald Johanson, a young anthropologist from the United States, arrived in Ethiopia to look for fossils. Working with two Frenchmen, Maurice Taieb and Yves Coppens, Johanson found a primitive primate knee joint that had washed out of a slope during a rain. It seldom rains in the Afar desert, but when it does, the rain is heavy, cutting gullies into the gravel and bare rock. On very rare occasions, these rains uncover ancient fossils. As luck would have it, such a rain occurred not long before Johanson's arrival. This fossil knee joint was intriguing because it was about 3 million years old. Its structure indicated that this individual had walked erect.

Because of this find, Johanson decided to return to the same ancient lake bed the following year to continue the search. If any more humanlike fossils were embedded in the ground, they might be of a similar age. Johanson and his colleague, Tom Gray, had been searching all fall with no success when they parked their vehicle on the slope of a gully on the morning of November 30, 1974.

The temperature had reached 43°C (110°F), and they were about to return to camp when Johanson noticed part of a fossilized arm bone lying on the ground partway up the slope of a gully. As he searched further, he found pieces of a skull, thighbone, and pelvis, along with other skeletal parts. Remarkably, the skeletal parts all seemed to be from one individual, and a very humanlike individual at that. The two scientists barely could contain their excitement. They had just made an amazing find.

5. Fill in your T-table with examples as you read the following information about how scientists use evidence and inference to gain insight into the past.

   Though Lucy's complete skeleton was not recovered, scientists are still able to infer many things about her. For example, the structure of Lucy's pelvis (the bones that make up the hips) has been carefully documented. Her pelvis was compared with those of modern chimpanzees, apes, and humans. Humans walk upright. Scientists call this bipedal. Apes and chimps are knuckle-walkers and are not bipedal. In knuckle-walking, animals use all four limbs. The fingers stay partially curled under so that these animals use their knuckles to help support them while they walk. Lucy's pelvis shared many features with the pelvis of modern humans. From the evidence of Lucy's fossilized pelvis and the pelvis of modern humans, scientists made the inference that Lucy walked upright. Evidence from Lucy's knee bone also supports this inference.

   Scientists agree that hominids, apes, and chimpanzees came from one species that existed in the past. This extinct species is called the common ancestor of all three species. Recent estimates suggest that this ancestral species lived most recently about 7.3 million years ago. Chimpanzees, apes, and all other primates are not exclusively bipedal. Based on this evidence, scientists infer that the common ancestor of apes, chimpanzees, and hominids was not bipedal. What does this mean about Lucy? In the lineage from the most recent common ancestor of apes, chimps, and hominids (which existed about 7.2 million years ago) to Lucy (who lived 3.2 million years ago), there was a change from not being bipedal to being bipedal.

6. Write 1 or 2 short paragraphs that summarize how you think a scientist would explain how the lineage that led to Lucy changed over time. Explain how the lineage changed from an ancestral population that was not bipedal to a population that was bipedal.

   In this course, you are learning to see the world the way that scientists do. It is important to remember that in science, explanations must be based on evidence from nature. Natural causes can be tested by other scientists. Scientists cannot make explanations based on forces that are outside of nature. Such explanations are not testable. It is OK not to know at this point how scientists describe the process of change over time. But it is important that you make your best attempt. You will revisit your answers later in the chapter.

Analysis

Use the information you recorded in your science notebook during this activity to answer the following questions as part of a class discussion.

1. Compare hominids from Lucy's lifetime with hominids living today (humans) by answering these 2 questions.
   a. Do you think there have been more changes in the physical characteristics of the body (such as hands, feet, head, and posture) or more changes in how hominids lived (types of shelter, ways of getting around, and ways of gathering food)? What are your reasons for your answer?
   b. Do you think the processes that cause change in physical characteristics are the same or different from the processes that cause change in how hominids build shelters? Give reasons for your answer.
2. Examine the answers you gave to *Process and Procedures* steps 2 and 3 and answer the following.
   a. Which aspects of your descriptions did you base on evidence?
   b. Which aspects of your descriptions were inferences related to evidence?
   c. Which aspects of your descriptions were guesses?

Exploring Change

EXPLORE

Lucy was an individual who was a part of a larger population. Biologists define a **population** as a group of individuals of the same species that live in the same place at the same time. Lucy's population had characteristics that were different from the populations of hominids before her time. Similarly, modern human populations are different from ancient populations of our ancestors. But what process might explain how this change over time can happen? In other words, how does evolution work?

In the Explore activity, *Exploring Change*, you will make observations similar to those made by naturalists in past centuries. You will make predictions from your observations by using a simple model involving mathematical calculations. These observations will lead to a logical conclusion that will be a model for how evolution can work. You will then perform a simulation and use your model of evolution to explore the results.

Materials

Part A (per team of 2)

1 section of a piece of fruit with seeds
1 scientific calculator
paper towels
1 plastic spoon or knife

Part B (per team of 3)

25–50 objects of 1 type (for example, shells from a local marine snail *Littorina littorea* or leaves from different individuals of the same type of tree)
1 metric ruler
different-colored pens or pencils

Part C (per team of 3–4)

1 box lined with artificial turf
1 pair of forceps
1 pair of pliers
small seeds
large seeds
4 paper or plastic cups
1 tablespoon
1 measuring cup
timer (optional)

PROCESS AND PROCEDURES

Part A Seeds of Change

In part A, *Seeds of Change*, you will make simple observations of something you see every day—pieces of fruit. You will learn about how models help scientists understand the natural world and use your observations in a mathematical model that has some basic assumptions. You will examine the model to determine if the assumptions seem realistic.

1. Obtain a whole fruit or a section of a fruit, a paper towel, and a plastic spoon or knife from your teacher.
2. Remove all the seeds from your fruit using the plastic spoon or knife and place the seeds on the paper towel.
3. Count the number of seeds in the fruit.
4. Write a definition for the term "model" in your science notebook, using your own words.
5. Read the need to know, *A Simple Mathematical Model*. As you read, write a scientific definition of a model in your science notebook. Describe how this definition is different from the definition you wrote in step 4.

NEED TO KNOW

ENGLISH SPANISH

A Simple Mathematical Model

What do you think about when you hear the word model? Some of you may think of a handsome man on a runway, showing off a pair of fashionable jeans. Others of you may remember the model of the solar system that you built in elementary school. Still others may think of model cars or model airplanes.

The natural world that scientists study is very complex. Some natural events and systems are too large, too small, or too complex to study directly. To learn about these natural events and systems, scientists use models. Your elementary school solar system model is an example of a simple scientific model.

A scientific model is a simplified system that mimics a more complex system. Scientists use models in their work to help them explain and predict phenomena. A model of the atom helps your teacher explain the tiny particles of the atom in a way that you can visualize. All models represent something. This could be a process, a physical object or being, a mathematical relationship, an event, a system, or even an idea. Models are created with certain assumptions in mind, and all models have limitations.

For example, if you used Styrofoam balls in your solar system model, your model would have the assumption that the planets were made of the same materials. This assumption isn't true. But if you were only trying to develop a useful picture of how planets are arranged relative to the Sun, this assumption would not affect your final project. An important part of being a critical thinker is to understand and analyze the assumptions in the models that you study.

You will use a simple mathematical model to explore the following assumption: *All the seeds from a fruit survive, become adults, and make their own fruits.* The simple mathematical model you will use requires four other assumptions.

1. The fruit you have represents the last fruit of its kind on Earth.
2. All plants will die at the end of each year.
3. Each plant needs a square of land 1 m long by 1 m wide (1 m^2) to grow.
4. Each plant makes the same number of fruits per year. A green pepper plant will make eight green peppers. An apple tree will make 850 apples. An orange tree will make 300 oranges. A tomato plant will make 13 tomatoes.

The output from your model will be the number of plants in the next generation. You will track the number of plants for seven generations.

Row no. 1	Generation	0	1	2	3	4	5	6	7
2	Number of plants	*							
3	Number of fruits per plant (This number will be the same each time.)	*							
4	Total number of fruits in this generation (= row 3 × row 2)	1							
5	Number of seeds per fruit (This number will be the same each time.)								
6	Total number of seeds in this generation (= row 4 × row 5)								
7	If all survive, how many will make it to the next generation? (Use this value in row 2 for the next generation.)								

Figure 5.2 Seeds of Change Data Table. Copy this table into your science notebook, making it specific to the type of fruit you studied. The asterisks show the assumption that the fruit that you started with represents the last fruit of its kind on Earth.

6. Place a copy of the *Seeds of Change Data Table* in your science notebook, or draw your own table similar to the one in figure 5.2, making it specific to the fruit you are using.
7. Listen as your teacher explains how to perform the calculations for the first generation.

   You may want to take notes as your teacher is explaining the calculations.
8. Carry out the calculations for 7 generations.

   HOW TO

   You will probably want to express some of your numbers with scientific notation. For practice using scientific notation, see *How to Use Very Large and Very Small Numbers* in appendix D1.
9. Look over your completed *Seeds of Change Data Table*. On your own, circle the data that you think are the most important.
   a. Write the phrase "What I see" near the data you circled and describe the pattern you notice.
   b. Write the phrase "What it means" under your "What I see" phrase. Then write 1 or 2 short phrases that summarize what you think caused the pattern you observed.

      HOW TO

      Remember, these two phrases together are called *highlight comments* because they represent the essence, or the highlights, of the data in the table. For more information, see *How to Write Highlight Comments and Captions* in appendix B6.

10. Use the data in the *Seeds of Change Data Table* to help you answer the following questions in your science notebook.
    a. If we assume that each plant takes up 1 m^2 of space, how many generations will it take to fill up all the usable farmland on Earth with this plant? *Good, farmable land surface area* = 4.5×10^{13} m^2.
    b. If all the seeds survive, how many generations will it take to completely cover the total surface area of Earth with this plant? *Total surface area of Earth* = 5.11×10^{14} m^2.
11. Write in your science notebook whether you think the following assumption is true: "All the seeds from a fruit survive, become adults, and make their own fruits." Also describe whether you think this assumption would be true for organisms other than plants, such as bacteria and animals.
12. Read the following paragraph. Then answer the question posed at the end in your science notebook.

    Biologists monitor the number of individuals in populations for many different types of organisms. Forestry officials may monitor the number of Douglas fir trees in a forest. Fishery biologists keep track of the number of salmon in major rivers. Observations of a large number of organisms in a large number of different parts of the world suggest that most populations normally stay about the same size.

    You created a model about the growth of populations if every seed survives. You also read the observation that most populations are stable in size. How can you reconcile these two pieces of information?

Part B Celebrate Diversity

In part A, you used a mathematical model to make an observation that the naturalist Charles Darwin made. The observation you made is an important observation about the behavior of populations in nature. What else can we learn about the individuals that make up populations? Take a look at the people around you (classmates, friends, and family). Do the individuals in other populations of organisms differ from one another as humans do? You will explore the answer to this question in part B, *Celebrate Diversity* as you take measurements from 25 samples of the same structure from different individuals in a population.

1. Choose 1 object from your pile of shells or other collected specimen. Using a metric ruler, measure the length of the object to the nearest millimeter. Then record its length in your science notebook.

   Littorina species, such as those seen in figure 5.3, are one of many types of marine shells that can be found locally in California. This is one of the sample organisms that your group might choose to measure.
2. Find the longest object and the shortest object in the pile.
3. Guess how many millimeters larger the longest object will be. Then guess how much smaller the shortest object will be. Record your 2 guesses in your science notebook.

   You are making an estimate of what you expect to find when you measure the remaining objects. Having a rough idea of what you expect

Figure 5.3 Count the millimeter markings on the ruler to determine the length of the object (in this case, a small shell).

to find will help you know if you make a measurement that is way too long or too short to be accurate.

4. In your science notebook, create a data table in which you will record the measurements of each of your objects. Make sure that you label your data table.
5. Choose 2 team members to measure the objects. The third member will record the data. Measure the length of the remaining objects to the nearest millimeter (see figure 5.3). Record the measurements in your data table.

   Round off your measurements to the nearest millimeter. If an object measures 11.5 mm, for example, round up to 12.0 mm. If it measures 11.2 mm, round down to 11.0 mm.
6. Work with your team to construct a frequency distribution table similar to the one in figure 5.4. Draw it in your science notebook by using the following steps.

   Creating a frequency distribution table is an important first step in summarizing your data.

   a. Determine the range of lengths in the sample by finding the difference between the largest and smallest objects.
   b. Divide the range into 5–8 intervals, selecting a convenient size.

      For example, if the range of the sample is 24 mm (the smallest, 6 mm; the largest, 30 mm), it could be divided into five intervals of 4 mm each, as follows: 6–10, 11–15, 16–20, 21–25, and 26–30.
   c. Assemble your data in a table with the interval listed in the first column and the number of individuals in each interval listed in the second column.

Figure 5.4 **Frequency distribution table.** This table summarizes how many measurements are placed in each interval. The table lets you know how frequently measurements from each interval occurred in your sample.

Intervals (mm)	Number of individuals in the interval
6–10	2
11–15	6
16–20	10
21–25	4
26–30	3

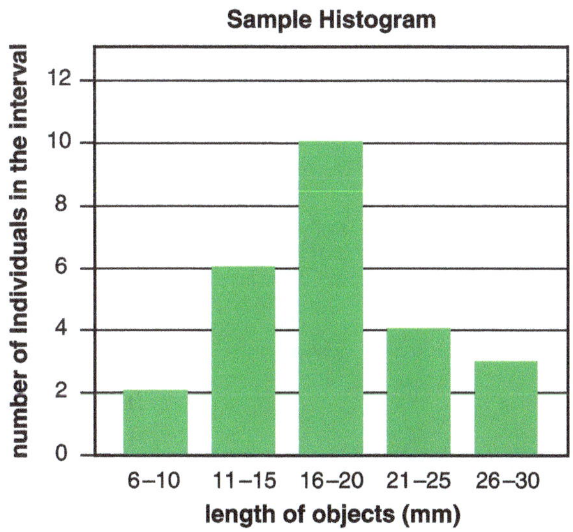

Figure 5.5 A histogram.

7. Construct a histogram similar to the one in figure 5.5. Use the data from your frequency distribution table.

   A histogram is simply a bar graph with the intervals on the horizontal axis and the frequency on the vertical axis. To learn more about making histograms, see *How to Create Graphs* in appendix D3.

8. Calculate the mean, or average, of your data. Draw a dashed vertical line on your graph to indicate the mean.

   The mean is the sum of all measurements divided by the number of individuals you measured.

9. Gather with your team and discuss the following questions. In your science notebook, write down the best ideas you hear.
   a. Are all your samples the same for the feature you measured, or do the samples vary?
   b. What features of the graph you developed help you answer the question about whether the samples are the same or vary?
   c. What might cause some individuals to be different from others for the characteristic you measured? List at least 3 different possible causes.
   d. Of the causes you listed, which ones would be passed on to the next generation?
   e. Imagine that only the individuals in the intervals to the right of the mean on your graph survived and reproduced. What would a graph that summarized the variation in length for the offspring from these surviving individuals look like? Draw the answer you would expect on your original graph, using a different-colored pen or pencil.

Part C Modeling Change

In parts A and B, you learned some important characteristics of populations. The observations you made about these populations occur in some way in every population of every different kind of organism on Earth. But what happens to populations that have these characteristics as they

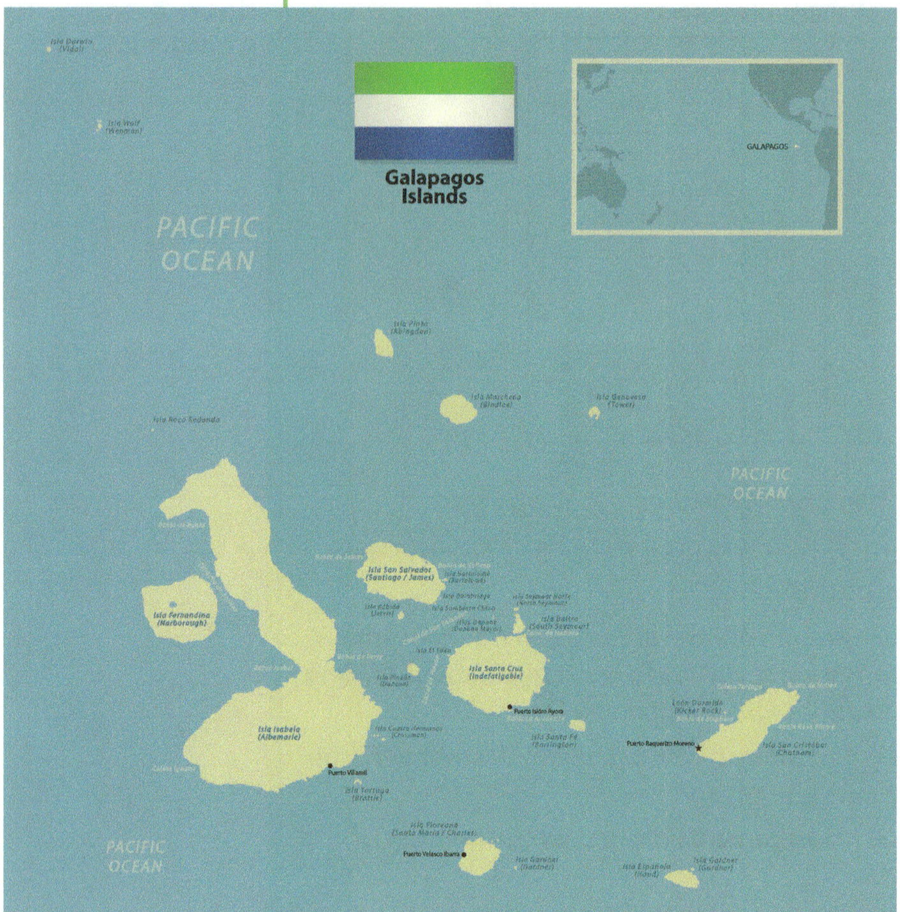

move forward in time? One way to answer this question is to develop a model. In part C, *Modeling Change*, you will complete an activity adapted from HHMI BioInteractive, called *Beaks as Tools*. You will model what happens to a population of Medium Ground Finches (*Geospiza fortis*) when faced with changes in resource availability due to drought. As with any population, there is variation and your task is to explore which organisms are successful.

The small island of Daphne Major is one the Galapagos Islands. They are located approximately 1000 kilometers off the coast of Ecuador in South America. The remote location has allowed for tremendous biodiversity in flora and fauna. The Medium Ground Finch is one species of bird that resides here. In non-drought years, food for these birds is plentiful and includes different seed sizes. In 1977, a major drought struck the island and the vegetation and available seeds changed considerably. Five years later, long-lasting rains allowed vines to overrun the island and again changed the vegetation drastically. This was followed by a second drought. Your team will play the role of Medium Ground Finches on the island of Daphne Major.

1. After gathering all materials, practice picking up large seeds (beans) and small seeds (rice) with the two types of beaks (pliers and forceps). *Note: The large seed must be cracked to count as eaten food in this simulation.*
2. Your team will be modeling the availability of food in three situations on an island in the Galápagos Islands. The availability of the food depends on the environmental conditions on the island of Daphne Major in the 1970s. Read about each situation in the table in figure 5.6. In your notebook, draw a table like the one shown. Then record a prediction for birds with each kind of beak in the last column. Use your observations from practicing in step 1, to help you make your predictions.

Environmental condition	Description of food available	Your prediction about the ability of each beak (bird) to pick up enough food
A: Land of Plenty	Plenty of both small and large seeds *Represented in simulation by* *$\frac{2}{3}$ cup large seeds (beans) and* *3 tablespoons of small seeds (rice)*	Large Beak: Small Beak:
B: Drought 1	Mostly large seeds *Represented in simulation by* *$\frac{1}{3}$ cup large seeds (beans)*	Large Beak: Small Beak:
C: Drought 2	Mostly small seeds *Represented in simulation by* *1.5 tablespoons of small seeds (rice)*	Large Beak: Small Beak:

Figure 5.6 Food availability on Daphne Major in the 1970s. The table shows environmental conditions during three periods on the island of Daphne Major. Notice that there is a description of the food available as well as how it will be represented in the simulation for each condition.

3. Prepare a data table to record the results of your investigation. To do this, attach a copy of the *Modeling Change Data Tables* in your science notebook or draw your own table similar to the one in figure 5.7.

Group Results	A: Land of Plenty		B: Drought 1		C: Drought 2	
	Small beak	Large beak	Small beak	Large beak	Small beak	Large beak
Trial 1	Small seeds: Large seeds: Total seeds:					
Trial 2						
Trial 3						
Trial 4						
Total						
Mean						

Figure 5.7 Modeling Change Group Results Data Table. Use this table to keep track of your simulation data. For each condition and beak size, you will need to record the number of small seeds, large seeds, and total seeds. You will do a total of four trials, then calculate the total number and mean for each category.

4. Decide on your roles for the activity. Two members of your team will be birds—one with a small beak (forceps) and one with a large beak (pliers). The remaining team members are observers and will assist with timing, counting seeds, recording data, and setting up the food availability conditions for *Land of Plenty, Drought 1, and Drought 2.*
5. To test your predictions, you will run a foraging simulation under food availability conditions that simulate the conditions before and during the two major droughts on Daphne Major. Read all steps in the Protocol and refer to the set up illustrated in figure 5.8 before beginning. Be sure to pay attention to what you will do based on your role for the activity.

PROTOCOL

SEP

PROTOCOL

Simulation Rules

1. Set up the food availability condition according to the description provided in figure 5.6. *For example, you will need to spread 2/3 cup large seeds (beans) and 3 tablespoons of small seeds (rice) throughout the artificial turf lined box lid for "Land of Plenty."* See figure 5.8 to see an example of how to set up the condition.

Figure 5.8 Example setup for the Land of Plenty condition in the simulation.

2. You team will run four 30-second feeding trials for each condition described.
3. Each finch will have two cups to place seeds in.
4. During each trial, finches should pick up as many seeds as possible and place them in separate cups for each bird—beans in one cup and rice in another. Large seeds (beans) must be cracked to count. Observers should count each large seed as it is cracked and small seeds (rice) can be counted at the end of each trial.
5. Remember to:
   a. Empty and count seeds in cups between each trial.
   b. Record the **total number of collected large seeds, total number of small seeds**, and **overall total number of seeds for each beak type** in your table at the end of each trial.
   c. DO NOT replenish the seed supply in your environment between trials.
   d. DO remove all seeds and set up for the next food availability condition (Land of Plenty, Drought 1, Drought 2) after 4 trials.
6. Once you have read these rules, set up the simulation as for "Land of Plenty" and begin your first of 4 trials. When finished with the 4 trials, repeat steps 1–5 for "Drought 1" and "Drought 2".

6. After the foraging simulation, work with your team to:
   a. Calculate totals for each beak type for each of the three food conditions.
   b. Calculate the mean (average) for each beak type of each of the three food conditions.
   c. Enter the **mean for the total number of seeds** for each beak type for each of the three food conditions into the class results data table.
7. Follow your teacher's directions on how to share and record the class data.
8. Study the class data and discuss the following questions as a team. Record your team's responses in your science notebook.
   a. Compare the means in your class table. Which beak collected the largest average number of all seeds? Identify one for each food condition.
      Land of Plenty:
      Drought 1:
      Drought 2:
   b. Which beak collected the smallest average number of seeds? Identify one for each food condition.
      Land of Plenty:
      Drought 1:
      Drought 2:
   c. Which beak collected most of the large seeds in Drought 1 and most of the small seeds in Drought 2? (You might have to refer to your group data to help you answer this question.)
      Large Seeds:
      Small Seeds:
   d. Did either of the two beak types consistently collect more seeds than the other under all conditions? If so, which one?
9. Calculate whether the finches were able to survive the two simulated drought conditions. To do this, first consider the total number of seeds each finch collected during Condition A: Land of Plenty. This is the amount that a bird would eat when there are unlimited seeds available. For a bird to survive, it generally needs to eat at least an average of 80% of the amount it would eat when there are unlimited seeds. You will need to calculate the amount each bird needs to survive, then compare the amount that the birds ate during each drought condition. Write your calculations in your science notebook.
   a. Find the food limit for small-beaked finches. To do this, use the class mean for beak type "S" under Condition A: Land of Plenty and calculate what 80 percent of the value would be. To do this, multiply the mean by 0.8. This is the amount of food that would allow small-beaked finches to survive
   b. Find the food limit for large-beaked finches. Follow the same steps as in 9a, but use the class mean for beak type "L" under Condition A: Land of Plenty.

c. Compare each finch's food limit with the average (mean) amounts it obtained in Conditions B and C (Droughts 1 and 2).
   - If the average is greater than or equal to the limit you calculated for that type of bird in 9a or 9b, the finch survived.
   - If the average is less than the limit you calculated for that type of bird in 9a or 9b, the finch died. Mark those that died with an "X".

10. Analyze the results of your calculations to determine how well the finches survived the droughts and factors that contributed to survival. Discuss the following questions as a team. Record your team's responses in your science notebook.
    a. Did any of the finches collect less food than its 80% limit during Drought 1 or 2 and starve to death as a consequence? List the ones that died during each drought.
       Drought 1:
       Drought 2:
    b. Was there any difference in the two beaks' ability to pick up small seeds? If so, what features made one beak more successful than the other?
    c. Was there any difference in the ability of the two beaks to pick up and crack large seeds? If so, what features made one beak more successful than the other?
    d. Consider the physical characteristics of the substrate (the artificial turf) in your model environment and the effort it took to collect large and small seeds under any of the food conditions.
       - What characteristics made seeds less accessible, and which seeds were most affected?
       - Would you say that the artificial turf made it hard for some finches to find enough food to survive? If so, which finch had the greatest trouble?
11. Participate in a class discussion about the simulation and what you learned.

Analysis

Complete the following tasks as a team. Record your team's best ideas in your science notebook.

1. In part B, you saw that variation is an important characteristic within all populations. How was this characteristic represented in your model with the Medium Ground Finches? What did variation allow for?
2. Based on your explorations, describe what might cause a population to change over time.

Explaining Adaptation

EXPLAIN

In the Explore activity, *Exploring Change,* you investigated a way that populations can change over time. You know about many types of change. Think about what you were like a year ago. What has changed? Your height, the shoes you wear, your hairstyle? These were all changes that you made as an individual. Do you think that populations change like you do as an individual, or is the process of change in populations different?

For hundreds of years, people wrestled with the question of how organisms come to "fit in" with their habitat. For example, there is a flower in the tropics that has a tube that is a half meter (18 in) long, with nectar at the bottom (figure 5.9b). In that same area of the tropics, there is a moth with mouthparts that are over a half meter (18 in) long (figure 5.9a). Inherited features of organisms that help them survive in their environment, like the moth's long mouthparts, are known as **adaptations.** Adaptations help organisms "fit" their environment.

How do these adaptations arise? Do individual organisms help form their own adaptations, by actions they take in their lifetimes? Or is adaptation a biological process that only emerges in a population over the course of generations? About 150 years ago, the answer to this was not very clear. Since then, the weight of scientific evidence has clarified how adaptations arise. In *Explaining Adaptation*, the Explain activity, you will examine scientific data about the adaptations of the Galapagos finches and you will learn about the biological process that affects populations and leads to adaptations.

Figure 5.9 (*a*) The nectar-feeding organ of this hummingbird hawk moth is longer than the moth itself. (*b*) It is adapted to the long tube of the red valerian flower.

Materials

different-colored pens or pencils

PROCESS AND PROCEDURES

1. Brainstorm a list of 3 *unusual* organisms that interest you. Write your list in your science notebook.

   If you are having trouble, think about animals that you have seen in the zoo or in a Biology textbook that you thought were interesting.
2. For each organism you listed, write down 1 characteristic that you think can be passed from one generation to the next. Explain how this characteristic might benefit the organism in its natural environment.
3. Contribute 1 idea from your brainstorming session to a class discussion about adaptations in organisms.
4. As you read the scenario, *Finches in a Drought*, think about the simulation you participated in during part C of the Explore activity, *Modeling Change*. The scenario describes how the beak of 1 population of finch changed due to a severe drought on the island where it lives. As you read and annotate the scenario using the Mark It Up! Strategy, pay particular attention to noting unfamiliar terms. Underline any information that helps you better understand the simulation.

   To learn more about the Galápagos Islands, read the sidebar, *The Finch Scientists of the Galápagos.*

SCENARIO

ENGLISH SPANISH

Finches in a Drought

The Galápagos Islands are dry places, receiving on average 25 cm (10 in) of rainfall a year. But in 1977, the Galápagos island of Daphne Major received nearly zero rainfall—only 24 mm, less than 1 in. This was the first of the two droughts that you simulated in part C of the Explore activity, *Exploring Change.* Using tweezers and nets, scientists on the island recorded how the number of seeds, insects, and cacti on the island declined sharply, as organisms struggled to grow and survive without enough water. Small, soft seeds were favored by several species of finches on the island. These were eaten first, leaving only the hardest, most difficult-to-open seeds (figure 5.10).

Figure 5.10 Caltrop seed and medium ground finch. Scientists found that the seeds of the caltrop plant can only be cracked by finches with beaks over 10.5 mm deep.

The medium ground finch was hard hit by the changes in its food supply. Its beak is not large enough to crack a whole caltrop seedpod. Instead, it has to go through a labor-intensive process of holding the seedpod to the ground while peeling back layers of woody material, finally revealing a few seeds nestled inside. In 1977, many medium ground finches did not have beaks large enough to accomplish this. All over the island, medium ground finches began to die. The struggle to survive was so severe that none of the medium ground finches mated in 1977.

The drought ended in the spring of 1978. Only 20 percent of the original population of medium ground finches remained. Scientists measured the beak depth of all of the finches on the island, before and after the drought (figure 5.11). As a group, those that had survived were different from those that had perished (figure 5.12).

Did *individual* finches change due to the drought? Scientists measured each finch before and after the drought. These finches did lose a lot of weight. But their beaks, made of a hard material like your fingernails, were each the same size before and after the drought.

Figure 5.11 Beak depth is measured at the widest point of the beak. Finches are caught when they fly into fine nets. They are then carefully untangled and measured.

The scientists knew from past studies that beak size is a heritable trait. **Heritable traits** are passed from one generation to the next. In other words, they are **inherited**. Because beak size is heritable, the scientists predicted that the offspring of the surviving finches would have beaks that resembled their parents' beaks.

In the spring of 1978, the surviving medium ground finches reproduced. There weren't many left, and there were many more males than females amongst the survivors. But the males sang their courtship songs, and the females chose their mates. The pairs made nests in the bushes and cactus plants, just as finches have been doing for millions of years. What do you think the offspring looked like? In particular, how deep do you think the offspring's beaks were?

To answer this question, study figure 5.13, which shows the beak depth of medium ground finch offspring born in 1976, before the drought, and in 1978, after the drought.

	No. of birds	Average beak depth (mm)
Original population	642	9.42
Drought survivors	85	9.96

Figure 5.12 Beak depth of medium ground finches before and after drought.
Source: Adapted from Boag, P. T., & Grant, P. R. (1981). Intense natural selection in a population of Darwin's finches (Geospizinae) in the Galápagos. *Science, 214*(4516), 82–85.

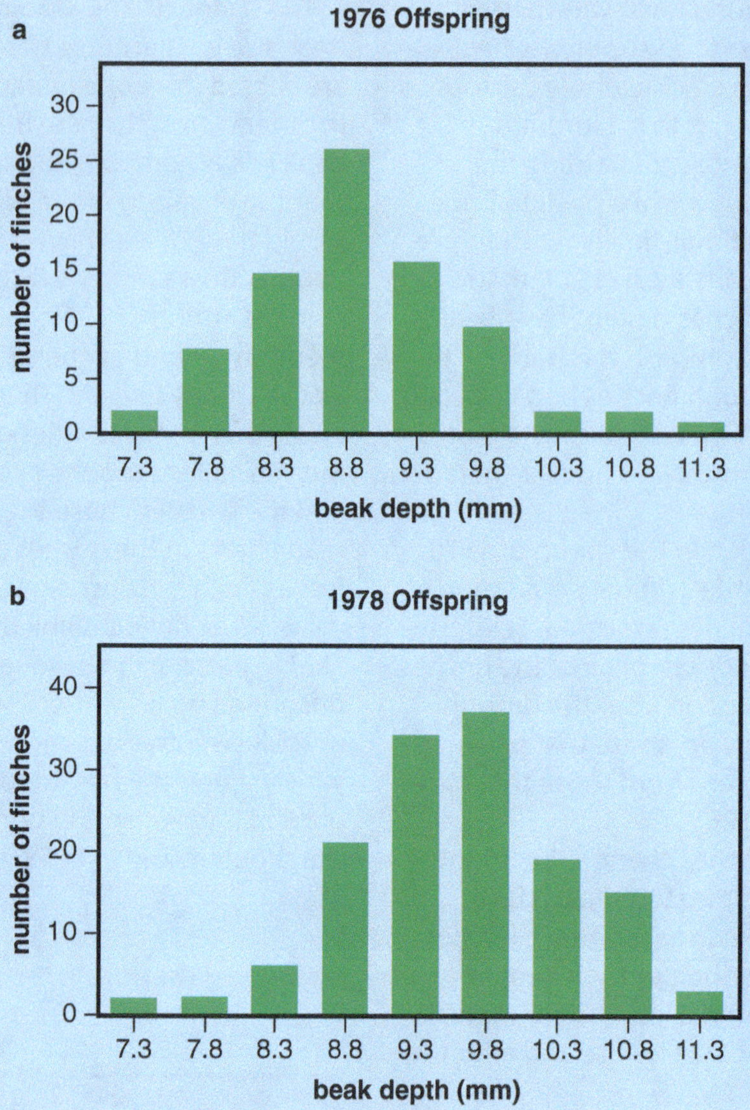

Figure 5.13 Medium ground finch offspring beak depth in (*a*) 1976 and (*b*) 1978. How were the postdrought offspring different from the predrought group?
Source: Adapted from Grant, B. R., & Grant, P. R. (2003). What Darwin's finches can teach us about the evolutionary origin and regulation of biodiversity. *BioScience, 53*(10): 969. Copyright, American Institute of Biological Sciences.

SIDEBAR

The Finch Scientists of the Galápagos

The Galápagos island of Daphne Major is not a typical tropical vacation destination, even though it is located close to the equator (figure 5.14). If you were visiting there, your packing list might look like this:

2 pairs of sunglasses
1 broad-brimmed hat
2 bottles of sunscreen
2 bottles of body lotion
canvas pants
long-sleeved shirts
gardening gloves
food and water for 2 weeks
1 stove
1 tent
pencils
1 ruler
1 scale
2 types of nets
science notebook
3 types of tweezers

You might have noticed that a swimsuit isn't on the list. This is because Daphne Major doesn't have a beach. Its coast is steep and rocky. Why would you need so much lotion to visit Daphne Major? In a normal year, the Galápagos Islands get only about 25 cm (10 in) of rain. That's less than an inch a month. You might wonder, can anything live in such a dry environment?

In fact, life thrives on Daphne Major. There are many species of plants, including shrubs and cacti, many species of insects, and several types of birds, tortoises, and iguanas. Ancestors of all these organisms originally arrived on the island after journeying over 950 km (600 mi) of ocean from South America or by moving from another of the islands in the group.

Figure 5.14 Daphne Major is one of the Galápagos Islands, located about 1,050 km (650 mi) west of South America. It was formed by an undersea volcano.

One group of bird species is known as Darwin's finches, because Charles Darwin collected the first specimens of these birds when he stopped on the Galápagos during his long voyage (see the sidebar, *The Voyage That Changed Biology,* for more about Darwin's historic trip). Although all the finches on the Galápagos are related to a single ancestor that migrated to the islands a little less than 3 million years ago, they display a wide variety of beak sizes and shapes (figure 5.15). Scientists who studied what these finches eat have found that their beaks are key to allowing each finch to specialize on a certain food during the leanest parts of the year—the dry season. Some specialize on very hard seeds, some on soft seeds, others eat cactus pads and fruits, and still others forage for insects.

You might have wondered why people visiting Daphne Major would need to bring all their food and water, and three pairs of tweezers. It's because this island is one of the uninhabited islands of the Galápagos. No bathrooms, no kitchens . . . no structures at all. If you go there, you must be prepared to survive its arid climate for weeks to months at a time.

Generally, people go to Daphne Major for only one reason: to collect data.

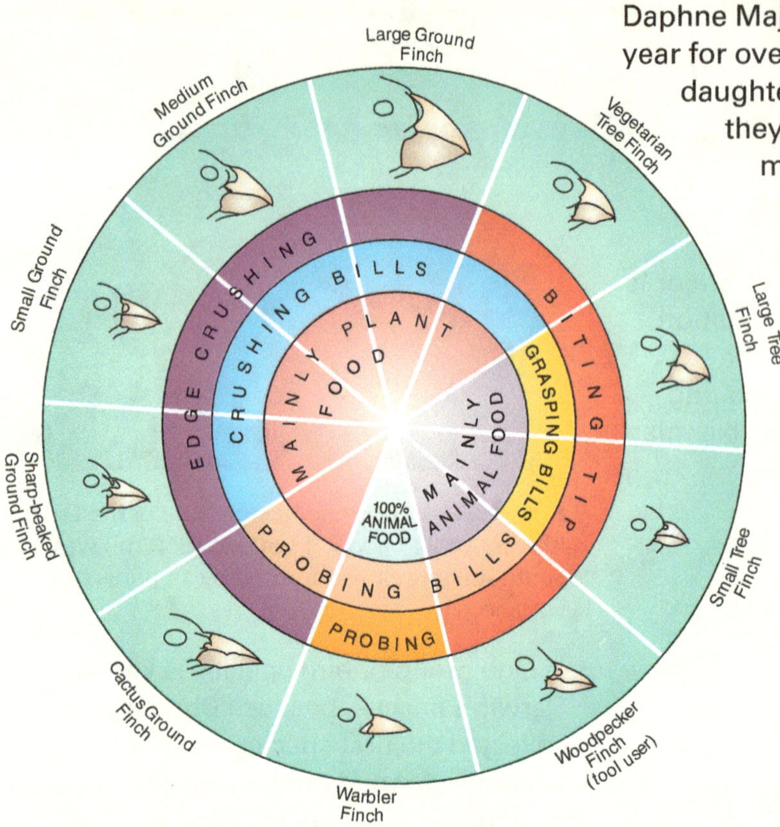

Figure 5.15 **Darwin's finches.** The different sizes and shapes of the beaks show how the beak of each finch species has adapted to gathering a specific food source.

Because Darwin's finches, and other interesting organisms on Daphne Major, have been evolving undisturbed for several million years, the island of Daphne Major is an ideal place to conduct biological studies. That's where the tweezers come in. Tweezers are said to be the most essential equipment for work on Daphne Major, because they allow a biologist to quickly pick up, sort, and count seeds . . . one of the finches' favorite foods.

Drs. Rosemary and Peter Grant are a pair of biologists who realized that the Galápagos Islands are a living laboratory of evolution (figure 5.16). In the early 1970s, they began taking their students on yearly trips to the Galápagos, working mostly on Daphne Major. They returned every single year for over 30 years, often with their young daughters, Nicola and Thalia. Sometimes they camped on Daphne Major for months at a time. With the help of their students, the Grants caught, measured, and banded every finch on the island of Daphne Major. They found every nest made by finches, measured the eggs, and followed the chicks after they hatched. They even tried to find the remains of dead finches, collecting the regurgitated pellets of owls on the island and dissecting them to recover their bird bands and measure finch bones.

The Grants' scientific team observed finches during each season of the year to record what they were eating. They counted each type of food to find out how abundant it was in the environment. Over time, their incredibly thorough and detailed work paid off, giving them a glimpse of evolution in action.

Figure 5.16 Drs. Rosemary and Peter Grant realized that the Galápagos Islands are a living laboratory of evolution.

SIDEBAR

ENGLISH SPANISH

The Voyage That Changed Biology

It was just after Christmas in 1831. The 22-year-old Charles Darwin (see figure 5.17) was nervous and excited as he climbed up the gangplank to the scientific research ship HMS *Beagle*. Darwin was hired to collect organisms as the ship sailed around the world. Just imagine how much more excited he might have been if he had known what his voyage would mean to modern science. Although young Darwin's journey would last only five years, what he learned would change his thinking about the living world. In turn, his ideas would change biology forever.

Figure 5.17 **Charles Darwin lived from 1809 to 1882.** These pictures were taken at different times in his life. On the left, he is 31 years old. In the middle, he is 72 years old. On the right, he is 45 years old.

The mission of the *Beagle* was to make a detailed map of the coast of South America. While the ship's crew completed that work, Darwin spent his time onshore, collecting thousands of specimens. Back on the ship, he made detailed observations of the interesting and exotic organisms he encountered. Darwin observed organisms in such diverse environments as the jungles of Brazil, the harsh plains ofTierra del Fuego (near Antarctica), the grasslands of Argentina, and the Andes Mountains. Figure 5.18 shows all the amazing places Darwin visited.

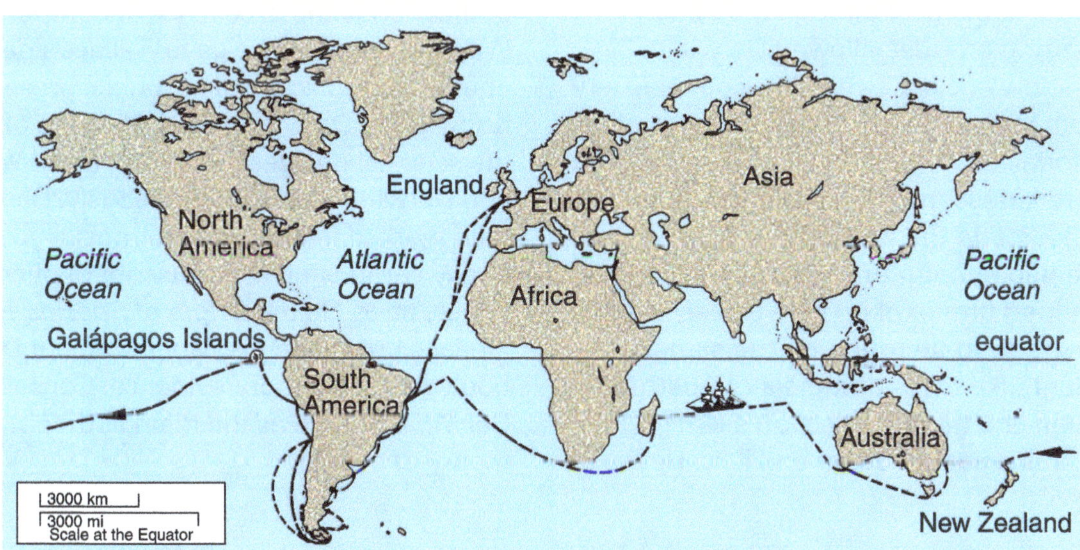

Figure 5.18 **Voyage of the HMS *Beagle*.** Darwin sailed around the world on the *Beagle*. The route included stops in South America and a stay in the Galápagos Islands. During the trip, Darwin collected evidence that he later used to support the theory of evolution.

During his travels, Darwin was puzzled by some of his observations. He found fossils of giant armadillos and wondered why the modern armadillos living in South America were much smaller. He noticed that the shells of giant tortoises varied a great deal between different Galápagos Islands. He wondered how those reptiles came to inhabit the islands in the first place. He also noticed that the mammals in South America were very different from the marsupial mammals he had read about living in Australia. He wondered if he could find a way to explain all of these patterns.

After he returned to England, Darwin spent nearly 20 years collecting more observations and testing his ideas. Then he published *On the Origin of Species*. You may be surprised to learn that Darwin did not use the word "evolution" in the first edition. Instead, he used the phrase "descent with modification." A "modification" is a change in a feature of an organism. Darwin proposed natural selection to explain how populations become modified, or changed, over time. In this way, he could account for how the shells of different giant tortoise populations may change over time (figure 5.19). "Descent" means that organisms are linked to the past through their ancestors. Recognizing this idea allowed Darwin to explain how ancient armadillos could be very large but still related to the smaller armadillos living today.

Overall, descent with modification helped Darwin explain the diversity of organisms that he had encountered on his travels. It also helped him explain why closely related species tend to be found on the same continent, like the marsupials of Australia.

Darwin and natural selection seem to go together like peanut butter and jelly. However,

Figure 5.19 A giant tortoise like the ones that Darwin studied.

Alfred Russel Wallace had the very same idea, at about the same time. Wallace made observations of organisms in the area now known as Malaysia. In fact, if Wallace had not made his ideas known, Darwin might not have published *On the Origin of Species* before he died. In a letter to Darwin, Wallace included a rough draft of his theory of evolution. His paper provided fewer examples of evidence than Darwin had collected. But Wallace clearly used natural selection to explain the evidence.

Darwin was shocked to learn that another scientist might publish the idea of natural selection before he did. Darwin had to acknowledge the excellence and importance of Wallace's work, so he gave Wallace's paper to another noted scientist for public presentation. Knowing of Darwin's nearly 20 years of study, the scientist decided to present both Wallace's and Darwin's work simultaneously. Darwin then worked feverishly for almost a year, finally completing a manuscript of over 500 pages, titled *On the Origin of Species*, in 1859. Wallace published his paper first, but Darwin's book contained more evidence. Consequently, Darwin is known as the main author of the theory of evolution.

5. Paste or sketch the graphs in figure 5.13 into your science notebook.
6. Write at least 3 highlight comments on the graphs in your science notebook to describe what you see and what you think it means.

   Focus on the highest points you see in each graph. For more information about how to use this strategy for steps 6 and 7, refer to *How to Write Highlight Comments and Captions* in appendix B6.

   HOW TO

7. For each graph, write a caption in your own words that summarizes the main point of the graph.
8. In your science notebook, record and answer this question: "How do you think a scientist would explain what caused the finches' beaks to change between 1976 and 1978?" Be as specific as you can about what changed, when, and why.

   Remember that scientists are limited to explanations based on observable evidence and natural causes.

9. Create a 2-column table in your science notebook with the title "two views of adaptation." The first column should be narrow and titled "ideas about how adaptations occur." The second column should be much wider and titled "what I learned." In the first column, write out the terms "inheritance of acquired characteristics" and "natural selection," leaving plenty of space in between them.
10. Read the essay, *Two Views of Adaptation*. As you read, use the Mark It Up! Strategy to annotate the reading. Use one symbol to mark information that you learn about inheritance of acquired characteristics. Use a different symbol to mark anything that you learn about natural selection. When you have finished reading, fill in the second column of the table you created in step 9, using your own words.

Two Views of Adaptation

ENGLISH SPANISH

How do changes across time occur to produce adaptations? In other words, what is the mechanism that causes populations to change over time? Two ideas have been important in the history of biology: acquired characteristics and natural selection.

Inheritance of Acquired Characteristics

The idea that organisms can change themselves was generally accepted by scientists in the 1800s. This is the idea that the changes that parents acquire during their lifetimes are inherited by their offspring. So individual parents can try to change, and the changes they make (or that they *acquire*) are passed on to the next generation (they are *inherited*). As a result, the population changes over time. This idea is called the inheritance of acquired characteristics. Another way of saying this is that individuals acquire adaptations because they want or need them.

This idea is still used by people today as a common-sense explanation for change. After all, humans change certain aspects about their bodies all the time. You can go to the gym and work very hard to make your muscles larger. It seems to make sense that other organisms make changes to themselves.

The inheritance of acquired characteristics was used to explain how specific adaptations came to be (figure 5.20). Think about trying to explain what caused the changes in the beak depth of the population of finches after the drought. An explanation based on the inheritance of acquired characteristics would go as follows:

- During the drought, soft-shelled seeds were quickly eaten and then were gone. The only source of food left for the finches was the hardest, most difficult-to-open seeds.
- Some finches tried to change their beaks to make them deeper and stronger.
- The finches that could change their beaks were able to eat seeds and survive.
- The surviving finches reproduced.
- The next generation had deeper and stronger beaks.

Figure 5.20 Jean-Baptiste Lamarck, a prominent French scientist born in 1744, made many important contributions to science. He invented the dichotomous key, helped found the science of paleontology, and described many species of animals. Importantly, he popularized the idea that all living things were related and had changed over time. He thought that change occurred as acquired characteristics were inherited.

Natural Selection: A Population View of Adaptation

In 1858, Charles Darwin and Alfred Russel Wallace (figure 5.21) made a joint presentation to the Linnean Society in London, England. They had each come up with the same explanation for how organisms change over time. Their explanation did not rely on an individual organism "willing" those changes to occur. The process they described is called natural selection. Natural selection is based on some critical observations and one logical conclusion that is the outcome of the observations:

- Observation 1: More offspring are born than can survive.
- Observation 2: Individual organisms within a population vary from one another in particular characteristics.
- Observation 3: Some variations among individuals can be passed from parents

Figure 5.21 **Charles Darwin and Alfred Russel Wallace.** Born in 1809, Charles Darwin first studied medicine and theology to please his father. Alfred Russel Wallace was born in 1823 and trained as a surveyor. Both men developed their ideas during, and after, international voyages. Read the sidebar, *The Voyage That Changed Biology,* for more about Darwin's voyage.

to offspring. Scientists call this heritable variation.

- Conclusion: Individuals that have inherited traits that better enable them to survive and reproduce will leave more offspring. The frequency of these traits will increase in future generations. In other words, the population will change over time. These beneficial traits are called adaptations (figure 5.22).

Let's use natural selection to explain the changes in the beak depth of the population of finches after the drought:

- More finch offspring are born than can survive. An important factor that limits the size of the finch population is the food supply—in this case, seeds.
- The finch population had individuals with different beak depths.
- Beak depth can be passed from parents to their offspring.
- During the drought, soft-shelled seeds were quickly eaten and then were gone. The only sources of food left for the finches were the hardest, most difficult-to-open seeds.
- More individuals with deep beaks survived.
- These individuals produced more surviving offspring than individuals with shallow beaks.
- The whole population changed because, on average, more individuals in the next generation had deeper beaks.

It is important to recognize a few facts about natural selection. First, natural selection describes how populations change over time. Individuals do not evolve. Second, the process of natural selection is not random. The variation present in a population is somewhat due to chance. So are changes in the environment. But the characteristics that become more frequent in a population are the very ones that help individuals survive and reproduce. Thus, the process is not random.

Figure 5.22 Examples of adaptations: (*a*) leaf insect, (*b*) flicker, (*c*) red Irish lord fish. Can you identify the adaptations possessed by these organisms? How might each adaptation benefit the organism? How do adaptations develop?

11. Which of the following statements represents an explanation based on the inheritance of acquired characteristics, and which represents an explanation based on natural selection? Record your answers in your science notebook and prepare to explain them during a class discussion.
    a. The beak size of parent finches increased during their lifetimes because they needed to crack hard seeds. Those changes were passed down to their offspring.
    b. The beak size of parent finches did not change during the drought, but parent finches that had larger beaks survived the drought to have offspring. Those offspring had beaks like their parents.
12. With your partner, review your work from step 8 by reading the question and the answer you wrote in your science notebook. After you review your answer, complete the following tasks.
    a. Was the answer you wrote in step 8 based on the inheritance of acquired characteristics, natural selection, a mix of the 2, or some other explanation? Write your response next to your answer to step 8, using a different-colored pen or pencil.
    b. Write a new answer to step 8 that uses only natural selection as an explanation. Briefly explain how this answer is different from your previous answer. Use the following words in your answer:
        - Variation
        - Over-reproduction
        - Natural selection

    Write your new explanation in your own words.
13. Participate in a class discussion of the ideas of inheriting acquired characteristics and natural selection.
14. Make a 2-column table in your science notebook. Title the left column "why scientists accept natural selection." Title the right column "why scientists reject the inheritance of acquired characteristics." As you read the essay, *Natural Selection: A Powerful Idea*, use the Mark It Up! Strategy to annotate the reading. Put a "1" next to any information that would fit with the first column of why scientists accept natural selection. Put a "2" next to any information that would go into the second column of why scientists reject the inheritance of acquired characteristics. When you finish reading, write 2 or 3 observations into each column of your table.

Natural Selection: A Powerful Idea

ENGLISH SPANISH

Scientists have collected large amounts of evidence that supports natural selection as an important way that populations can change over time. Scientists have documented the following three facts:

- If there are no limits to population growth, organisms would reproduce very rapidly.
- Even so, our observations tell us that most populations in the wild stay fairly constant in size.
- Resources available in any natural environment are limited.

In other words, there is ample evidence from the natural world that most populations produce many offspring and, at most times, more offspring are born than can survive.

What about variation? In every population of living organisms that has ever been examined, scientists find some form of variation. And scientists know how to test to see if specific variations can be inherited. Many traits that vary among individuals in a population can be inherited. In fact, in a growing number of cases, scientists can now point to specific changes in an organism's DNA that cause specific variations. On the other hand, variations that appear during an organism's lifetime are often not heritable and cannot affect the course of evolution.

Just like the finch study, scientists have collected evidence of natural selection in a wide range of organisms. Natural selection is indeed a powerful mechanism of evolution. But maybe you're not convinced. Have you heard the saying that the "proof is in the pudding"? You might think you have good ideas about how to make pudding, but the true test is in how it tastes. Similarly, a strong test of scientific ideas is when they are put to practical use. Let's explore the results from attempts of people to put both the inheritance of acquired characteristics and natural selection to practical tests.

From the 1930s to the 1960s, the Soviet agricultural minister Trofim Lysenko wanted to change crop plants to grow better in cold regions of the Soviet Union. Winter wheat could grow successfully in cold regions. But a different type of wheat called spring wheat matured more quickly and could result in larger yields. Lysenko used the idea of acquired characteristics to try to change spring wheat so it could thrive in cold regions. To do this, he told farmers to soak and bury spring wheat seeds in snow the winter before planting. This grand experiment failed miserably. Lysenko's ideas kept Soviet scientists from finding effective answers to agricultural problems for decades.

In contrast, the idea of natural selection has been very important for agriculture. One major issue in agriculture today is the damage caused by insects and other pests. Scientists in a field of science called resistance management use models of natural selection to help keep our crops safe from pests. Understanding natural selection also helps us understand a related process called artificial selection. In **artificial selection**, a human chooses the individuals that will reproduce, based on certain traits. In natural selection, some part of the environment favors certain variations. Artificial selection can cause rapid changes in a group of organisms. For example, figure 5.23 illustrates six different types of vegetables that humans eat. All six plants are from the same species. Humans selected different traits in the ancestors of these species to suit different human tastes. Artificial selection has been important in shaping every agricultural product on which your life depends.

The idea of natural selection is the foundation for the rapidly growing field of biotechnology. Biotechnology is a $30 billion a year industry. This industry uses ideas of natural selection and artificial selection to help develop new drugs, vaccines, and other important industrial products. Scientists use an understanding of natural selection to develop a new

Figure 5.23 **Six types of kale.** These kale plants differ primarily in the part of the plant that stores the most starch.

vaccine against the virus that causes the flu every year.

Aside from biotechnology, an understanding of natural selection has influenced medicine in other ways. An understanding of natural selection helped doctors develop new strategies for treating people infected with HIV. They use similar strategies to keep us safe from disease-causing bacteria and viruses that become resistant to specific drugs.

Finally, natural selection's explanation of change that occurs within populations, not individuals, explains some truly bizarre facts of life. For example, some female spiders, like this golden-orb web spider, attack and eat their mates (after mating). Natural selection provides a way to understand how the male spider becomes the nutrition that will help his offspring survive.

Analysis

Answer the following questions in your science notebook. Be prepared to contribute your ideas to a class discussion.

1. Review the list of 3 organisms you wrote for step 1 of part A, *Two Views of Adaptation*. From your list, choose 1 adaptation for an organism and explain how it may have changed from an ancestor that had a different form of the trait. Include the terms "heritable," "variation," "population," "over-reproduction," and "natural selection."
2. During step 6 of the Engage activity, *Lucy*, you recorded your initial ideas about how Lucy's lineage may have changed over time from an ancestral population that was not bipedal to a population that did walk upright. Now that you are more familiar with the concepts of natural selection and adaptation, revise your answer to this question to incorporate these concepts.

   When you revise a previous answer, use a different-colored pen or pencil. Draw a single line through sentences you are revising to indicate that you have changed your thinking. Revisions are important to help you monitor your own understanding. Documenting how your ideas have changed makes learning easier.
3. How did the information in this activity help you understand the simulation you did with the bird beaks and seeds in part C of the Explore activity, *Exploring Change*? How did the simulation help you understand more about what you read in this activity?

Further Challenges

1. Charles Darwin had a big impact on the way scientists view changes in populations. To learn more about his work, read the sidebar, *The Voyage that Changed Biology*. Choose one of the stops on Darwin's journey and research more about the area and the organisms that live there.
2. To learn more about Peter and Rosemary Grant, read *The Beak of the Finch* by Jonathan Weiner, a Pulitzer Prize–winning account of their work with Darwin's finches.

ELABORATE

Evidence for Evolution

In the Explore activity, *Exploring Change*, you saw how variation, over-reproduction, and selection can lead to change in a population. In the Explain activity, *Explaining Adaptation*, you practiced using the idea of natural selection to explain how these changes produce adaptations in finches. You may wonder, though, how we know that changes in living organisms have happened across time. When scientists are researching a topic, how do they convince themselves that their ideas are valid? As you have already learned, they use evidence.

In this Elaborate activity, *Evidence for Evolution*, you will study some of the evidence that different scientists have accumulated about evolutionary change. To do this, you will assume the role of a specialist and work to collect evidence about biological changes across time. Your team of specialists will use this collection of evidence to develop a presentation. As you learn from the presentation from another team of specialists in the class, you will begin to see how combinations of evidence from different branches of science can support the same explanation.

PROCESS AND PROCEDURES

In this activity, you will examine evidence that supports the idea that biological changes are the result of evolutionary processes. You will see some of the changes brought about by the processes you have studied during the previous activities. You will also learn about how scientists study these changes.

1. Work with your team of 4 to decide how to split into 2 pairs. One pair will take on the role of a paleobiologist, and the other pair will take on the role of a physical anthropologist:
   - Paleobiologist: A scientist who investigates biological change by studying fossils and the history of Earth. Skim the sidebar, *Interview with a Paleobiologist,* for more information. You will read it more carefully later in the activity if you take on this role.
   - Physical anthropologist: A scientist who investigates biological change by studying ancient hominids and living primates, including humans. Skim the sidebar, *Interview with a Physical Anthropologist,* for more information. You will read it more carefully later in the activity if you take on this role.
2. With your partner, read and study the information assigned to your role. Complete the procedures outlined for your role. As you work with your partner, make certain you both contribute ideas.

   The information and procedures for both the paleobiologists and the physical anthropologists can be found under the heading Procedures for the Specialists. After everyone has had time to study his or her special field of science, you will meet again with your team of four so that you can teach each other what you have learned.

SIDEBAR

ENGLISH SPANISH

Interview with a Paleobiologist

INTERVIEWER: Why did you go into the field of paleobiology?

MS. SCHUTZ: Because I love solving puzzles. Initially, I studied forensics. However, I realized that I wanted to ask questions that were more broad and included nonhumans. In essence, I try to understand how mammals lived by looking at their skeletons. I study both living and extinct mammals. I am interested in how the shapes of skeletons evolve and how this affects how mammals move.

INTERVIEWER: You're a graduate student. What does that mean, and how did you get to this place in your career?

MS. SCHUTZ: A graduate student is like an apprentice. Graduate school is a place where you learn your craft with the assistance of scientists who are already established, after you have completed college.

So, as you can imagine, it is very important to choose the right mentor to follow. You need mentors to be good at their craft, but you also need them to be effective teachers, so they can help you become a master yourself.

How did I get to where I am? Well, I am a curious person. I love finding out how things work. This ranges from how airplanes fly to how the bones of birds are structured so they hold up to all the abuse that comes with flying. Have you ever seen an albatross land, or a cormorant dive from a great height into the ocean? The forces involved are incredibly strong, strong enough to break human bones. It is all about form and function.

Growing up, I was very lucky to have a very adventurous grandfather who encouraged me to go after whatever made me curious. I spent the first nine years of my life growing up in Colombia, South America, being exposed to a great deal of nature. Once I caught an absolutely stunning little frog that had colors so bright they looked like they came out of a crayon box. My grandfather took one look and slapped it out of my hand. Then he washed my hands and arms maniacally. I later realized that he had probably saved my life because I'd caught a poison dart frog.

Heidi Schutz. Heidi Schutz is a graduate student in the Department of Ecology and Evolutionary Biology, University of Colorado at Boulder.

My family was also not into "gender" roles or at least less so than the families of other girls my age. I played with any kind of toy I wanted—regardless of whether or not it was a "boy" or "girl" toy. My greatest possessions as a kid were my telescope, my radio-controlled car, and my three-foot-tall stuffed pony.

INTERVIEWER: What do you find most interesting or exciting about your work?

MS. SCHUTZ: Digging fossils can be exciting. Once I was working in a cave in Colorado that had very little room. The roof of that particular room collapsed and a poor camel was standing there at the time. Yes, camels once were prevalent in North America. We also had cheetahs . . . ever wonder why pronghorn are so fast? Because I'm not very tall, I was able to spend about nine hours a day in that hole to recover the camel fossil.

I like learning something new every day and the challenge of working on puzzles that are really hard. I also really enjoy interacting with other people who do what I do. A common misconception about scientists is that they sit around all by themselves in their offices or their labs being "brilliant." There is also the idea that scientists are so smart that they can barely function in the social world. That is just not true. To me, science is about teamwork. The best scientific discoveries, in my mind, are made when people from different fields work together and combine their knowledge.

Interviewer: What do you find most challenging or frustrating about your work?

Ms. Schutz: Sometimes . . . even now, I am treated differently because I am a woman or because I am Latina—or both. That really frustrates me. However, it also motivates me to change those attitudes in my profession. It motivates me to make it easier for people after me who want to get into this line of work. I am beginning a new job next year as a postdoctoral fellow, and it will involve mentoring minority and female graduate students as I conduct further research.

Interview with a Physical Anthropologist

Interviewer: Why did you decide to become a scientist, and how did you choose this field for your specialty?

Dr. Covert: As a child, I was quite interested in nature. I enjoyed hiking, camping, and fishing with my family. When I entered college in 1972, I was excited about studying the natural world. I took classes in chemistry, physics, biology, and anthropology in pursuit of knowledge on this subject. As the semesters passed, I realized what I was most interested in was the evolutionary history of humans and our closest relatives, the nonhuman primates, so I decided to specialize in physical anthropology. I was fortunate to study this subject at the University of Massachusetts for my BA, Arizona State for my MA, and Duke University for my PhD. During the past 20 years, I have participated in paleoanthropological research in Egypt on four occasions, in Vietnam on five occasions, and in North America on 15 occasions. I have

Herbert H. Covert. Herbert Covert is a professor in the Department of Anthropology, University of Colorado at Boulder.

been a professor of anthropology at the University of Colorado at Boulder since 1986.

Interviewer: What is the most interesting or unusual thing that has ever happened while you were doing research?

Dr. Covert: In 1981, I had the good fortune of discovering the facial skeleton of an early monkeylike primate called *Aegyptopithecus*, a creature that is thought to be a good model for the last common ancestor of humans, apes, and Old World monkeys. It was quite exciting to carefully excavate it from the ancient sands of the Sahara desert. With each sweep of the paintbrush I was using to uncover it, I was able to get a more complete view of the face of my great-great grandparent removed by 32 million years or so!

Interviewer: What do you find to be the most fun or most challenging aspect of your work?

Dr. Covert: I find traveling and meeting people that share my interests in the natural history of life is the most rewarding aspect of my job. Knowing that scholars in Vietnam and Egypt have the same intellectual aspirations and interests as I have is a wonderful reminder that people from around the world share the same wonderful curiosity about the natural history of our planet.

3. Return to your team of 4 and prepare to share what you learned with your team and to learn from your other team members.
4. On a new page in your science notebook, create a data table or diagram like the one shown in figure 5.24 to help you track the evidence from your team. Label the table "evidence for evolution." Leave at least 2 more pages blank in your science notebook to add more evidence for evolution from your studies later in this chapter and the next.
5. Work with your partner to complete these tasks.
   a. Describe your specialist role to the other members of the team. Make sure your description includes the following:
      - Briefly describe the procedures you followed to gather your evidence.
      - Explain how the poster you made shows evidence to support the idea that organisms change across time.

Type of study	Evidence
Paleobiology	
Physical anthropology	
...	

Figure 5.24 Evidence for Evolution table. Draw a table like this in your science notebook to help you keep track of each specialist's evidence for change across time.

b. When the other team of 2 is describing its procedures and evidence, take notes in your *Evidence for Evolution* table. Be sure to ask the team to clarify any evidence you do not understand.

6. Discuss with your partner how having more than 1 type of evidence is stronger than any of the separate pieces. In your science notebook, record the main ideas from your discussion.

You have learned about evidence that supports evolution that comes from the work of two different types of scientists. Think about why a combination of evidence from both scientists would be stronger than evidence from only one.

Procedures for the Specialists

Paleobiologists

Paleobiologists are scientists who specialize in the study of fossils and the history of life. Much as detectives reconstruct a crime from the evidence left behind, paleobiologists search for clues to biological change by studying fossils. A fossil is any preserved part or impression of an organism that lived in the past. Plants, insects, fish, plankton, and mammals are just a few examples of organisms that leave fossils. One method used in research is a process of dating rocks and fossils called stratigraphy. **Stratigraphy** is the study of how the layers on Earth are organized. "**Strata**" is another name for "layers." The singular form to refer to one layer is **stratum**. For this task, you will study and make a model of Earth's strata. In your role as a paleobiologist, you will examine both geologic and fossil evidence because understanding biological change requires an understanding of geologic change.

Materials (per 2 students)

1 glass beaker with strata (prepared by your teacher)
1 empty glass beaker
strata materials
1 set of colored envelopes
online resource
poster board or butcher paper
felt-tipped markers

1. Find out more about the type of work a paleobiologist does. You may wish to read the sidebar, *Interview with a Paleobiologist*. Think about how you will explain the job of a paleobiologist to your teammates.
2. Complete the following activities with your paleobiologist partner.
   a. Observe the beaker of strata provided by your teacher. Work with 1 other team member to make a copy of each stratum and its marker in your own beaker.

Look closely at the layers. Make your beaker's layers match as closely as possible.

b. Answer the following questions about your beaker and its strata. Record the answers in your science notebook.
   i. Which layer of materials is the oldest (has been in the beaker the longest)?
   ii. If these strata were layers in Earth, what inferences might you make about the relationship between the depth of the layer and the amount of time that has passed?

      Consider how the strata might be similar to a stack of papers and books in your locker, as shown in figure 5.25.

c. Consider the locations of the 3 colored markers that you placed in the strata. If you found these colored markers in Earth's strata, which would you infer to be the oldest?
d. Obtain 3 colored envelopes from your teacher. Each envelope corresponds to 1 of the colored markers in your strata.
e. Imagine that each envelope represents a fossil discovery. Which fossil would be the oldest? Which would have formed most recently?
f. Open the envelopes and carefully observe the fossil pictures in each.
g. In your science notebook, compare each fossil with each of the others and with modern-day organisms. What similarities and differences do you observe?

   Consider how the fossils look and their relative ages.

Figure 5.25
Books and papers stacked in a locker. How old are the books and papers below the pink memo? Above the memo?

3. Create a poster with your partner that you can use to explain to your teammates how fossils form and how they can be dated. Include an illustration of your strata model.

   The essay, *Fossils: Traces of Life Gone By,* and the sidebar, *Technologies for Dating Fossil Evidence,* are resources that will help you understand fossil formation and provide more information for your poster. If you choose to read the essay and the sidebar, determine the specific information you need for your poster. Then choose appropriate symbols to represent that information and use them to annotate the readings using the Mark It Up! Strategy.

4. Watch the video segment, "Plate Tectonics," which is about continental drift, and think about the changes that the continents have undergone over time.
   a. In your science notebook, describe how you think the movement and separation of continents affected organisms living on land in Earth's distant past. Describe the types of evidence you would expect to see to support the idea that continents have separated.
   b. Add information to your poster about how the evidence for plate tectonics supports the theory of evolution.
5. Check that you have completed all the steps for the paleobiologists. Finish preparing your poster to use as a visual aid when you teach your teammates about paleontology. Return to step 3 in the general procedure.

Fossils: Traces of Life Gone By

ENGLISH SPANISH

Glancing around nervously, a whalelike creature waddles awkwardly on its very short legs down the sandy slope. Once it enters the water, however, it moves gracefully, swimming with its powerful tail.

The mammal described is an extinct *Rodhocetus.* Scientists infer that this animal was closely related to an ancestor in the lineage that led to modern whales. As a result, it shared many features with an ancestor of modern whales. *Rodhocetus* also has something very special in common with some dinosaurs, bacteria, ferns, horses, and even Lucy. The bodies of some members of these species became fossils after they died. In each case, a series of unusual events occurred that preserved their bodies in the form of fossils. Scientists search for these fossils and use them in their studies of biological change.

Fossils do not form easily. You may have seen scavengers eating dead animals or seen how microorganisms can decompose dead bodies. Sometimes, though, all or some of the body parts of a dead organism such as a plant or animal fossilize. Usually, the hard parts of organisms (such as bones, teeth, shells, or wood) fossilize, not the soft parts (such as flowers, organs, or fat). This is because hard parts decompose slowly, while soft parts decompose quickly. There are several ways that fossils are formed. These include petrification, preservation, and trace records.

Although fossilization is rare, it can happen in several ways. Some fossils form when minerals dissolved in groundwater seep into the tissues of a dead organism and turn it into stone. This is called petrification. Others form when organisms are buried in the mud at the bottom of lakes and rivers. Before the organisms decompose, minerals in the water replace atoms and molecules in body tissues and form a cast in the shape of the organisms. Occasionally, the impressions organisms make in mud or sand dry out, are covered, and turn to stone. These impressions are called imprints.

Preservation can occur in a variety of ways (figure 5.26). Some fossils form when

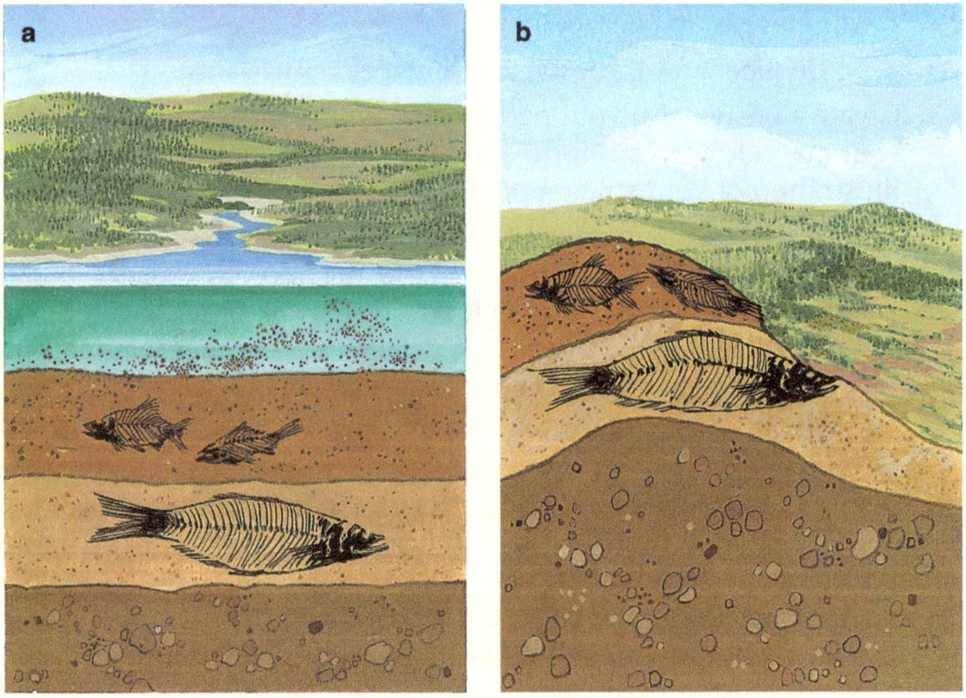

Figure 5.26 Fossil formation. (*a*) Dead organisms or their skeletons may remain intact in underwater sediments through long periods of time. Eventually, minerals circulating in underground water replace the bone of the skeletons, forming fossils. (*b*) Buried fossils may be brought to the surface by any of the forces that uplift segments of Earth's crust. Once near the surface, the fossil-containing rock layers are subject to erosion, which can expose the fossil from the rock.

organisms become trapped and preserved in amber or tar. Amber is a sticky substance produced by evergreen trees that hardens when it dries. Tar is a gooey, oily substance that sometimes forms in the ground when plant material decays. Finally, fossils sometimes form when organisms are covered by snow or ice and freeze before they start to decay.

In rare cases, scientists have found soft parts of fossilized organisms. Examples include entire woolly mammoths frozen in arctic ice and prehistoric insects fossilized in amber. In 2007, a high school student found a mummified dinosaur with intact skin and muscles in South Dakota.

We are most familiar with the fossilized remains of the parts of plants and animals. However, organisms with soft bodies can leave fossilized traces of their activity, such as burrows. Impressions such as footprints, feather prints, and leaf prints can also fossilize (figure 5.27). These types of fossils allow scientists to make inferences about an organism's life and behavior. For example, footprints of dinosaurs that walked in soft sand or mud sometimes survive as recognizable tracks in solid rock. These trace fossils suggest how the animal walked, including the length of its stride and where it placed its feet.

Figure 5.27 Fossil footprints. Trace fossils include footprints, like these made by hominids around 3.6 million years ago.

The fossil record not only provides information about once-living organisms, it also shows how Earth itself changed across very long periods of time. For example, if we find fossils of marine animals on what now are mountaintops, we may infer that the rock in that area formed under a sea. This may be because mountain-building geologic forces lifted up an ancient seabed. In this way, the rocks in which fossils are found give clues to physical changes in climate and land conditions.

Sometimes fossils are destroyed by animals, weather, or other changes in Earth's surface. If a preserved fossil escapes destruction, it may become exposed on the surface of Earth by erosion. Often, this is how fossils are discovered. Even if a fossil survives and becomes exposed at Earth's surface, it cannot serve as evidence of biological change unless someone discovers it and realizes its importance.

Another way the geologic record can show changes in Earth is through the study of fossils that are similar but are found in different places on Earth. About 230 million years ago, all the land on the planet was grouped together in one gigantic continent called Pangaea. Since that time, the land has broken apart and moved into the current arrangement of continents. The scientific theory that is related to the movement of the continents is called the theory of plate tectonics.

The movement of the landmasses can be tied to changes in Earth's biological makeup. Sometimes the changes in land connections allowed organisms to migrate to a new area. Other changes in the land caused connections to be lost, so some populations of organisms were isolated. For example, scientists have found identical plant fossils along the coasts of western Africa and eastern South America. This is strong evidence that these continents were once connected and have moved apart due to plate tectonics. Have you ever wondered why kangaroos are only found in Australia? This continent has been isolated from other landmasses for about 80 million years. Over this time, mammals in Australia have evolved into a number of unique groups including the kangaroo, the echidna, and the extinct Tasmanian wolf (figure 5.28). A combination of what we know about fossils and about the physical structure of Earth serves as a geologic record of change.

By looking at fossil evidence and comparing fossils with the organisms living today, a fundamental feature of living organisms becomes obvious. Throughout Earth's history, the different species of organisms living on Earth have changed. New species form, branching off from existing species. Other species die off. The most dramatic change possible for a species—its complete disappearance—is called extinction. Figure 5.26 shows several examples of species that have become extinct. The fossil record is full of many now-extinct species. Of all the species that ever existed on Earth, 99.9 percent are now extinct. Species continue to become extinct due to both natural and human-made changes in the environment. But new species continue to emerge as well, often occupying roles in the environment left vacant by extinct species. You will learn more about the formation of new species in chapter 6, *Products of Evolution: Unity and Diversity*. The fossil record provides a powerful line of evidence for evolutionary change.

Figure 5.28 Australian mammals. All of the mammals native to Australia are marsupials. After birth, their young develop fully within their mother's pouch. They include (*a*) kangaroo, (*b*) echidna, and (*c*) the extinct Tasmanian wolf.

Figure 5.29 **Extinct animals.** (*a*) xerces blue butterfly, (*b*) dodo bird, (*c*) Steller's sea cow, (*d*) golden toad.

SIDEBAR

Technologies for Dating Fossil Evidence

From sundials to atomic clocks, humans have invented many tools to measure time as it flows from seconds to minutes to hours. Certain natural processes also measure time. For example, a cut through the trunk of a fallen tree reveals growth rings. Counting those rings is a way to measure the number of years the tree was alive and growing. The rings, however, do not tell how long it has been since the tree fell. Scientists often want to know when a rock was formed or how long ago an organism lived. Clues often exist that paleobiologists can use to calculate how long ago a fossilized organism lived, with the help of technology.

The age of a fossil can be calculated in two ways. First, the *relative* age of the fossil can be determined by studying how the layers of rock containing the fossil were deposited. Another word for "layers" is "strata." So this method is called stratigraphy. Second, the age can be calculated directly, from the age of the rock containing the fossil or from the fossil itself.

In stratigraphy, each rock layer provides a record of past events and past environments (see figure 5.30). The relationship between layers of rock and the occurrence of certain fossils was noted over 200 years ago. Around 1800, William Smith, an English civil engineer, became interested in rock strata because of their relationship to the structural success of canals he was building. He noticed that certain layers contained fossils and that throughout England there was a match between the type of rock layer, its placement between other layers, and the fossils it contained.

About the same time, two geologists, Georges Cuvier and Alexandre Brongniart, were studying fossils in rock strata in France. When they compared the fossils they found with modern life-forms, they discovered that the modern forms were more similar to the fossils from the higher rock layers than those from the lower layers. In fact, no modern species were found in the oldest and deepest layers. The reason is related to the way fossils are formed. Sediment that settles on top of a dead organism is more recent than the sediment under the dead organism. Thus, scientists infer that any fossils that they find in a particular rock layer are older than any fossils found above that layer and younger than fossils found below that layer. Using this knowledge and other techniques, geologists can develop estimates of the time period for each layer of rock. They can then determine the approximate age of a fossil.

Direct dating methods can determine the age of fossils with great accuracy. Direct dating methods depend on naturally occurring radioactive elements. These elements are unstable forms of matter that break down, or decay, into other chemical elements at a steady rate. Different radioactive elements decay at different rates. For a particular element, however, decay occurs at a constant rate. Elements that undergo decay are called **radioactive isotopes.**

Figure 5.30 Stratigraphy is a relative dating method. The stratification of rock in this cliff was exposed when a highway was built. Fossils of older organisms are found in the deepest layers. Organisms that lived more recently are found in the youngest layers.

a Living organisms incorporate carbon-14.

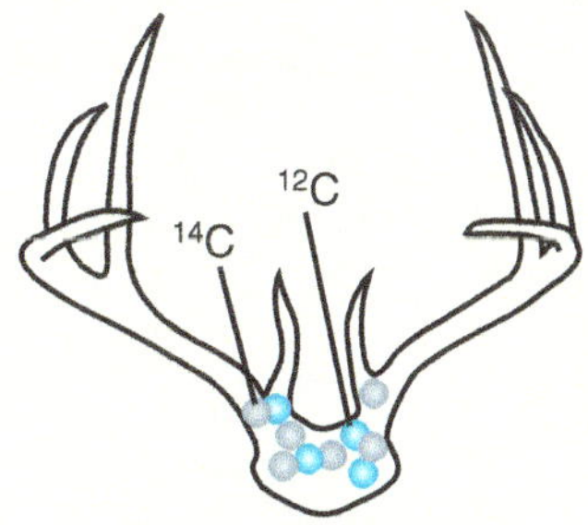

b The proportion of carbon-14 to carbon-12 in an organism's remains indicates how long it has been dead.

Figure 5.31 Carbon-14 becomes part of plants and animals. Small amounts of carbon-14 exist in the atmosphere as carbon dioxide, which plants incorporate into their tissues. This carbon-14 enters animals when they eat plants. Scientists can measure the amount of carbon-14 in fossils. Because they know that the half-life of carbon-14 is 5,730 years, they can calculate the age of the organism.

One method that is useful for dating organic material is carbon-14 (^{14}C) dating. Scientists can determine the age of material that is between 500 and 47,000 years old using this method. Normally, carbon atoms contain six protons and six neutrons. The unstable carbon-14, however, has six protons and eight neutrons. Carbon-14 atoms are naturally a part of some carbon dioxide molecules ($^{14}CO_2$) in the atmosphere. Plants take in this carbon dioxide and use the carbon-14 to make sugars during photosynthesis. Animals take in carbon-14 when they eat plants.

While plants and animals are living, they contain stable amounts of carbon-14 and carbon-12. The amount of carbon-12 in an organism does not change, even after death. After the organism dies, however, the level of carbon-14 in their tissues begins to go down as it loses its "extra" neutrons. Scientists can determine the proportion of carbon-14 to carbon-12 to estimate the age of a fossil. Older fossils have a lower amount of carbon-14 compared with living things, because some amount of the carbon-14 has been lost by radioactive decay. Fossils that are not as old have proportions of carbon-14 to carbon-12 that are more similar to living things today. This is because there has not been as much time for the decay to happen. Figure 5.31 helps illustrate this process.

Other types of radioactive dating can be used to measure material that is much older. These techniques are used to determine the age of the rocks in which the fossils are found. One example uses the radioactive element potassium-40. Potassium-40 decays at a much slower rate than carbon-14. Because of this slow rate of decay, geologists have used it to date rocks that are up to 4 billion years old. By choosing the appropriate radioactive element, scientists can date all kinds of rocks and fossils, whether they are a few hundred years old or a few billion years old.

Scientists often combine radioactive-isotope dating techniques with other methods for dating material, such as measuring magnetic properties. This improves the accuracy of their estimates of fossil ages because the combination of techniques lets them check for the consistency of the dates. These multiple lines of evidence, which support scientific reasoning, strengthen scientists' confidence in the accuracy of their fossil dates.

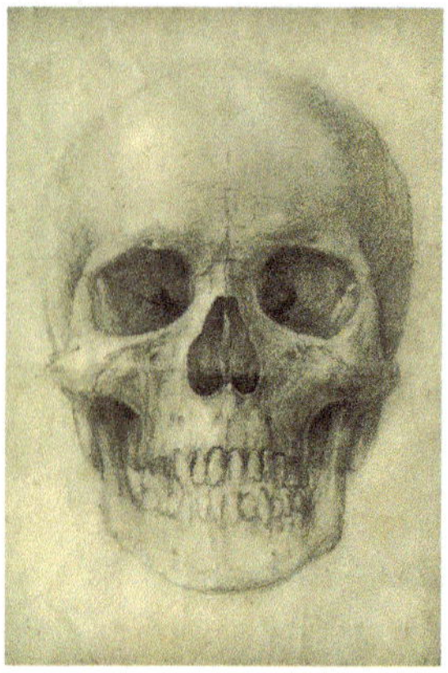

Physical Anthropologists

Physical anthropologists are scientists who compare ancient hominids with living humans and other primates to understand how humans have changed over time.

Materials (per 2 students)

- 2 small, clear, flexible metric rulers
- 2 protractors (jumbo sized, if available)
- online resource
- glue
- poster board or butcher paper
- scissors
- felt-tipped markers
- different-colored pens and pencils
- 1 set of selected human, chimpanzee, and mystery fossil bone pictures

1. Find out more about the type of work a physical anthropologist does. You may wish to read the sidebar, *Interview with a Physical Anthropologist*. Think about how you will explain the job of a physical anthropologist to your teammates.
2. Complete the following activities with the other physical anthropologist on your team.
   - **a.** Obtain an envelope that contains pictures of selected human, chimpanzee, and mystery bones. Then work with a partner to do the following tasks.
     - **i.** Carefully observe the human and chimpanzee bones for similarities and differences.
     - **ii.** Make a list in your science notebook of all the similarities and differences you can find.
     - **iii.** Compare each of the mystery fossil bones with the human and chimpanzee bones. In your science notebook, make a 3-column table to record your observations. In the first column, list each mystery bone. In the second column, record whether each mystery bone is either more like human bones, more like chimpanzee bones, or not like either human or chimpanzee bones. In the last column of your table, briefly describe the evidence that led to your decision in column 2.
     - **iv.** Based on your observations, write an explanation in your science notebook for how the mystery fossil bones might be related to humans and chimpanzees. Use the evidence you have from your observations to support your ideas.
   - **b.** View the image sequence "Hominid Skulls" on the video and make observations. As you compare the skulls, focus on the following features, make measurements where appropriate, and record the information in your science notebook:
     - Size of lower jaw
     - Prominence of brow ridges
     - Slope of face
     - Width of face
     - Size of forehead
     - Size of brain case
     - Size of molars

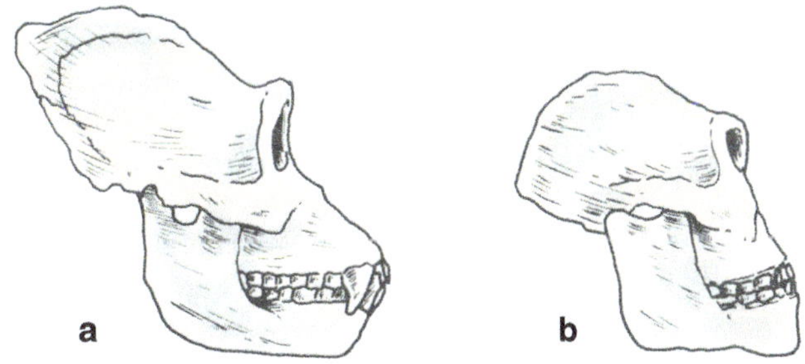

Figure 5.32 Compare the features listed in step 4b on these (*a*) gorilla and (*b*) early hominid skulls.

Fossil name	Trait						
	Jaw size	Brow ridge	Slope of face	Face width	Forehead size	Brain case size	Molar size

Figure 5.33 **Record of fossil measurements.**

Use the sketches of skulls in figure 5.32 to help you determine which measurements to make. Create a data table like the one in figure 5.33 to record your results. Review the sequence of skulls or individual frames as many times as necessary to make your comparisons.

c. Read the essay, *Primates Show Change Across Time,* and use the Mark It Up! Strategy to annotate it. Choose symbols that will help you answer the questions below. When you have finished reading, discuss the questions in your group. Record your ideas about each question in your science notebook.
   i. What evidence did you find in the essay and in your observations of the hominid skulls that indicates change occurred between early and modern hominids?
   ii. What evidence indicates similarities between early and modern hominids?

d. Summarize the pattern of changes that you observe in the hominid skulls from *Australopithecus afarensis* to *Homo sapiens.*

3. Create a poster with your partner about the evidence you have studied.
   a. Glue the human, chimpanzee, and mystery bones onto a poster board.
   b. Add a summary, in your own words, for how humans have changed over time.

   You will use this information to teach your teammates about the evidence that physical anthropologists contribute to our understanding of how species change over time.

4. Check that you have completed all the steps for the physical anthropologists. Finish preparing your poster to use as a visual aid when you teach your teammates about physical anthropology. Return to step 3 in the general procedure.

Primates Show Change Across Time

ENGLISH SPANISH

A carpenter walks up to a cabinet she is building and places a metal handle in a spot she has marked. Holding the handle with her left hand, she turns a screwdriver with her right hand to attach the handle to the wood. Somewhere, 5,000 years ago, a hunter stepped quietly through a thin forest, keeping his sight fixed on a small deer a short distance away. As he walked, the hunter pulled an arrow from his quiver and set it to the bow, ready for a shot that may win his family several good meals and a hide for clothing.

Walking upright is an advantage for humans because it allows us to save great amounts of energy when we move long distances. It also allows us to use tools while we move about. While all primates are able to grasp objects with their opposable thumbs to some degree, humans have very nimble fingers and hands. The biological changes that made the refined use of our hands possible took place over a long period of time.

The fossil record of primates provides evidence for the idea that biological changes have occurred at various rates in primate history. Living primates include lemurs, chimpanzees, gorillas, orangutans, and humans. All the species of primates alive today descended from one ancestral species that lived millions of years ago. Primates that are most closely related to each other share additional, more recent common ancestors.

Early hominids like Lucy had brain cases more similar in size to those of apes than to those of humans. Still, hominid skulls show that their brains were larger for their body size than those of modern apes. Lucy belongs to a group of hominids called *Australopithecus*. Compared with modern primates, individuals in this group had front teeth that were more humanlike, being considerably smaller than those in modern apes. Lucy's pelvis and knee bone support the idea that she walked upright (figure 5.34). Taken together, Lucy's skeletal remains suggest that her species had characteristics that fall in between more ape-like ancestors and modern humans.

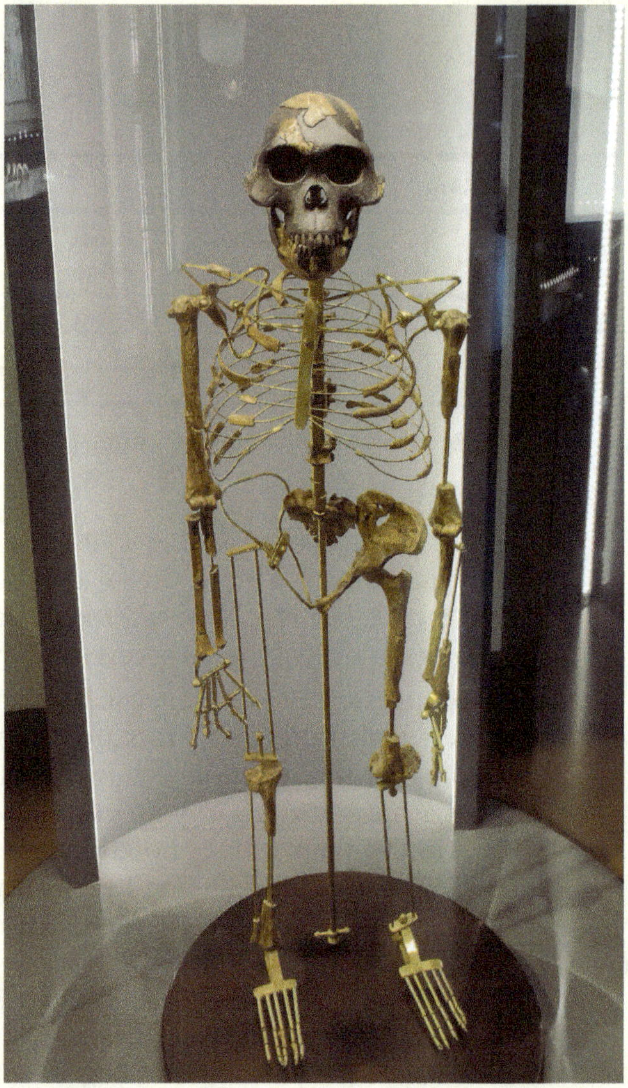

Figure 5.34 Lucy's skeleton on display at the Natural History Museum in Vienna, Austria. Note that parts of the skeleton are filled in with wire and other materials to give a more complete view.

In 1995, a new species of hominid was uncovered in the Lake Turkana region of Kenya, in East Africa. It had many of the same features found in Lucy. In particular, its leg bones suggest that it stood upright and walked upright. It was named *Australopithecus anamensis*, and it was scientifically determined to be around 4.1 million years old. This date makes *A. anamensis* more than half a million years older than Lucy's species.

Scientists continue to discover fossils that provide evidence for the idea that hominids have changed over time. Three fossils were found in Africa between 2001 and 2004 that are several million years older than Lucy. One of these, *Sahelanthropus tchadensis*, dates to between 6 and 7 million years old. It was first thought to be a hominid. Later, it was classified as an ape, partly because of its small brain size. The two other fossils, *Orrorin tugenensis* (around 6 million years old) and *Ardipithecus ramidus* (around 4.4 million years old), were also discovered. They walked upright and had teeth that were more similar to modern humans' teeth than to apes'. These two new fossils are now considered to be the oldest known hominids.

These fossils have helped scientists understand how hominids evolved. Using these fossils, scientists now think that hominids split from the common ancestor shared with chimpanzees between 7 and 6 million years ago.

Other hominids that lived on Earth around 2 million years ago, such as *Homo habilis* and *Homo erectus*, show a more humanlike appearance than Lucy and other hominids. They have a more vertical face, smaller front teeth, a smaller lower jaw, and a significantly larger brain case, like the *Homo erectus* skull in figure 5.35. Though these past species were probably not the direct ancestors of humans, these specimens suggest that the hominid lineage that led to humans gradually changed in form over time. These fossils are consistent with the scientific explanation that humans evolved from earlier hominids.

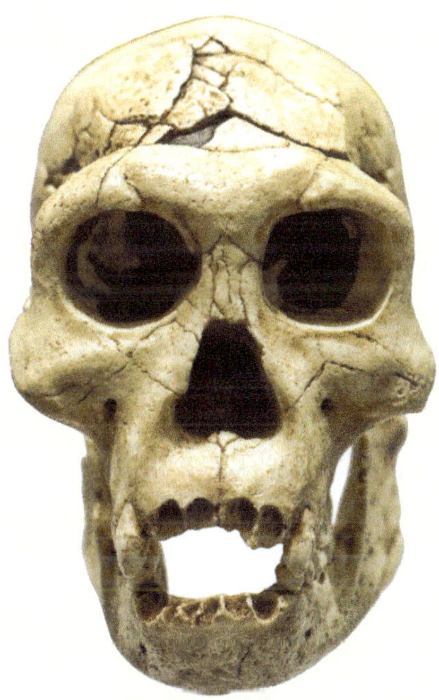

Figure 5.35 This skull from *Homo erectus* shows more humanlike characteristics than Lucy's. What do you notice when you compare the two?

Analysis

Complete the following tasks individually and record your response in your science notebook.

1. Review the *Evidence for Evolution* table you made to record information from paleontology and physical anthropology. With different-colored pens or pencils, circle groups of evidence that you consider to be different lines of evidence supporting evolution.

   A line of evidence is generally generated by scientists of one type, like paleobiologists. Scientists using different methods generate different, independent lines of evidence.

2. Add a new line of evidence to your *Evidence for Evolution* table called "natural selection in living populations." Fill in this evidence column with the most-relevant information you learned about finches in the Explain activity, *Explaining Adaptation*.

Further Challenges

Plate tectonics, the movement of continents, and the ice ages have had a tremendous effect on the distribution of living organisms throughout Earth's history. The movement of continents and the rising and falling of sea levels created and destroyed many land bridges. These land bridges connected one continent to another. Think about these concepts as you use the fossil data in figure 5.36 to complete the following tasks.

1. Propose a place of origin and a migration route for marsupial mammals.

   Mammals are animals characterized, in part, by hair, sweat glands, and the nursing of young with milk secreted by mammary glands. Marsupials are mammals that produce embryos that spend only a short time in the mother's uterus. At birth, the immature infants must crawl into the mother's pouch, where they continue to mature while nursing from the mother.

2. Consider what the fossil record indicates about ancient distributions of marsupial mammals (figure 5.36). Then propose an explanation for the present marsupial mammal distribution.

Figure 5.36 Distribution of marsupials.

Continent	Oldest marsupial fossil (mya = million years ago)	Present marsupial distribution (number of families)
North America	110 mya	1
South America	75 mya	2
Antarctica	55 mya	0
Europe/Asia	52–58 mya	1
Africa	50 mya	0
Australia	30 mya	10

Modeling Earth's History

ELABORATE

In the previous activities, you read about how biological changes can take place over vast periods of time. The Lucy fossil is extremely old—more than 3 million years old. In the Elaborate activity, *Evidence for Evolution*, you learned that the split between the chimpanzee and hominid lineages occurred between 7 and 6 million years ago. This is a time span that is difficult to comprehend because time for humans generally means tens and hundreds of years.

In chapter 4, The Cellular Basis of Activity, you briefly explored **geologic time**, the time span of Earth's history. In that chapter, you learned about geologic time when you studied how oxygen concentrations in the atmosphere have changed over time. One way to grasp the immensity of Earth's history is to make a model of a timeline. In this Elaborate activity, you will build on your knowledge of geologic time by developing a model of a timeline of Earth's history to help you better understand when certain human events and major geologic and biological events occurred. You will also begin to see patterns in how events took place. This timeline should help you appreciate geologic time and the changes that have taken place since the formation of Earth.

Materials (per team of 4)

clothespins or paper clips (optional)
chart paper
event cards that your teacher provides

PROCESS AND PROCEDURES

1. Discuss the following question with your teammates: "How long ago do you think each of the following events took place, and in what sequence?"
   - First evidence of dinosaurs
   - Breakup of Pangaea
   - First evidence of hominids
   - First evidence of life (bacteria)
   - Formation of Rocky Mountains
   - Anatomically modern humans becoming common
   - Significant amounts of oxygen in atmosphere
   - First evidence of land plants
   - Lucy

   At one time, all the continents were joined in one gigantic continent called Pangaea. It is OK if you are not confident about the sequence of some of these events. Try to place the events you know something about first, and then make your best educated guess about the sequence of the remaining events.

2. Make a list on chart paper of the events in the order that your team thinks they happened. You may have some ideas about when certain events occurred. If so, next to those events write your best ideas about how many years ago you think the event took place.

   List the most recent event at the top of the paper. Write large enough that the class will be able to read your list when you post it at the front of the class.

3. Post your team's list of events at the front of the class as your teacher directs.
4. Compare your team's estimated times and sequence with those of other teams. Then discuss the following questions with your teammates.
   a. Consider your estimates. Explain whether they were guesses or inferences based on evidence.
   b. Why did your team's estimates differ from those of other teams?
5. Examine the *Major Events in Earth's History* table that your teacher provides. Then answer the following question in your science notebook: "Which times or sequences of occurrence surprised you?"

   Scientists use several tools to understand geologic time. One of these tools is the use of evidence and inference. This table is based on inferences from evidence that scientists have gathered about the history of Earth and its living organisms, much as you did in the Elaborate activity, *Evidence for Evolution.* It also is based on scientific theories that geologists and paleobiologists have developed about the time spans and patterns of change in Earth's history. For these reasons, the dates in this table are generally considered to be accurate. For a review of evidence and inference, read the chapter 1 need to know, *Evidence and Inference,* in the Explain activity, More Evidence.

Dinosaurs lived on Earth millions of years ago. Birds are thought to have evolved from dinosaurs.

6. Study the marked clothesline that your teacher has prepared. Discuss with your teammates how it might be used to represent when events occurred in Earth's history.
7. Study the event cards that your teacher provides. Be sure you understand what the event is and when it occurred. Fold each card in half, crosswise, to form a tent.
8. Calculate where on the timeline each event card should go and place your cards in the appropriate locations.

   The distance between the red marks on the clothesline represents 1 billion years. The distance between the short black marks represents 100 million years.

9. Locate the correct time on the timeline for each of your events. Fasten your event cards to the timeline as your teacher directs.
10. Create your own version of the timeline in your science notebook. Be sure to include information on the length of time between markings as well as the sequence of events.

Analysis

Discuss the following questions with your teammates, and then answer them in your science notebook. This will prepare you to participate in a class discussion.

1. Think about how much time is required for biological changes to occur.
   a. How many years passed between the appearance of the first hominids and modern humans?
   b. Imagine that the generation time of hominids was 20 years. (In fact, this is an estimate, as we do not know the actual generation time of hominids.) How many generations passed between the appearance of hominids and modern humans?
   c. What does this tell you about the rate of biological changes in organisms with long generation times?
2. What did the timeline help you understand about Earth's history?
3. Look at the events that represent changes in Earth's surface or atmosphere. Compare the location of these events on the timeline with the events showing the first evidence for certain plant or animal groups. How do you think these events might be related? Describe at least 2 patterns.
4. Describe in 1 or 2 sentences how the timeline you developed is a model. Add to your description by including the following:
   - Two strengths of your model
   - Two limitations of your model
   - One assumption in your model

   You may wish to review the need to know, *A Simple Mathematical Model*, in the Explore activity, *Exploring Change*, as you answer questions about your model.

Further Challenges

Sometimes, words people use in everyday language have different meanings when we use them in science. For example, you might give your friend an explanation of why you were late. As you have learned, a scientific explanation includes a claim, evidence, and reasoning, or science ideas. There are other words that are used in slightly different ways in science.

a. In your notebook, write the words "hypothesis," "theory," and "law," leaving a few lines of space between each one.

b. Write a definition and example of how you might use each word in a conversation with friends or family.

c. Read the sidebar, *Just a Theory?* As you read, use the Mark It Up! Strategy and highlight any information that relates to how these words are used in science.

d. Using a different-colored pen or pencil, add a definition and information for each word to describe what it means in science.

SIDEBAR

Just a Theory?

What do we mean when we say that we have a theory about something? Usually, we mean that we have an idea about how to explain something (figure 5.37). We may suggest a theory about who killed the butler in the late-night thriller. Or we may have a theory about why Ms. Figueroa always looks so tired when she delivers the mail on Monday morning. Or we may have a theory about what the coach of the local baseball team said to his pitcher after the team lost last night's big game.

When we use the word theory in this way, we're talking about ideas we have that are tentative. That is, when we say that we have a theory about something, we may mean that we don't have much evidence. If we don't have much evidence, other people don't usually agree with or share our theories. Nevertheless, it is fun to suggest clever ideas.

Figure 5.37 The word "theory" often means different things to different people.

In contrast, scientists use the word *theory* in a very different way than we just described. A **scientific theory** is a comprehensive explanation of a set of natural phenomena. Every scientific theory is supported by vast amounts of evidence from multiple sources. Evidence for a theory is collected by many different scientists, not just one. Scientists rely on scientific theories because they explain so much.

A good example of a scientific theory is the atomic theory. It was proposed in 1803 by the British chemist John Dalton. One of Dalton's important ideas is this: *All matter (or things) in the universe is composed of tiny particles called **atoms**.* Physicists have collected an enormous amount of evidence for Dalton's idea. Some evidence came from technological breakthroughs. They allowed physicists to detect and subdivide atoms. Even though Dalton did not predict that atoms could be broken into smaller pieces, like electrons and protons, the general idea of his theory still stands. This example helps show that useful theories in science gain explanatory power as more evidence is discovered. New evidence leads to improvements on the details of the theory, and the theories we have today in science have stood the test of time. However, science is always open to revising theories, based on new evidence.

Another example is the germ theory. This theory describes how some diseases are caused by microorganisms, like bacteria. It used to be that people thought supernatural forces or invisible vapors caused diseases. But in 1876, Robert Koch found that bacteria caused the disease anthrax. Scientists then continued to look for other bacteria that caused diseases. Now we know that many diseases are caused by bacteria or other microorganisms. All this evidence strengthened the germ theory of diseases. Today, no scientists doubt that microorganisms can cause disease.

From both of these examples, we can see that scientific theories have many pieces of evidence that support one broad idea. In this way, a scientific theory is similar to a bicycle. A bicycle is made of many different parts, which all support one another. As people work with bicycles, the details of the parts change, producing a better bicycle. But the overall purpose of the bicycle (to provide fast, human-powered transportation) does not change. What is the purpose of a scientific theory? To explain how something in nature works. The role of theories in science has not changed.

Clearly, scientists use the word theory very differently from the way most nonscientists do. To understand how an explanation becomes known as a scientific theory, we need to know the term hypothesis. You may have already learned that a **hypothesis** is a trial idea about something. Hypotheses are not guesses; they are based on some observation and logic. However, they have not been thoroughly tested. Testing our hypotheses allows us to decide which trial ideas adequately explain what we study and which ideas do not.

You probably can see, then, that the everyday ideas we call "theories" are usually *hypotheses*. These trial ideas or explanations can be tested, but they aren't scientific theories. In scientific terms, the police officers who investigate a murder do so by testing a set of hypotheses.

In contrast, a scientific theory is built upon evidence from a large number of tested hypotheses. So a single hypothesis does not turn into a theory. As new hypotheses are tested, the theory helps explain the new information. Scientific theories can even help scientists think of new hypotheses to test (figure 5.38). This is because they often predict undiscovered phenomena. For example, when a new disease is discovered today, germ theory allows scientists to predict that they may find a bacteria or virus that causes it.

Why do scientists call Darwin's explanation the *theory* of evolution instead of the *hypothesis* of evolution? Why is evolution almost universally accepted by scientists around the world?

First, evolution is considered a theory because of the enormous amount of evidence that supports it. All of this evidence—not just some of it—supports evolution. The evidence comes from many different areas of science. Evolution also successfully predicts new phenomena. For example, evolutionary theory predicts that fossils should exist that have characteristics that are in between those of fish and reptiles. In 2006, scientists reported that they had discovered

Figure 5.38 When used by scientists, the word theory means an explanation that is supported by evidence and well accepted by other scientists.

such a fossil, which they named *Tiktaalik* (figure 5.39).

Did Darwin explain everything there is to know about evolution? Absolutely not. As one example, Darwin was not able to explain *how* characteristics are passed from one generation to the next. In fact, the ideas he had about inheritance were wrong. This is because he lived before the concept of genes was widely known. However, this does not make Darwin's theory incorrect. Scientists have added information about genetic inheritance to the theory of evolution. And because scientists have been able to add to Darwin's ideas, it shows that evolution is a sound scientific theory. When they found new information, their findings supported Darwin's theory, rather than contradicted it.

You are now probably wondering, if scientific theories are so important, how are they related to scientific laws? Social laws are very important for human societies. If you break the law, you are likely to get in a lot of trouble. Fortunately for you, scientific laws are virtually impossible to break. They describe relationships between things in the natural world. Often, you can recognize a law because it has its own mathematical formula. Scientific laws always produce a predicted result given a specific set of circumstances. You may have tested some of them in school. For example, Newton's law of universal gravitation predicts how fast gravity will pull a falling object toward Earth.

Scientific laws and scientific theories are both important. But laws are more limited than theories in what they explain. In fact, many scientific laws are a part of a larger scientific theory. Gravitational theory includes Newton's law of universal gravitation. But it also includes an explanation for why gravity works the way it does.

Keep in mind that the ideas (hypotheses) you think of are important, too. While not being the same as scientific theories, your ideas can often be tested. If you find evidence for your hypothesis and other people see its importance, it has the potential to make a lasting impression on science.

Figure 5.39 *Tiktaalik* is an extinct animal that has many characteristics of a fish, such as scales, fins, and gills. But it is the oldest fossil known to have a movable neck and wrist bones. These characteristics are found in modern amphibians. For these reasons, scientists call *Tiktaalik* a transitional fossil.

Evolution in Action

In this chapter, you studied a mechanism for evolution in living organisms—natural selection. You also examined evidence and inferences that support the theory of evolution. If you have developed a deep understanding of evolution, you can use and apply the concepts you have studied to analyze and explain new situations. This is much more powerful than simply memorizing information.

In this Evaluate activity, *Evolution in Action*, you will apply your knowledge to explain three possible endings to a story about a teenage girl who had surgery. Use the vocabulary and concepts that you have learned to write well-reasoned explanations for the three possible outcomes to the scenario. By successfully completing this activity, you can demonstrate to yourself and your teacher that you truly understand.

PROCESS AND PROCEDURES

1. Read the scenario, *A Turn for the Worse*.

SCENARIO

ENGLISH SPANISH

A Turn for the Worse

The attendants wheel the teenage girl into the operating room. Her mother waits anxiously in the sitting room at the end of the hall. The doctor worries that the girl's inflamed appendix might burst before she can perform the operation. In spite of the danger, she successfully removes the appendix from the girl's abdomen. After surgery, nurses take the patient to the recovery room. After about 30 minutes, she regains consciousness and speaks to her mother.

All seems to be going well in the first 24 hours after surgery. However, on the following day, the girl begins to run a fever, which quickly rises to 41°C (105°F). The doctor realizes that her patient is again at risk of dying.

The girl in this story has a bacterial infection. A strain of *Staphylococcus* bacteria entered her abdomen during surgery. It is now rapidly multiplying inside her body. Will she survive this infection? Use your knowledge of evolution and your scientific thinking skills to propose an explanation for what happens next. Write a claim, or answer, to the questions—Why did the girl die after getting the infection? or Why did the girl survive the infection?—depending on the year. Then, using information from the need to know *Antibiotics* as evidence, write your reasoning for your answer to the questions. As you complete this activity, you will evaluate what you have learned about the way living organisms change across time.

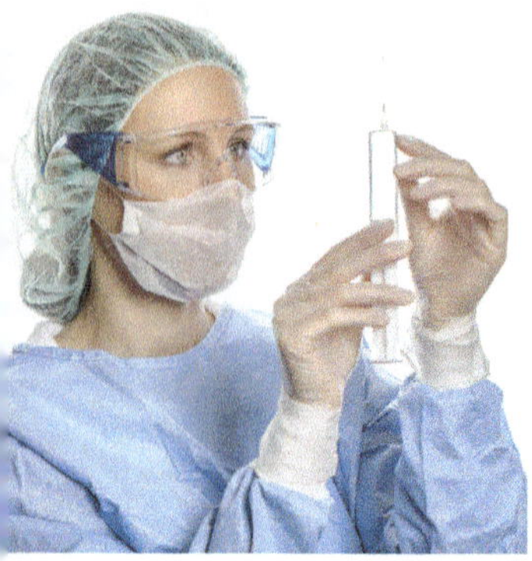

2. Read the following 3 possible outcomes for the opening scenario. Each outcome takes place in a different time period in the history of Western medicine.

Possible Outcome 1

The year is 1925: The girl becomes delirious from fever. After a few days, she dies.

Possible Outcome 2

The year is 1945: The girl receives an injection of the antibiotic penicillin, followed by repeated doses. Within 24 hours, her fever is reduced. In a week, she is released from the hospital, well on her way to recovery.

Possible Outcome 3

The year is 1965: The girl receives an injection of the antibiotic penicillin, followed by repeated doses. At first, she seems to be getting better. But after two days, her fever rises rapidly again and she becomes delirious. After a few days, she dies.

3. Read the background information in the need to know, *Antibiotics*. As you read, write in your science notebook any facts that may help explain the 3 scenarios you read.

NEED TO KNOW

ENGLISH SPANISH

Antibiotics

An antibiotic is a medicine that kills certain types of bacteria. Fortunately, antibiotics do not affect human cells. You may have heard of antibiotics such as penicillin and tetracycline. Researchers discovered a few antibiotics in the 1920s, but these drugs were not widely available until the 1940s. Penicillin was the first widely available antibiotic. It quickly became an important tool for fighting bacterial infections. Before antibiotics, bacterial infections killed many people.

Today, doctors choose from a wide range of antibiotics. They use these to treat a variety of bacterial infections such as strep throat, pneumonia, and even acne. The dependence of modern medicine on antibiotics has changed the way we think about health. This is a good example of how technology leads to cultural change. Many patients now demand antibiotics, even when an antibiotic will not cure their illnesses. For instance, antibiotics do not kill viruses. The common cold is most often caused by a virus. So taking antibiotics for a cold is not useful. Also, not every antibiotic is effective against all bacteria. Doctors must know which bacteria are causing the illness. This is why doctors usually take a sample, like a throat swab, before prescribing an antibiotic.

When antibiotics are not used correctly, they may not kill all of the problem bacteria. The bacteria that survive have inherited differences in their DNA that allow them to survive exposure to the antibiotic. These bacteria are said to be *resistant* to the antibiotic. The survivors may then reproduce and become more abundant. They pass the genetic differences on to their offspring.

The evolution of resistant bacteria is most likely to occur when individuals do not take all of their antibiotics. When all of an antibiotic is taken, all of the bacteria are usually killed. When a patient stops taking antibiotics early, though, the most-resistant bacteria remain. The person may feel better, but the resistant bacteria then start to reproduce. When many people misuse antibiotics over many years, the antibiotic may eventually become ineffective. This is because all of the problem bacteria have inherited the differences that allow them to be resistant.

In recent years, scientists have found bacteria that are resistant to several antibiotics at once. Just one example is the bacteria that cause tuberculosis. Tuberculosis is a lung infection that causes death if not treated. It has become more common around the world due to multidrug resistance. The problem is currently most severe in Russia, China, and India. Fortunately, tuberculosis is not nearly as easy to catch as the common cold or the flu. However, tuberculosis can be spread through the air when people are in close quarters. Because of this, doctors suggest that patients with multidrug-resistant infections should not travel by airplane.

4. Sketch the graph in figure 5.40 in your science notebook. Then apply the highlight comments and captions strategy to your sketch.

   Use this strategy to refer to parts of the graph that you think are important (what you see) and explain why you think it is important (what you think it means). Then write a caption for the graph that summarizes how the graph may help you explain the three scenarios you read. You can review *How to Write Highlight Comments and Captions* in appendix B6.

5. For each possible outcome, write 1 paragraph that explains why it was possible at that time in history.

   Base your explanations on your experiences in this chapter. Give a basic overview of how these outcomes could have happened.

SEP

HOW TO

Figure 5.40
Increase in resistance of hospital strains of *Staphylococcus aureus*.

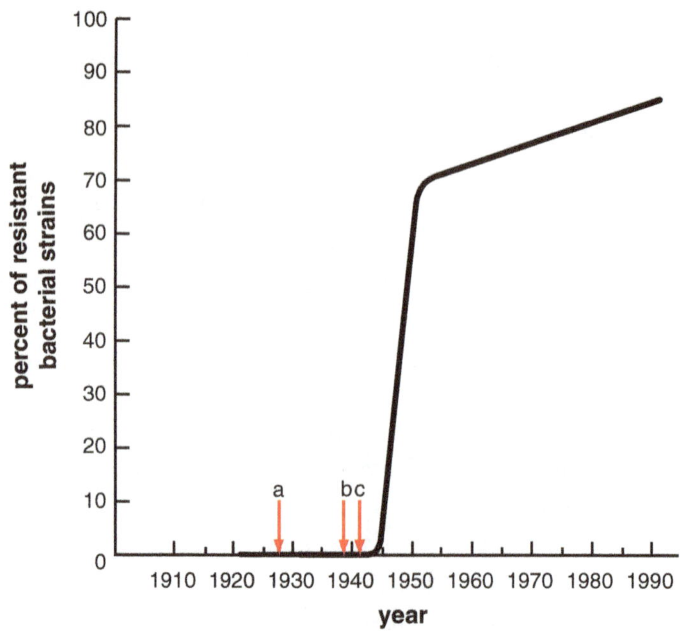

a. 1928: In the laboratory, Sir Alexander Fleming observed that the mold *Penicillium* kills the bacterium *Staphylococcus*.

b. 1939: Ernest Chain, Sir Howard Florey, and researchers at Oxford University isolated the antibiotic penicillin from the mold *Penicillium*.

c. 1941: Large-scale production of penicillin begins.

Analysis

1. In a short essay, explain how these 3 possible outcomes illustrate evolution in action in modern times. Your written response should explain what changed over time and how it happened. A good response or answer will
   - describe the evolutionary change that occurred,
   - identify the selective pressure in the bacteria's environment,
   - explain the role of variation in characteristics of the bacteria,
   - explain the role that new generations of bacteria played,
   - use the terms "adaptation" and "natural selection" in the context of the 3 possible outcomes, and
   - use specific examples and evidence to support your claims.
2. The frequency at which a species reproduces (generation time) affects how quickly evolution occurs. Explain why the trait of penicillin resistance could become common in populations of *Staphylococcus* bacteria in less than 24 years, while changes in other populations may take millions of years.

   You may want to consider the following: The generation time for humans is about 28 years. The generation time for bacteria is about 20 minutes.

3. Consider the observation that bacteria evolve resistance to antibiotics. Is this observation a new line of evidence for evolution or is it an example of one of the lines of evidence that you learned about in this chapter?

   To prepare to answer this question, first review your answer to *Analysis* question 2 in the Elaborate activity, *Evidence for Evolution.* In your table, add any new lines of evidence you learned about from the Elaborate activity, *Modeling Earth's History.* Then think about whether the observation that bacteria evolve resistance is a line of evidence or an example; explain why.

Further Challenges

1. Farmers sometimes add antibiotics to their livestock feed. This has been shown to enhance the growth of the livestock. What effect might this practice have on the evolution of bacteria if it is used frequently by farmers in a community? Read the *KQED News* article, "California Now Has Strictest Limits in U.S. on Livestock Antibiotics" (October 2015). How might these limitations address the spread of antibiotic resistance?
2. According to the Centers for Disease Control and Prevention, more than 2 million Americans get sick with antibiotic-resistant infections each year. Read the *KQED Science* article, "U.N. Takes 'Historic' Action Against Antibiotic Resistance" (September 2016). Describe how the overuse of antibiotics can lead to extreme consequences, including death. What is the U.N. declaration trying to do to combat this problem?

CHAPTER 6

Products of Evolution: Unity and Diversity

. . . the Tree of Life, which fills with its dead and broken branches the crust of the earth, and covers the surface with its ever branching and beautiful ramifications.

Charles Darwin

How many different types of life can you find in this coral reef? How many different types do you think you would find if you went snorkeling among the soft coral, fish, and other organisms of this coral reef?

One of the most remarkable aspects of life is its enormous diversity. Think of the differences between the microorganisms in the ocean and the blue whales that feed on them. In chapter 5, *Evolution: Change Across Time*, you learned how populations of organisms become different from one another over time through natural selection. In this chapter, you will consider how differences that accumulate in populations can lead to the formation of new species.

Your challenge in this chapter is to look at life's diversity as a scientist. Though living organisms are diverse, they also have many features in common. This is because all living things are linked to one another through ancestors in the past. Living things are descended from a common ancestor, but lineages change over time. This is why Charles Darwin described evolution as "descent with modification." Have you ever seen a diagram of your family's tree? In this chapter, you will learn how scientists use diagrams similar to a family tree to represent the relationships among organisms. You will examine how the unity that underlies all of life helps scientists organize and explain life's diversity. In addition, you should find, as many scientists do, that a close study of all the different kinds of life on Earth will increase your appreciation for life's diversity and complexity.

GOALS FOR THE CHAPTER

Throughout chapter 6, *Products of Evolution: Unity and Diversity*, you will work to gain an understanding of the following concepts:

- ✔ Common ancestry helps explain why organisms share so many features.
- ✔ Over vast periods of time, life on Earth is shaped by natural selection, speciation, and extinction.
- ✔ Biological classification is based on the natural nested groups in the tree of life.
- ✔ Many different sources of evidence support the theory of biological evolution.
- ✔ Scientific theories are supported by independent sources of evidence.
- ✔ Scientific explanations are tentative and change over time as more evidence is discovered.

To help you develop these understandings, you will complete the following activities.

ENGAGE

How Many Kinds of Zebras?

Key Idea: Because of life's diversity, determining which groups of organisms represent distinct species can be complex.

Linking Question:
How can we use unifying features to explore diverse forms of life?

CHAPTER 6

Products of Evolution: Unity and Diversity

EXPLORE

Using Unity to Explore Diversity

Key Idea: We can use characteristics that are shared among groups of living organisms as evidence of evolutionary relationships.

Linking Question:
How do scientists represent the unity and diversity of life on Earth?

EXPLAIN

Evolutionary Trees—the Pattern of Evolution

Key Idea: Evolutionary trees are claims about relationships and are based on data that scientists use to understand the patterns of evolution.

Linking Question:
How do new species develop from common ancestors, as depicted in evolutionary trees?

EVALUATE

Your Critter's Adaptations and Ancestry

Key Idea: Evolution is the process that explains both the unity and the diversity of life.

MAJOR CONCEPTS

- ✔ Common ancestry helps explain why organisms share so many features.
- ✔ Over vast periods of time, life on Earth is shaped by natural selection, speciation, and extinction.
- ✔ Biological classification is based on the natural nested groups in the tree of life.
- ✔ Many different sources of evidence support the theory of biological evolution.
- ✔ Scientific theories are supported by independent sources of evidence.
- ✔ Scientific explanations are tentative and change over time as more evidence is discovered.

Linking Question:
How can a specific organism be understood through the unifying principle of evolution?

ELABORATE

Evidence for Common Ancestry

Key Idea: Evidence for descent with modification comes from studies of comparative anatomy, development, and biological molecules such as DNA.

Linking Question:
What is the evidence that organisms share common ancestors in 1 great tree of life?

EXPLAIN

Descent with Modification

Key Idea: Natural selection can result in changes in characters that affect reproductive isolation among different populations. This causes speciation. The branching process of speciation results in life's diversity.

ENGAGE

How Many Kinds of Zebras?

In chapter 5, you studied adaptation and considered several explanations for the adaptations in the Galapagos finches. In *How Many Kinds of Zebras?*, you will focus on different kinds of zebras. If you were to travel across Africa, you would notice that in certain places, the zebras look different. Are these differences just variation within a species, or are there different species of zebras?

Take a look at the zebras in figure 6.1. Is each different type a different species? This question may seem easy to answer, but in fact, the identification of species can be difficult for scientists, especially if only a little is known about the organism. Sometimes scientists discover new information about an organism that leads them to change their minds about what they call a species. In this Engage activity, you will participate in a process to learn about the difficulties that scientists face as they sort through life's diversity. Can you tell a zebra by its stripes?

Materials (per team of 4)

online resource (optional)

Figure 6.1 Zebras. All these zebras live in Africa but each group has a different common name. What similarities and differences do you observe among these four zebras?

PROCESS AND PROCEDURES

1. Make a table in your science notebook similar to the one shown in figure 6.2.

Characteristic	Zebra			
	Chapman's	Grant's	Grevy's	Mountain
Overall size				
Shape of head				
Stripes on belly				
Width of black stripes				

Figure 6.2 Sample observation table. Record your observations in a table similar to this one.

2. Fill in your table using what you observe in figure 6.1. Discuss your observations with your team.
3. Write down an initial hypothesis about how many species of zebras you think there are. Make sure to record the reasoning you used to make your decision.

   The fundamental question here is whether each group of zebras represents a separate species or whether some are members of the same species. Provide reasons for your answer, using the information you listed in your table.
4. Label a new row in your table as "range overlap." Examine the map in figure 6.3. Estimate how much each zebra's range overlaps with the range of the other types of zebras. Fill in this new information in your table. If your ideas about the number of zebra species have changed, write a revised hypothesis under your original. Include the reasons you revised your hypothesis.

   "Range" describes where scientists have found an organism. For example, the area in which the mountain zebra is found is in southwestern Africa, so its range is southwestern Africa.
5. Read aloud the information on the *African Zebras* handout, taking turns with your teammates. Add to new rows in your table information you think is important. Write down a final hypothesis about how many zebra species there are.

SEP

SEP

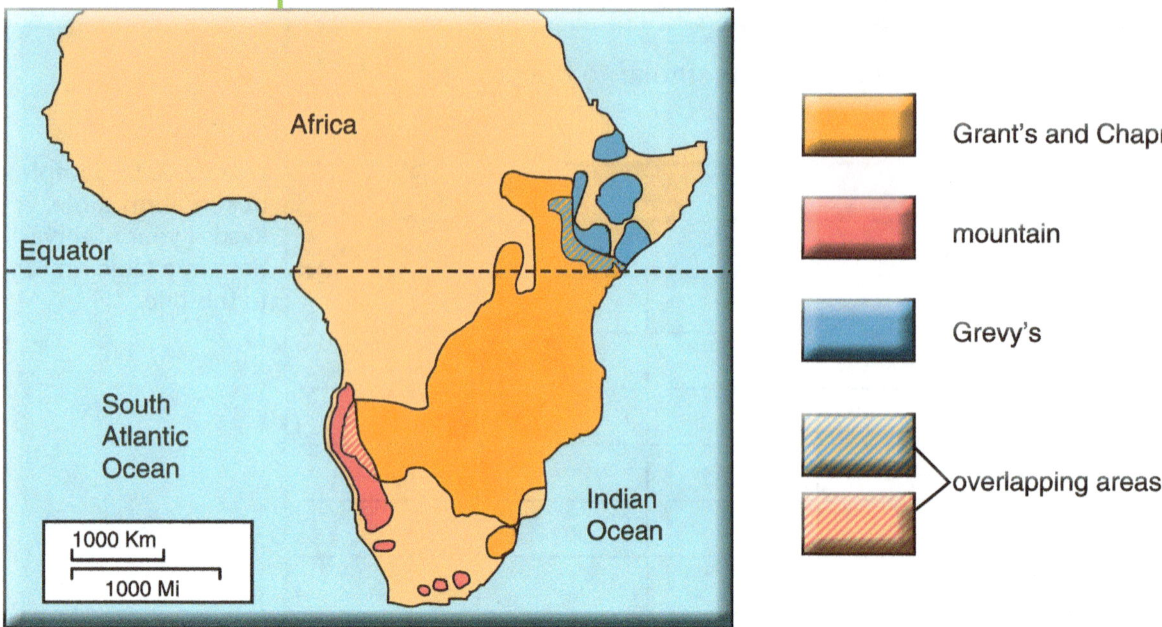

Figure 6.3 How do the ranges of the four zebras compare?

6. Discuss the number of zebra species with your teammates. Revise your hypothesis of the number of species if your teammates convince you of their reasoning. Record the reasons you revised your hypothesis.

Analysis

Answer the following questions in your science notebook.

1. Review your hypotheses about the number of zebra species. As you gathered more information, did your ideas change about how many species of zebras exist? Explain why or why not.
2. Write down an initial definition of "species." Use your own words and leave some space in your science notebook to revise your answer later.
3. Write down your ideas for why the 4 types of zebras have so many features in common, such as their striping, their diet, and their behavior.
4. Imagine 2 zebras that are different species. There was a time in the past when these 2 species did not exist. Instead, 1 zebralike species existed. Write a short story explaining how you think these 2 distinct species might have arisen from a common ancestor that was a single zebralike species.

It is OK if you aren't sure how a new species forms. What is important is that you write out any ideas you already have about this scenario. Feel free to use your imagination to compose a logical scenario.

Using Unity to Explore Diversity

EXPLORE

In the Engage activity, *How Many Kinds of Zebras?*, you thought about the definition of the word species. You compared various characteristics found in different kinds of zebras to help you refine your thoughts. You also thought about why the different types of zebras have a lot in common.

Every species has peculiarities of structure, function, and behavior that make it distinctly different from all of the others. In chapter 5, you learned that natural selection is one very powerful way that populations of organisms change over time. Yet all living organisms share certain features. For example, if you read the essay *A Closer Look at Cellular Respiration* in chapter 4 you learned the process for converting food into usable energy is very similar in all organisms. In this Explore activity, *Using Unity to Explore Diversity*, you will focus on these shared features in order to explore how different species are related. Your focus question is, "How can similarities, or unifying features, be used to explore diversity?" To do this, you will make observations of several different kinds of plants and an alga (figure 6.4).

Figure 6.4 A scientist makes observations about plant growth and appearance.

Materials (per team of 4)

plant and alga cards (1 set of 5 cards)
large sheets of paper
felt-tipped markers
different-colored pens or pencils

PROCESS AND PROCEDURES

1. Read the scenario, *Exploring Nature,* to help you get ready for your exploration.
2. Gather as a team of 4. Examine the plant and alga cards that your teacher provides your team. While you compare the pictures of each card, think about how these organisms may be related to one another.
3. Discuss the following question with your team and record your answer in your science notebook. One way to begin exploring relationships is to think about characteristics (or characters, for short) that organisms share with other organisms. How can the number of shared characteristics help scientists determine how closely related 2 organisms are?
4. Copy figure 6.6 onto a new page in your science notebook.
5. As a team, make observations of your plant and alga cards as you fill in your *Plant Exploration* table. If you think the specimen has the character listed in the table, place an *X* in the appropriate cell in the table. Otherwise, leave the cell blank. Discuss any patterns you notice.
6. Compare your team's table with the *Plant and Alga* table that your teacher will show you. Ask questions about any characters that you do not understand.

SCENARIO

ENGLISH SPANISH

Exploring Nature

You wake up excited about the opportunity to get out of the city and into nature! Today, you and your friend Alayah get the chance to visit Muir Woods and the surrounding area. This will be Alayah's first visit to the area. Looking outside you notice that the fog in your Tenderloin neighborhood has cleared and the Sun is shining. Today will be a great day for getting outside.

While exploring, you and Alayah come upon some very interesting organisms. Along the trail is a big-leaf maple. But it is not alone—growing on the tree is a lot of moss and a licorice fern (figure 6.5a). Alayah is surprised when you take out your pocketknife and slice off a small piece of the fern for each of you to taste. To her surprise, it tastes like licorice. You continue on your hike into a large grove of massive evergreen trees. Some of these Douglas fir trees (figure 6.5b) are up to 90 m (295 ft) tall. After taking a break for water, you continue to your destination, a creek bank with a surrounding meadow. The meadow around this part of the creek is filled with red columbine plants. You choose to eat your lunch on some rocks near the edge of the creek. As you look down, you notice large clumps of green algae in the water.

As you are walking back to the parking lot, you reflect on the various organisms you have seen. They look so different, but your biology teacher has said that algae and plants are related. You and Alayah decide it would be interesting to try to collect evidence to study how these organisms are related. You decide it is best to represent these relationships with a diagram.

Figure 6.5 Some of the plants seen on the hike: (*a*) licorice fern, (*b*) Douglas fir, and (*c*) red columbine.

Character	Alga	Moss	Fern	Conifer	Flowering plant
Parts of the organism are green					
Has cells					
All species have more than 1 cell					
Grows > 10 cm tall					
Makes seeds					
Makes cones					
Makes flowers					

Figure 6.6 Plant Exploration table. Copy this table into your science notebook.

7. As a team, examine the row of data for "makes seeds." What does the pattern of data for this character say about the relationships among the 5 plants and alga? In your science notebook, draw a diagram that represents your team's ideas. Repeat this process with the character "grows > 10 cm tall."
8. Divide your team into groups of 2. With your partner, use all of the data in your *Plant Exploration* table to create a diagram in your science notebook that summarizes how the 5 plants and alga are related.

   You may want to try arranging the plant and alga cards on your table as you discuss how to draw your ideas. Continue examining your data row by row, as you did in step 6. Once you have made a diagram for each character, look to see how you might combine those diagrams for all five types of plants and algae.

   Another way to approach this task is to start by thinking about which two plants you think are most closely related, and why. Then ask, "How could we represent this in a diagram?" Then try adding the next-most-related plant to your diagram. Continue until all of the specimens are represented.
9. Trade the diagram that you and a partner created with the diagram of the other pair on your team. Do not tell the other pair how to read your diagram. Try to determine the pattern of relationships on your teammates' diagram.
10. Explain to your teammates how you came up with your diagram. Discuss the similarities and differences between your diagrams.
11. Ask your teammates for advice on ways to improve your diagram. If something doesn't make sense to you about your teammates' diagram, provide advice on ways to improve their diagram. Revise your original diagram in a different-colored pen or pencil.
12. Work with your partner to draw your final diagram on a large sheet of paper. Your diagram should summarize the relationships between the 5 plants and alga, based on data in your *Plant Exploration* table.

Analysis

Work with the members of your team to answer the following questions about how scientists study the relationships among living organisms. Record your answers in your science notebook.

1. During the activity, you wrote an answer to the question, "How can the number of shared characteristics help scientists determine how closely related 2 organisms are?" Using a different-colored pen or pencil, revise your answer to this question, based on your exploration and discussion with your team.
2. Select 1 of the diagrams from your class that you think most clearly represents the relationships among the specimens. Write down the features of the diagram that make it easy to interpret.
3. Imagine that you collect more data on your specimens using different characters. Do you think your ideas about the relationships among the plants and alga could change based on these additional data? Explain your answer.

Further Challenges

To expand your knowledge of the diversity of plants and algae, use the plant and alga challenge cards to add data to your *Plant Exploration* table. Then draw an expanded diagram of plant and alga relationships that includes all 10 plants and algae.

EXPLAIN

Evolutionary Trees—the Pattern of Evolution

Every living organism on Earth is related. Mountain lions and house cats are an example of species that share a common ancestor somewhat recently. Bacteria and corn plants, on the other hand, share a common ancestor that lived a very long time ago. In this Explain activity, *Evolutionary Trees—the Pattern of Evolution*, you will learn to interpret the diagrams that biologists use to summarize patterns of relationships among living things. As you do so, focus on the question, "How do scientists represent the unity and diversity of life on Earth?"

Materials (per team of 2)

various materials for constructing evolutionary trees
Who Is More Closely Related? animation on the online resource
different-colored pens or pencils

PROCESS AND PROCEDURES

1. Copy or paste into your science notebook the diagram from the *Scientific Relationships* handout that your teacher will show you. This diagram summarizes scientific claims about the relationships among the plants and alga you studied in the Explore activity, *Using Unity to Explore Diversity*.

   Does the shape of this diagram remind you of any particular living organism?

2. Compare and contrast the diagram you see on the *Scientific Relationships* handout to the diagram you and your partner developed in step 12 of the Explore activity, *Using Unity to Explore Diversity*.

   Make a T-table in your science notebook with the columns "similarities in the diagrams" and "differences between the diagrams." Think about whether or not the two diagrams show the same pattern of relationships.

3. Place the following characteristics onto the evolutionary tree shown in the *Scientific Relationships* handout.
   a. Has chloroplasts (green color).
   b. Has cells.
   c. All species have more than 1 cell.
   d. Grows > 10 cm tall.
   e. Makes seeds.
   f. Makes cones.
   g. Makes flowers.

      In the Explore activity, *Using Unity to Explore Diversity*, you used these characters to make a claim about evolutionary relationships. Now you are trying to understand when and in which groups these characters evolved. You will revisit your answers later in the activity.

4. Read the essay, *From Cell to Seed,* for a short description of the evolution of modern plants and algae. As you read, examine where you placed the characters on the evolutionary tree in step 3.
5. Copy or paste into your science notebook the diagrams from the *Comparing Evolutionary Trees* handout that your teacher will show you. Answer the questions associated with each image.
   a. According to the diagram in image 1, is the whale more closely related to the dog or the fish? Explain your answer.
   b. The evolutionary tree shown in image 2 has similarities and differences to the tree shown in image 1. According to this diagram, is the whale more closely related to the dog or the fish? Explain your answer.
   c. Explain whether the relationships shown on the two evolutionary trees in image 3 are the same or different.

From Cell to Seed

Have you thanked a green plant today? Plants play a critical role in our existence on Earth. As you know from chapter 4, they produce most of the oxygen that we breathe, the food that we eat, and many materials that we use—from rubber, to lumber, to medicines, to coffee. Perhaps we should ask, "Have you thanked a 3.5 billion-year-old single-celled organism today?" Such an organism likely was the ancestor of all modern plants. To understand how that could be, we need to trace the history of plant evolution. One of the ways we can begin to understand this history is to recognize that each of the major events in the evolution of different plant groups involved the evolution of a major adaptation. These events led to the emergence of the hundreds of thousands of different species of plants that currently inhabit every imaginable place on Earth, from the frozen arctic tundra to lush tropical rain forests.

The Ancient Seas

We begin our survey at a point about 3.5 billion (3,500 million) years ago (see the timeline in figure 6.7). Evidence indicates that plants, like all other modern species, evolved from single-celled organisms that first lived in ancient seas and resembled modern bacteria in many ways. The atmosphere above these seas is thought to have consisted of a mixture of gases, largely water vapor, carbon dioxide and carbon monoxide, nitrogen, hydrogen sulfide (the gas that makes rotten eggs smell), and hydrogen. This mixture of gases probably contained little or no oxygen, as you know from the chapter 4 Elaborate-Evaluate activity *Coevolution of Earth's Systems and Life on Earth.*

The first single-celled organisms that lived in these seas most likely used complex molecules from their environment as their source of energy. At some point, however, the growing population of living cells started

era	millions of years ago	major evolutionary event
Cenozoic	7	Most recent common ancestor of gorillas, chimpanzees, and humans exists.
	60	Evolution of primates.
Mesozoic	66	Extinction of dinosaurs.
	130	Evolution of flowering plants.
	210	Evolution of mammals.
Paleozoic	360	Evolution of amphibians.
	430	Evolution of land plants.
	500	Evolution of vertebrates (jawless fishes).
Precambrian	1,200	Evolution of multicellular organisms.
	2,400	Free oxygen builds up in the atmosphere.
	3,500	First single-celled organisms evolve.
	4,100	Oldest Earth rocks form.
	4,600	Earth originates.

Figure 6.7 Timeline of major evolutionary events. Why do you think each of these events is significant?

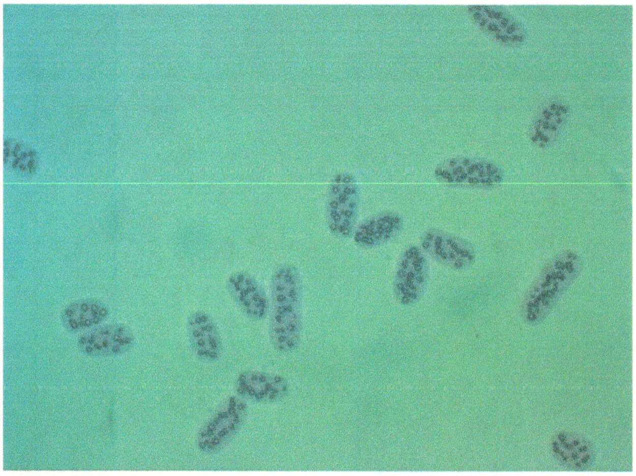

Figure 6.8 Phototrophic bacteria, *Chromatium okenii*, though only 15 μm long, can build complex molecules. The globules you see are sulfur particles.

using these complex molecules faster than they were being formed. Scientists think that the limitation of resources favored the survival of occasional cells that were able to make their own molecules. These cells probably used sulfur compounds, carbon dioxide, and the energy in sunlight to build their own complex molecules. Bacteria that use these kinds of chemical reactions, such as those in figure 6.8, still exist today.

The Accumulation of Oxygen in the Atmosphere

Bacteria that could use light, water, and carbon dioxide instead of sulfur compounds to make their own food were favored by natural selection. As you know, this process, called **photosynthesis**, releases oxygen gas. By about 2.1 billion years ago, so much oxygen had been released by these tiny photosynthesizers that the composition of Earth's atmosphere was fundamentally different.

Scientists think that photosynthetic organisms like these were the ancestors to modern plants. Today, photosynthesis is the major method of supplying energy either directly to living systems or indirectly to organisms that prey on others.

Modern photosynthetic cells, such as the blue-green bacteria shown in figure 6.9 and including those cells found in plants, contain a green pigment called chlorophyll that absorbs energy from sunlight.

The Evolution of Eukaryotic Cells

The fossil record indicates that at just about this time, another key evolutionary event occurred. Somewhere between 2.1 and 1.6 billion years ago, more-complex cells show up in the fossil record. These cells had a structure called a nucleus. A nucleus is a compartment for the storage of genetic material. Organisms with cells that have a nucleus are called **eukaryotes**. Amazingly, there is evidence that some of these early eukaryotes incorporated photosynthetic bacteria within their cells. As a result, they were able to carry out photosynthesis. The descendants of these bacteria are known today as chloroplasts. Remember from chapter 4 that photosynthesis occurs in chloroplasts. Algae and all plants contain chloroplasts.

Figure 6.9 **Photosynthetic bacteria.** Bacteria such as these *Gomphosphaeria*, which are each 1 μm in diameter, contributed to the production of oxygen in Earth's early atmosphere.

The Evolution of Multicellular Organisms

About 1.2 billion years ago, eukaryotic organisms made up of more than one cell first evolved (they were multicellular). Within these multicellular organisms, some cells were able to perform special functions that were different from the functions of other cells. This enabled multicellular organisms to exploit more resources from their environment. These benefits of multicellularity may have enabled the movement of organisms from the ocean to land (see figure 6.10). A multicellular organism would have had a better chance of surviving on land than would a one-celled organism. The outer layer of cells may have protected the inner cells from drying out rapidly. The inner cells may have been efficient at photosynthesis.

Two plant groups apparently evolved from such multicellular green algae. One group is represented today by mosses. These organisms possess few adaptations to life on land and require a moist environment in order to live and reproduce. The other group, which includes fossils of the oldest land plants, has many adaptations to life on land. This group includes modern ferns, conifers, and flowering plants.

Adaptations That Enhanced Survival on Land

One adaptation to life on land was the development of a waxy material that reduced water loss by providing a protective covering over the outer plant cells. The development of vascular tissue was another important adaptation (figure 6.11). Part of this "plumbing system" carries water from the ground up through all the parts of the plant. A vascular system enables plants to grow to large sizes (far greater than 10 cm).

Figure 6.10 ***Chara.*** *Chara* is a multicellular alga. The first land dweller may have looked like this alga. Its ancestor was only able to live and reproduce on land if ocean spray or tides kept it moist.

The oldest common ancestor of modern vascular plants probably existed about 420 million years ago. The earliest vascular plants required a moist environment to

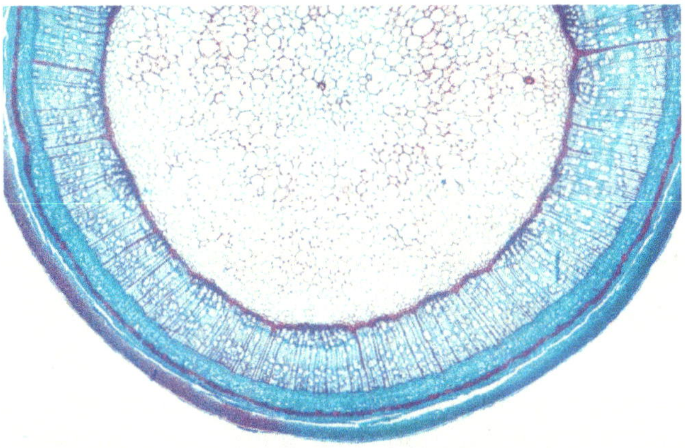

Figure 6.11 This magnified cross section of a young maple tree stem (photographed at 7×) shows the vascular tissues as rings around the outside.

reproduce. As a consequence, these plants were more successful in environments that contained at least a moderate amount of moisture.

Eventually, however, new species evolved that carried out sexual reproduction internally. This eliminated the need for a very moist environment. These changes probably occurred about 400 million years ago and enabled these new species to inhabit drier areas. In the most complex modern vascular plants, the reproductive structures are located in cones or flowers.

Another adaptation that evolved about 350 million years ago was the production of seeds. Seeds are products of sexual reproduction. They contain an inactive, tiny plant embryo packaged in material that provides food when the seed germinates and begins to grow. This system allows wide dispersal of new organisms because the seeds are spread by various means (wind, animals, or water) and begin to grow in new locations.

In cone-bearing plants such as pines and firs (refer to figure 6.12a), seeds are dispersed from open structures. But in flowering plants, seeds develop inside a specialized structure that becomes the fruit of the plant (see figure 6.12b). Fruits are made from structures in flowers. These flowering plants, products of millions of years of evolution, are the most successful type of living plants. They are represented by more than 250,000 species and have adapted to every continent on Earth except Antarctica. Your life depends on them.

Figure 6.12 How are these seeds likely to be dispersed? (*a*) The female cone in this piñon pine will mature and have seeds on the inside of the scales. (*b*) A papaya has seeds inside the fruit.

d. Explain whether the relationships shown on the two evolutionary trees in image 4 are the same or different.

When you first learned about graphs, it was important for you to practice interpreting what the information in a graph means. Similarly, you need to be able to interpret evolutionary trees to understand the unity and diversity of life. Recording your initial ideas makes your learning easier.

6. Use the TSAR strategy to compare your answers for step 5 with those of a partner.

Remember that in the think-share-advise-revise (TSAR) strategy, you first *think* about the answer to the question. Your thinking should be recorded in your science notebook. You then *share* your answer with your partner and listen to your partner's answer. In the next step, you *advise* your partner on ways to improve his or her answer and listen closely to the advice from your partner. Finally, you *revise* your answer using a different-colored pen or pencil if you decide to change parts of your answer. For more information about using the TSAR strategy, see *How to Use the Think-Share-Advise-Revise (TSAR) Strategy* in appendix B3.

HOW TO

7. Read the essay, *Interpreting the Tree of Life,* to help you explain how to interpret an evolutionary tree and how biologists use evolutionary trees to explain evolution. Use the Mark It Up! Strategy as you read. Underline words or phrases that relate to how to interpret evolutionary trees. Also label the following on the evolutionary tree from the *Scientific Relationships* handout: node, branch, taxa, and root.

Interpreting the Tree of Life

ENGLISH SPANISH

All the organisms on Earth are linked together through a long line of ancestors. A diagram that summarizes these relationships is called an **evolutionary tree**. The tree of life is an evolutionary tree that summarizes the patterns of relationships among every living organism. Scientists' understanding of the tree of life is growing rapidly as improvements in technology enable them to study the relationships among many different types of organisms.

Claims Based on Shared Characters

Claims about relationships are based on characters that are shared with some living things and not shared with other living things. More closely related organisms are expected to share a larger number of characteristics. Evolutionary trees, then, are claims based on evidence about who is more closely related to whom. They also summarize claims about when different characters evolved.

Not all characters are significant when considering questions of relatedness. Some characters are acquired during an organism's lifetime, such as bigger muscles built up by weight lifting. Because these characteristics cannot be inherited, they cannot be used as clues to evolutionary relationships. The characteristics that represent biological relatedness are those that are heritable. These characteristics can be passed on from parents to offspring by way of DNA. Biologists use heritable characteristics to make claims about relationships.

Today, biologists most frequently use patterns in DNA itself as a source of evidence for evolutionary relationships. Just as with other characteristics, the greater the similarities in DNA sequences, the more closely related two organisms are thought to be. (Refer to figure 6.28, which compares DNA across primates.) As is true in any scientific investigation, claims about relationships are more likely to reflect true evolutionary history when they are supported by more than one line of evidence. Thus, most modern studies of relationships involve multiple lines of evidence.

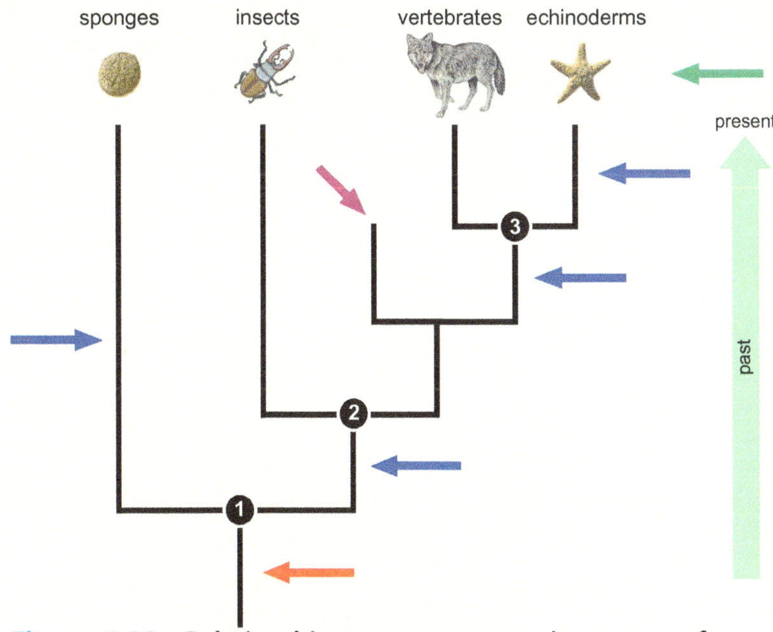

Figure 6.13 Relationships among some major groups of animals. Vertebrates are animals with a backbone. Echinoderms are a group of animals that include sea stars and sand dollars.

Interpreting Evolutionary Trees

To learn how to interpret evolutionary trees, first identify what each part of the tree represents. The evolutionary tree in figure 6.13 shows the relationships among four major groups of animals. Refer to figure 6.13 as you read the following descriptions:

- The thin green arrow points to the specific organisms that scientists studied. The organism may represent one species or a larger group of species. For example, the sea star represents a larger group of organisms known as echinoderms. Notice that in this diagram the top of the *y*-axis is labeled as present time. This means that all the groups in the study are alive today.
- Each of the blue arrows points to a different lineage of organisms. Each line, or lineage, is called a branch. Notice that all the branches on the tree are connected at some point.
- The purple arrow points to a lineage that went extinct. Though they are not shown, many extinct lineages would exist on this evolutionary tree.
- The red arrow points to the common ancestor of all the animals on the diagram.
- The numbered circles show the splitting of lineages.
    - Point 1 is the most recent common ancestor of all the animals.
    - Point 2 is the most recent common ancestor of insects, vertebrates, and echinoderms.
    - Point 3 is the most recent common ancestor of vertebrates and echinoderms.

Online Resource

8. Work with a partner and view the *Who Is More Closely Related?* animation on the online resource. As you view the animation, write down the steps for using an evolutionary tree to determine who is more closely related to whom. Also write the answers to each question posed on the screen.
9. Revisit your answers to the questions posed in step 5. Revise your answers, using a different-colored pen or pencil.
10. Examine figure 6.14, which shows the relationships among groups of vertebrates. Using the materials your teacher provides, apply what you learned in the previous steps to make at least 1 more evolutionary tree that shows the same patterns of relationships in an alternative way. Make a 3-dimensional model and draw your model in your science notebook.
11. Exchange the evolutionary tree you developed in step 10 with a partner's. Examine his or her tree and determine if your partner's tree depicts the same patterns of relationships as those shown in figure 6.13. Give your partner feedback to improve his or her tree. Use your partner's feedback to improve your evolutionary tree.

    It will be helpful to review the steps for using an evolutionary tree to determine who is more closely related to whom. You recorded these steps in your science notebook during step 8.
12. With your partner, brainstorm a list that includes at least 3 practical benefits of knowing evolutionary relationships.

    You have seen that knowledge of evolutionary relationships helped you interpret patterns of plant evolution. In what other ways would

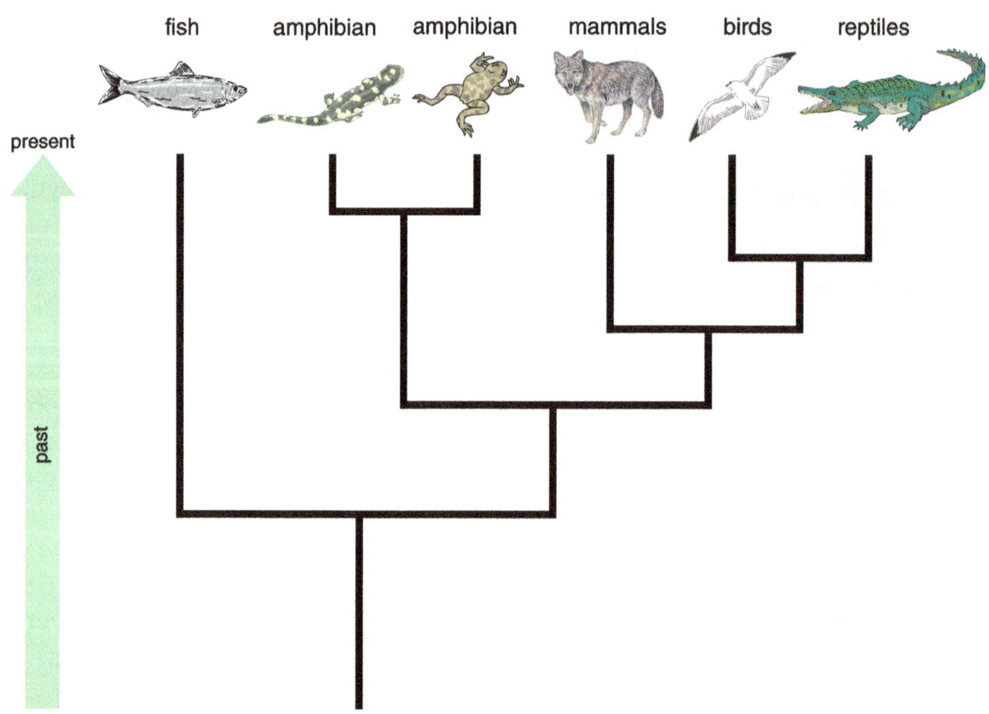

Figure 6.14 A larger evolutionary tree of some vertebrates. Can you show these same relationships in another way?

knowledge of relationships be helpful? Could this knowledge help you interpret fossils that you find? Help you identify an unknown organism, such as one that causes disease? Help solve a crime? Think broadly.

13. Make a list in your science notebook titled "uses for evolutionary trees." Add ideas from your brainstorming session that you think should go on the list. Then read the essay, *Using Evolutionary Trees*. As you read the essay, add more items to the list.

    To learn more about how evolutionary trees have helped us understand human evolution, read the essay, *Human Evolution*.

Using Evolutionary Trees

ENGLISH SPANISH

Scientists use data to explore relationships among organisms. They often look for patterns in the relationships they uncover.

Clades

Natural groups, also called **clades**, are one pattern that scientists try to identify. A clade includes an ancestor and all the groups that are descended from that ancestor. In this case, ancestors are lineages, not individuals. Look at the evolutionary tree in figure 6.15. Insects and crustaceans form a clade. The ancestral group and the two lineages that split from that group are shown in red.

Use the evolutionary tree in figure 6.15 to practice identifying clades. Do insects, crustaceans, and echinoderms form a clade? In fact, they do not form a clade because the vertebrates are not included. All the groups that came from an ancestor must be included. Vertebrates, insects, crustaceans, and echinoderms do form a clade. These groups of organisms belong to a clade that includes all the animals that have bilateral symmetry, at least in early developmental stages. This means that the animal has two halves that are

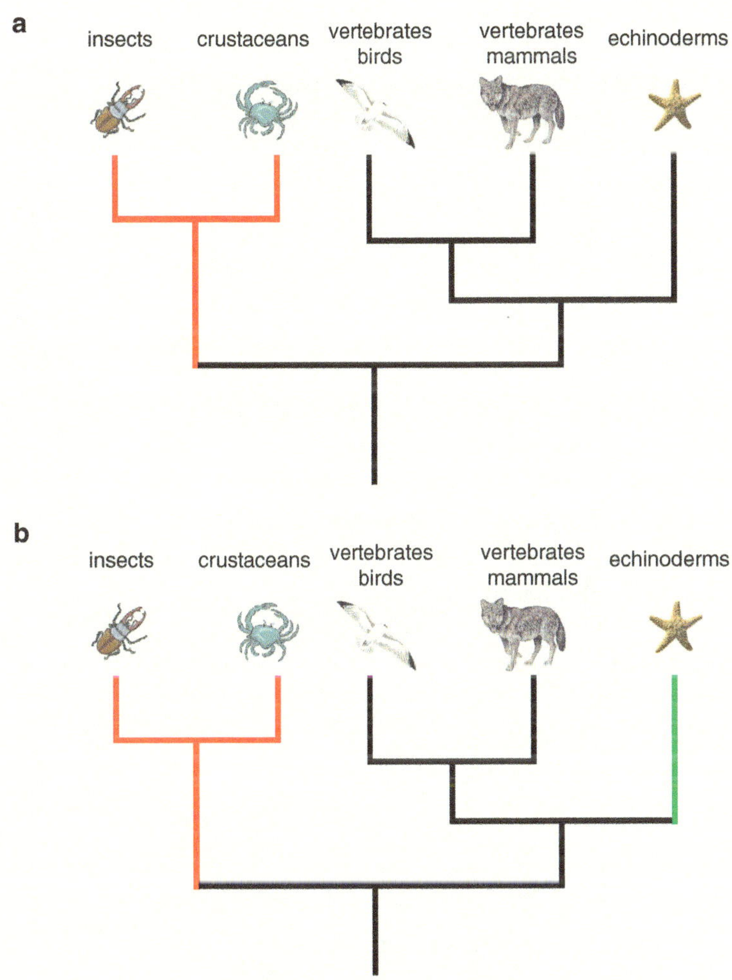

Figure 6.15 Identifying clades. (*a*) Insects and crustaceans form a clade. (*b*) Insects, crustaceans, and echinoderms do not form a clade. Overall, how many different clades can you identify on this evolutionary tree?

nearly mirror images. There are four clades on the evolutionary tree in figure 6.15. Can you find them all?

Benefits of Tree-Thinking Skills

The tree of life helps biologists classify all of life's organisms. But identifying the clades to which an organism belongs provides other useful benefits to our society. It is very important for medical personnel to use "tree-thinking" skills to better understand disease-causing agents. For example, scientists have constructed evolutionary trees to examine the patterns of evolution of the HIV virus responsible for AIDS. Some argue that AIDS is one of the most serious infectious diseases to have ever affected humankind. From these studies, scientists identified that there are two main types of HIV. They determined this because the two types formed two different clades. One type evolved from a virus in chimps, and the other type evolved from a similar virus found in a type of monkey called the sooty mangabey monkey. Scientists also determined that HIV moved from chimps and these monkeys into humans multiple times.

Understanding evolutionary trees can also help scientists track the illegal selling of meat from endangered species. For example, many species of whales are endangered, and special laws prohibit hunting and the marketing of their meat. But there are other closely related whales that are more abundant. Their meat is legal to sell. Scientists can analyze the DNA of meat sold in suspect products. If the analysis shows that the meat came from a clade that only contains the endangered species, then investigators have strong evidence that the meat was obtained illegally.

Evolution of Shared Characteristics

Scientists can use evolutionary trees to better understand how characteristics evolved. Figure 6.16 is the same evolutionary tree that you saw earlier in figure 6.15. Vertebrates, insects, crustaceans, and echinoderms all have bilateral symmetry. Scientists infer that the common ancestor of all the animals shown on this diagram also had bilateral symmetry. This observation explains why bilateral symmetry is found in all the groups that evolved from this ancestor.

Birds and mammals are part of a clade called the vertebrates. Fish, amphibians, and reptiles are also vertebrates but are not shown on this tree. The presence of a backbone is a shared characteristic of the vertebrates. Studying charactersistics of organisms using an evolutionary tree helps scientists infer when these features evolved. For scientists, evolutionary trees are a type of model that helps them make claims using the evidence they have about when features evolved. The tree in figure 6.16 shows that bilateral symmetry evolved early in this group of animals. The vertebral column evolved later, and only in some animals. Where do you think feathers evolved on this evolutionary tree?

The fact that clades of organisms share certain features has many practical benefits. Almost all new medical treatments are tried out first on laboratory animals such as rats and mice, instead of experimenting directly on humans. Because they share a common ancestor in the past, mice and rats share many biochemical and physiological features with humans. As a result, experiments on these animals give scientists an indication of how treatments may affect humans.

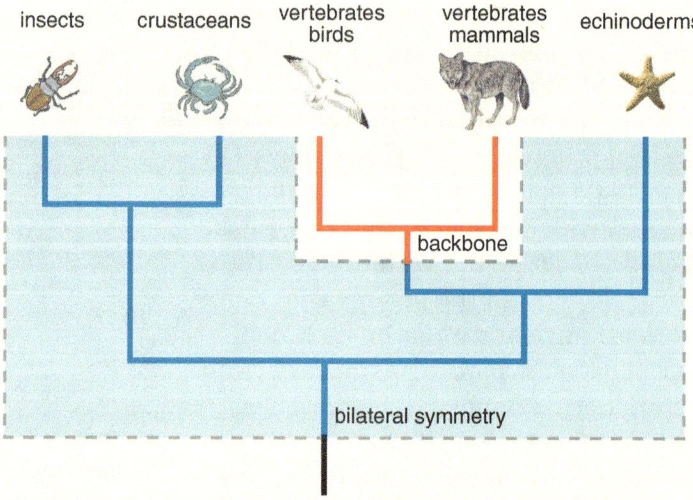

Figure 6.16 Character evolution. Evolutionary trees can be used to identify when and in what group certain characteristics evolved.

Human Evolution

ENGLISH SPANISH

An evolutionary perspective gives us the opportunity to reconsider and revise some previous ideas about how living things change. This perspective also sheds valuable light on what it means to be human. Figure 6.17 depicts evolution by using the metaphor of a ladder. In this model, more-complex groups replace simpler groups through time, with the evolution of humans at the end point of the line. People have this model of evolution in mind when they ask questions such as, "If humans evolved from chimpanzees, why are there still chimpanzees?" There are two errors in the figure 6.17 model. Examine the evolutionary tree for living primates in figure 6.18 as you use your tree-thinking skills to help you recognize the errors. First, humans did not evolve from chimpanzees.

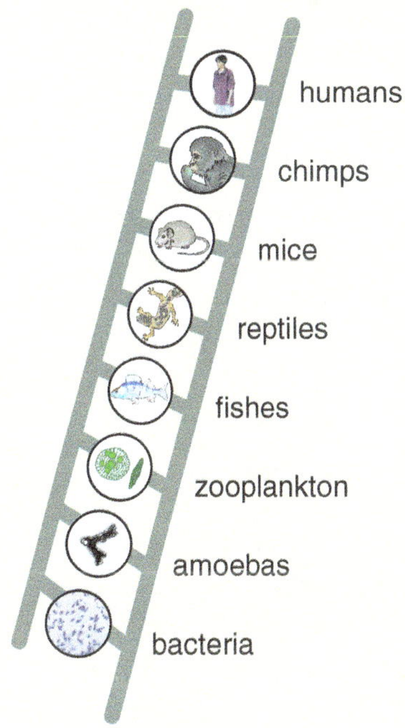

Figure 6.17 Ladder of evolution. Ancient philosopher Aristotle proposed that living things followed a progression, with more complex beings replacing simpler beings. What evidence do you have that this model is incorrect?

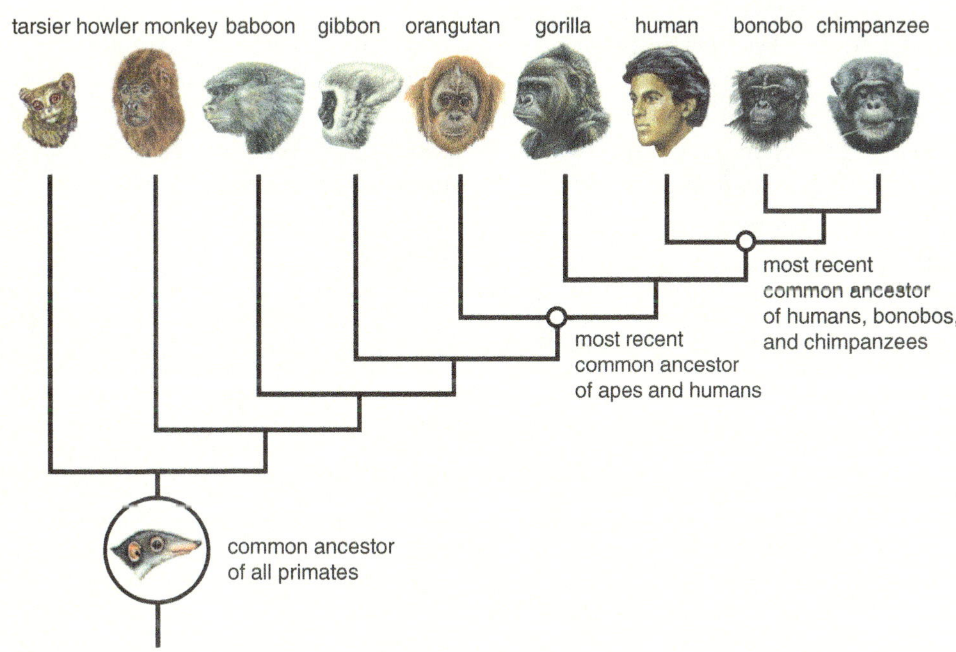

Figure 6.18 Primate evolutionary tree. This tree shows the relationships among the groups of living primates. The common ancestor of all primates lived about 60 million years ago.

Instead, chimpanzees and humans share a common ancestor. Both the lineage that led to humans and the lineage that led to chimpanzees have been evolving separately for about 6 million years. Second, evolution does not simply involve one group replacing another. Through the process of speciation, two species can evolve from what was previously one species. This helps us understand how the planet is covered with millions of species. Many speciation events result in a bushlike pattern of relationships, not a ladder of replacements.

Humanity's Place among the Animals

What does an understanding of evolutionary relationships teach us about humans? Humans share certain characteristics with certain other animals (for example, a backbone and mammary glands) that identify us as vertebrates and as mammals. Within the mammals, we humans and our close relatives are classified as primates. An important feature that distinguishes primates from their close relatives is having eyes that are located in the front of the face.

Scientists have concluded that the common ancestor of all primates was a mouselike animal that lived about 60 million years ago. The primates eventually split into three major lineages. The *hominoid* group includes apes (gibbons, orangutans, gorillas, and chimpanzees) and humans.

The Hominid Family Bush

The lineage that produced human beings is called the *hominid* family. Unlike apes, which walk on all fours, hominids are bipedal. Scientists have used fossilized bones and teeth to reconstruct our family tree, but it isn't easy. Since hominids have been evolving as a separate group for only 6 million years, the fossil record is not very long and the differences among fossils are not always obvious. This means that different scientists sometimes interpret the fossil record differently. Like the rest of science, our knowledge of human evolution is tentative and subject to change. As new fossils are found, our understanding of human evolution improves.

Despite uncertainties about the status and classification of some fossils, scientists have established many facts about hominids. For example, over the past 6 million years, the hominid lineage diversified. By looking at the fossil record, scientists have concluded that the hominid tree contains many species (perhaps 16 or more). The time frame for the existence of these species is shown in figure 6.19. You may notice that figure 6.19 does not show the relationships among the different hominid species. Reconstructing these relationships is an area of active research and debate. It is interesting to note that some of these different species of hominids overlapped in time and lived alongside one another. Figure 6.19 indicates that as many as five hominid species may have coexisted in Africa about 1.8 million years ago. All of these lineages have become extinct except one—the one that led to our own species, *Homo sapiens.*

It is also interesting to note that the oldest hominid fossils, about 6 million years old, were discovered in Africa. Hominid fossils first appear outside of Africa nearly 2 million years ago. The early presence of hominids only in Africa is consistent with the hypothesis that humans originated in Africa. Studies of patterns of diversity among modern peoples of the world also support this conclusion.

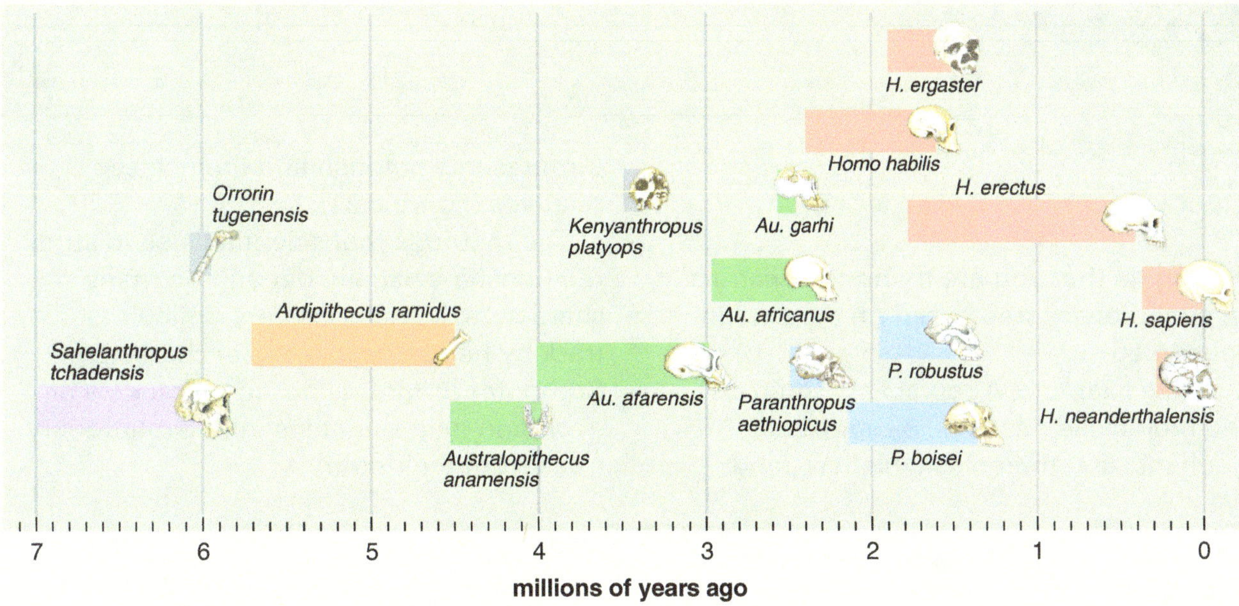

Figure 6.19 Timeline of hominid evolution. The fossil record of hominids shows that several species of hominids existed at the same time during the later stages of human evolution. Although scientists are not certain about the specific evolutionary pathway that led to modern humans, new discoveries will undoubtedly lead to new ideas.

Analysis

1. Describe how evolutionary trees are claims based on evidence that may be revised as scientists gather additional data.
2. How do evolutionary trees help us understand the patterns related to unity and diversity of life on Earth? Include the terms "common ancestor" and "characteristics."
3. Revisit and revise your answer to *Analysis* question 3 from the Engage activity, *How Many Kinds of Zebras?* Make your revisions in a different-colored pen or pencil.

Further Challenges

To practice using data on characters to build evolutionary trees, read the example in the sidebar, *Constructing an Evolutionary Tree.*

SIDEBAR

Constructing an Evolutionary Tree

Imagine that you are trying to construct an evolutionary tree based on the data in figure 6.20.

In this example, A represents squirrels and their close relatives, B represents elephants and their close relatives, and C represents goldfish and their close relatives (figure 6.21).

The methods that scientists use to study relationships usually depend on many characters. It becomes very difficult to track by hand the patterns in characters. Computer programs based on models of evolution help scientists infer evolutionary trees from their data.

Characters / Specimens	Squirrel	Elephant	Goldfish
1. Has tusks.		X	
2. Climbs trees.	X		
3. Has mammary glands.	X	X	
4. Has a backbone.	X	X	X

Figure 6.20 Character Data Table.

Instructions	Example
1. Identify which 2 species or groups have the most characters in common. In this case, squirrels and elephants share 2 characters. Goldfish only share 1 character with each of the other 2 groups. Thus, we infer that squirrels and elephants probably share a more recent common ancestor than either share with goldfish. To construct the evolutionary tree, first write each species (or group) from left to right. Place groups that you think are closely related next to each other. This row represents the present time. Next to this line, write the word "present."	A B C present
2. Each species represents a lineage that extends into the past. To represent these separate lineages, draw a vertical line straight down from each species.	A B C present | | |

Figure 6.21 Constructing an evolutionary tree.

Instructions	Example
3. Because squirrels and elephants share 2 characters and both only share 1 with goldfish, you can infer that squirrels and elephants may be more closely related to each other than either is to the goldfish. Represent this by drawing a horizontal line to connect the lines from the two closely related species. This shows that the 2 lineages used to be part of the same lineage.	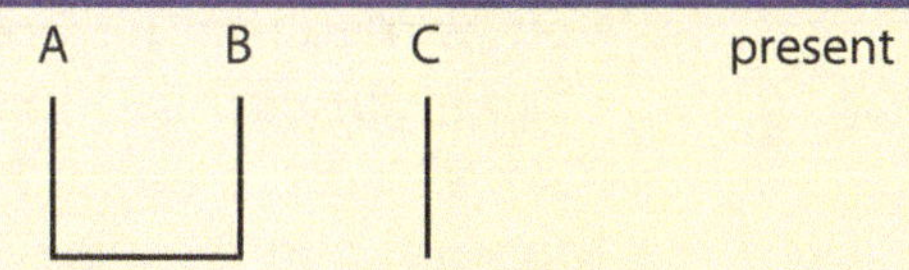
4. Once a union between 2 lineages is made, make a new vertical line in the middle of the horizontal line. Continue to extend any lineages that are not connected to other lines.	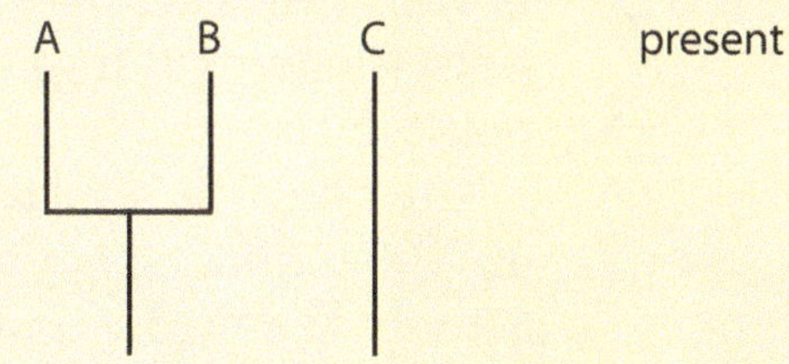
5. Represent the common ancestor of the next 2 lineages by drawing a horizontal line to connect the vertical lines. Repeat steps 4 and 5 until all the species are united.	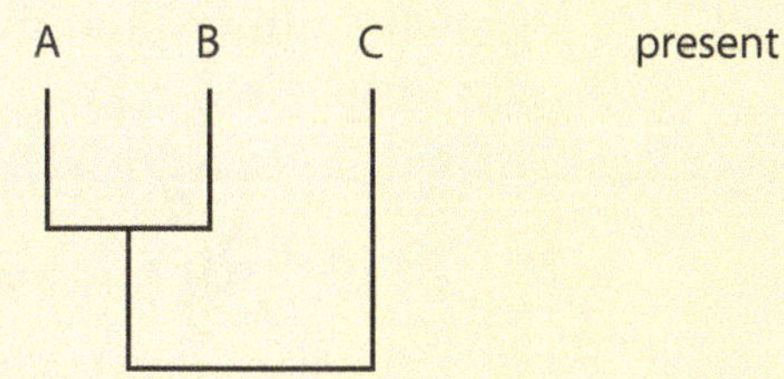
6. After all the species have been connected, draw in 1 last vertical line to represent the common ancestral lineage of all the species. Place the word "past" at the bottom of the diagram and put an arrow leading from the past to the present.	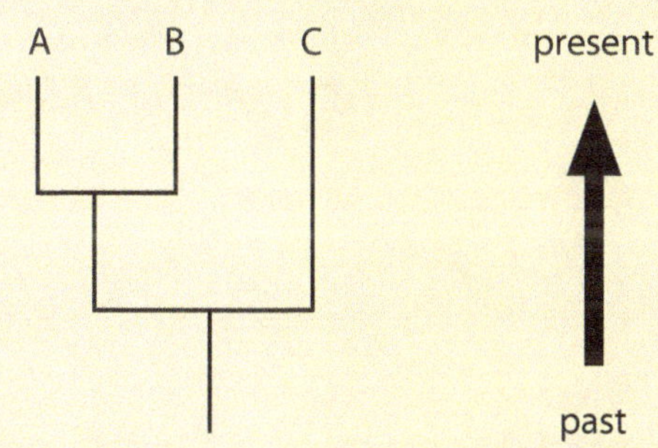
7. Place the characters from the study on their appropriate places on vertical lines. Notice that characters can change anywhere along branches, not just at nodes.	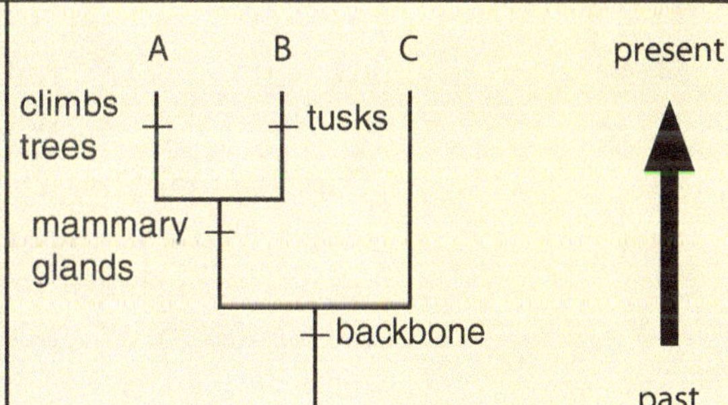

EXPLAIN

Descent with Modification

It is helpful to frequently reflect on what you have learned. In chapter 5, you learned that natural selection is a powerful mechanism that results in changes in populations over time. In the Engage activity, *How Many Kinds of Zebras?*, you examined some of the products of natural selection, namely four different groups of zebras. These groups of zebras have many similarities and also some differences. You worked with your classmates to decide which groups of zebras represent different species. You also started thinking about how these different kinds of zebras are related. In the Explore activity, *Using Unity to Explore Diversity*, you investigated how living things are related.

In the previous Explain activity, *Evolutionary Trees*, you learned how evolutionary trees summarize the patterns of relationships among different organisms. All living things share a common ancestor with one another at some point in time.

Life on Earth is incredibly diverse—scientists estimate that there are 40–100 million species. But how do new species come about? In the Engage activity, you wrote down your ideas about how new zebra species may have formed. In this Explain activity, *Descent with Modification*, can you use your growing understanding of evolution to add to your explanation for the origin of species?

Materials (per team of 2)

Descent with Modification animation on the online resource
Speciation animation on the online resource
Reproductive Isolation animation on the online resource
different-colored pens or pencils

PROCESS AND PROCEDURES

1. Work with a partner to create an initial concept map of some important terms you have studied so far. Use the following terms in your map: "evolution," "natural selection," "species," "evolutionary tree," "populations," "individuals," and "geologic time."

   Concept maps are a way for you to explore and represent how different concepts are connected. You will draw arrows to link terms and then use verbs or short phrases to describe the connections among the different terms. Concept maps help you organize information, which helps you learn. For more information about developing a concept map, see *How to Construct a Concept Map* in appendix B7.

HOW TO

http://www Online Resource

2. Continue to work with a partner and open the *Descent with Modification* animation on the online resource. Follow the instructions on the first 2 panels.

3. Write a few sentences in your science notebook to explain how the image on the second panel is a summary of descent through time for the population of plants (figure 6.22).

   To help you see this, start with a plant from generation 6 (top of the figure). Follow the lineage of this plant: identify its 2 parent plants; then identify the 2 parents for each of them. Continue this back through several generations. You will quickly see that your generation 6 plant is related to plants throughout the population. A lot of mixing occurs in this population.

   Remember the difference between a population and a species. A population is a group of organisms of the same species living in one area. A species usually is made of many populations.

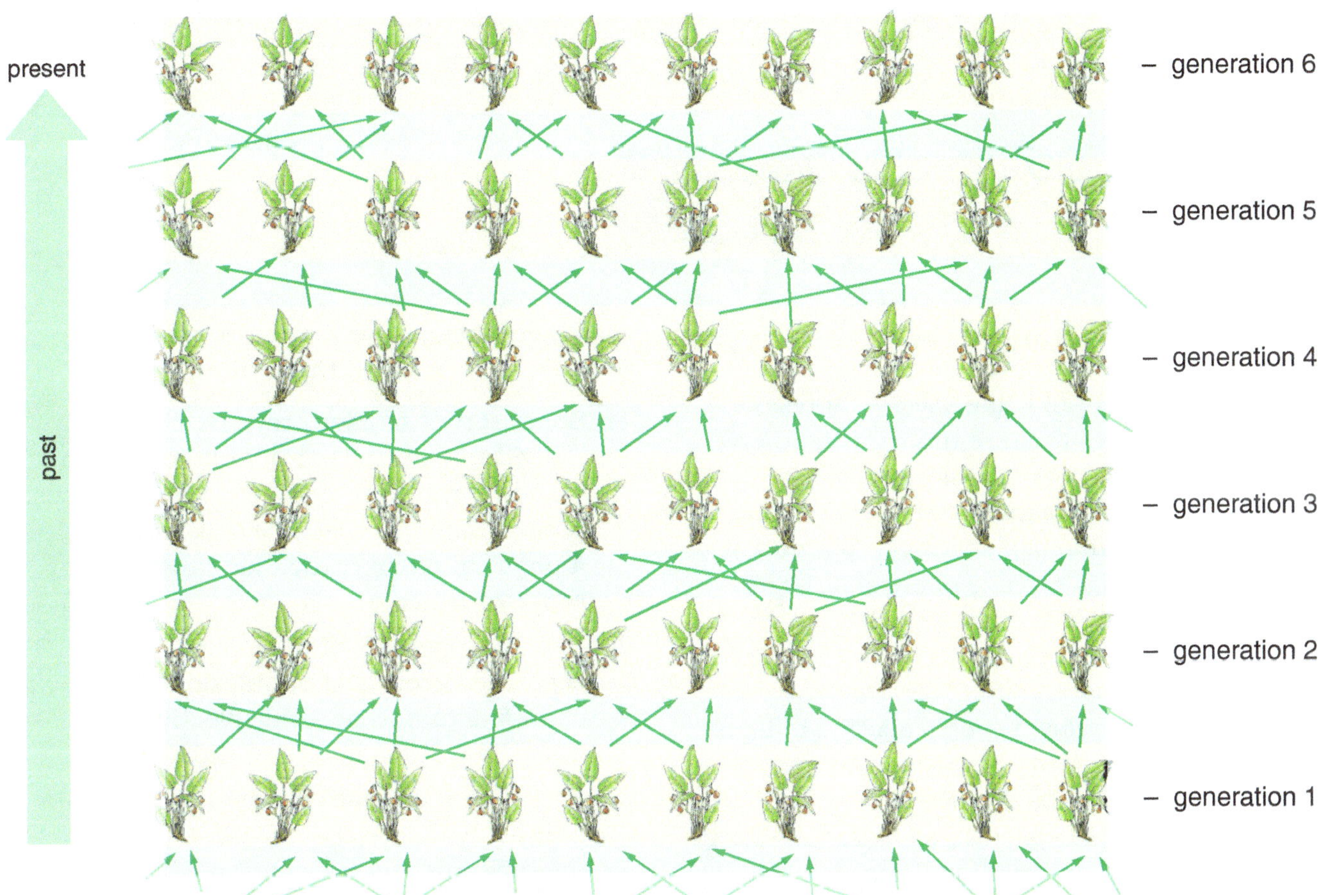

Figure 6.22 Plant populations through time. This figure represents a population of plants going forward in time. Most populations would have many more individuals, but the number was kept small in this example for clarity. The green lines show the parent-offspring relationships over several generations. Over several generations, this population is mixed. Less mixing occurs among different populations, and little to no mixing occurs among species.

4. Work with your partner to answer the following questions in your science notebook.
   a. Would you expect that many different physical traits would vary among individuals in 1 population of these plants? Why or why not?
   b. Think of just 1 trait, for example, petal length on a flower. If you compared the average length in 2 different populations, do you think the averages would be the same or different? Explain your answer.

   You might want to look over your notes from the chapter 5 Explore activity, *Exploring Change*, and Explain activity, *Explaining Adaptation*, to help you answer these questions and think about variation.
5. Click on the third panel of the animation and read the information on the screen to help you answer this question: "How are populations and species represented in evolutionary trees?"
6. As you read the essay, *A Closer Look at Species*, enter terms that are new to you into your personal glossary. Make sure to include drawings and other connections to help you remember these terms.

A Closer Look at Species

ENGLISH SPANISH

The long line of species and populations from which modern humans emerged is called a lineage. Today, all living species represent different lineages that are united with other lineages at some point in the past. We call each distinct lineage a **species**. Perhaps the most common definition of a species is "a group of organisms that is capable of breeding or mating to produce fertile offspring." Species are the fundamental unit of biological classification. Scientists estimate that 40–100 million species exist on Earth today, yet only about 2 million have been formally named. There are plenty of new species left to be discovered.

As lineages of species go forward over long periods of time, interesting things can happen. Lineages of species can split and form new lineages. This results in the formation of new species. This process is called **speciation**. Some lineages leave no descendants. This is called **extinction**.

When lineages permanently split in the process of speciation, three things happen:

1. Some sort of barrier prevents the individuals in the two populations from breeding with one another. In some cases, populations may become separated by a physical barrier, such as a mountain range or a canyon.
2. Over long periods of time, the two separated groups of populations keep changing due to natural selection and other mechanisms of evolution.
3. Eventually, changes develop in the two groups that prevent individuals from the different groups from breeding successfully with each other. These two groups are now **reproductively isolated**.

Once groups of populations have lost the ability to interbreed, they are forever on different evolutionary paths. After reproductive isolation evolves, many scientists consider the two lineages separate species.

In some ways, it seems that the definition of a species is straightforward. However, the diversity in different types of organisms poses some problems with a simple definition. Defining species based on their ability to interbreed is not always possible. For example, many organisms that reproduce sexually do not interbreed in natural conditions. Consider strawberry plants. Strawberries reproduce mainly by means of shoots or roots that grow into new plants, as shown in figure 6.23a. Also, many organisms, such as bacteria and other microbes, rarely or never reproduce sexually. Instead, they divide into two cells or a new cell buds off (depicted in figures 6.23b and 6.23c). In these cases, biologists must use other data to define the boundaries between species.

Though the term species is not easy to define, the process of speciation is incredibly important. As you observe all the organisms around you, it is astonishing to think that each of the millions of species, from bacteria to pine trees, came about through speciation.

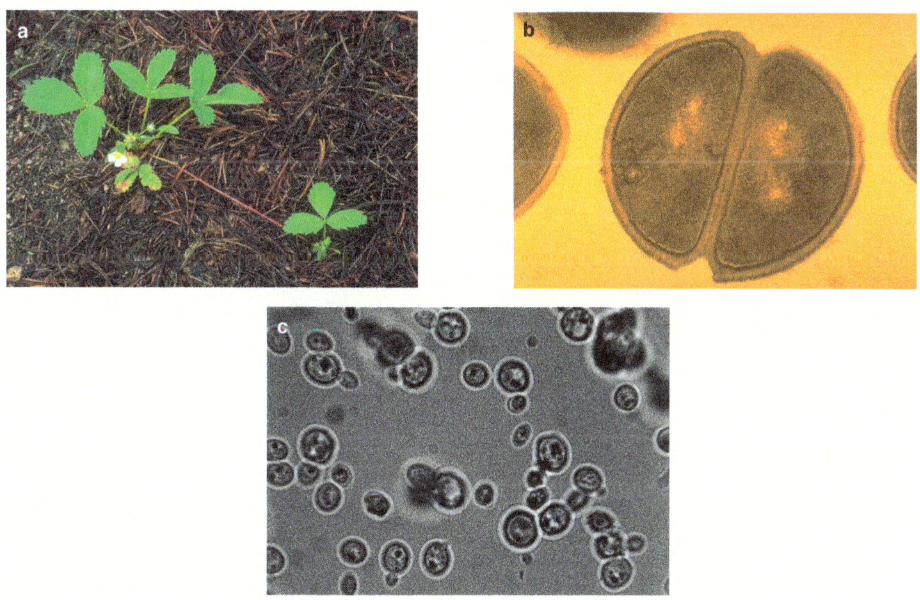

Figure 6.23 **Not all organisms reproduce sexually or interbreed.** (*a*) Strawberries develop new plants at the ends of runners. (*b*) Bacteria (*Staphylococcus*), photographed at 10,000×, divide into two new bacteria. (*c*) Yeast bud off new cells (photographed at 600×).

7. Revisit your answers from the Engage activity, *How Many Kinds of Zebras?*, and complete the following tasks.
   a. Revise your definition of a species given in Engage *Analysis* question 2, using a different-colored pen or pencil.
   b. With your partner, brainstorm at least 2 factors that may keep individual zebras from breeding and making fertile offspring in the groups that you identified as different species.

      Think about all the factors that may affect whether two individuals breed and form fertile offspring. Could changes in behavior, timing of breeding, or physical factors be important? Can you come up with other ideas?

Online Resource

8. Work with a partner and view the *Speciation* animation on the online resource.
9. Place a copy in your science notebook of the *Squirrel Speciation* handout that your teacher will give you; it is shown in figure 6.24. Use highlight comments to help you interpret the series of lines on the right side of the image.

   It may be helpful to reread portions of the essay, *A Closer Look at Species*. Remember, you want to describe what you see and what you think it means. Appendix B6, *How to Write Highlight Comments and Captions* may be a helpful resource.

HOW TO

Online Resource

10. Work with a partner to complete tasks on the *Reproductive Isolation* animation on the online resource.
11. With your partner, decide which reproductive isolating mechanism you think may have played a role in the isolation of 2 of the different zebra species that you learned about in the Engage activity, *How Many Kinds of Zebras?* Write a short paragraph summarizing the reproductive isolating mechanism and how you think it affected the 2 zebra groups.

    To learn more about how scientists observe and study speciation, read the sidebar, *Speciation*.

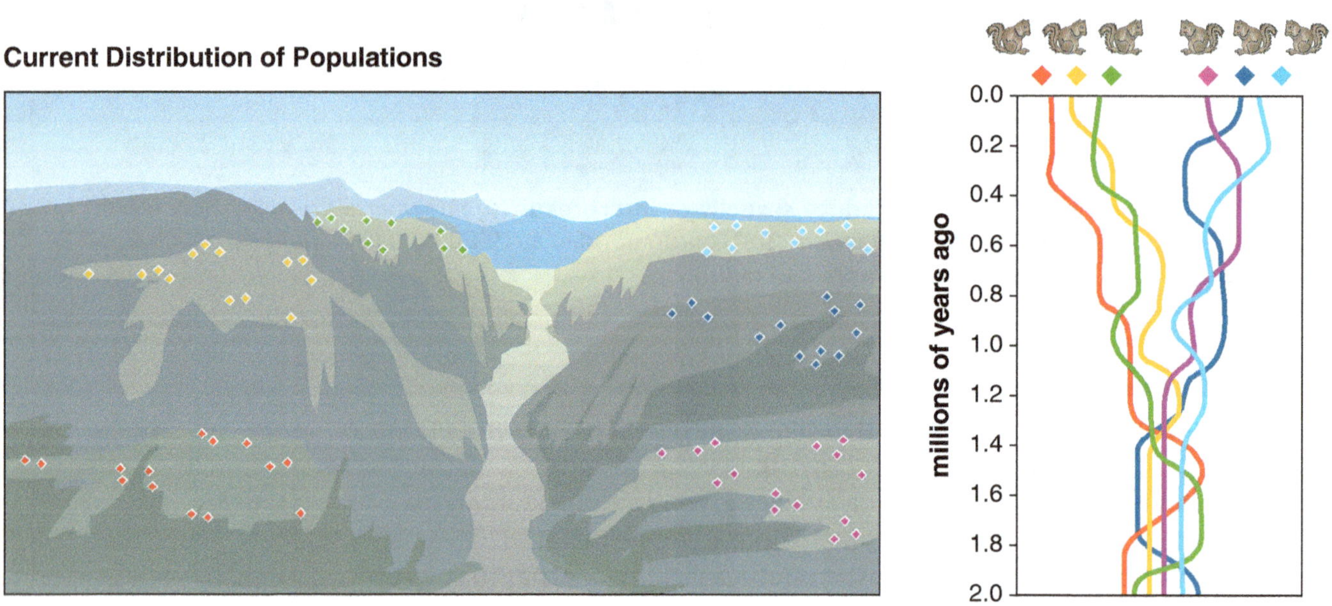

Figure 6.24 **Squirrel speciation.** The *y*-axis for the graphic on the right represents the relative degree of isolation among the different populations. Why do some of the lines start to diverge?

SIDEBAR

ENGLISH SPANISH

Speciation

It is amazing to think about the millions of species on Earth and realize that they all came about through the process of speciation. But how do biologists study speciation? Biologists study the formation of new species by exploring the evolution of reproductive isolation. Remember, once two lineages can no longer interbreed, they are forever on different evolutionary paths. However, it may take hundreds of thousands of years, or even over a million years, for reproductive isolation to evolve between two new species. To overcome the challenge of time, scientists try to understand speciation by developing models. Scientists then study those models in laboratory experiments and experiments in more natural settings.

One exciting group of studies involves a type of fly called a melon fly (*Bactrocera cucurbitae*). Melon flies are found all over the world and cause damage to more than 81 plant species. Scientists like to study these flies because they can produce a new generation in as little as 11 days. This means scientists can form many generations each year. Like many insects, melon flies develop from eggs into larvae. Larvae then develop into adults that can mate, and females can form new eggs.

In one experiment in the lab, scientists noticed that the larvae vary in how long they take to develop into adults. The scientists artificially selected the slowest-growing larvae to survive and then allowed the individuals to reproduce. They did the same thing with the fastest-growing larvae. After only 40–50 generations (about five years), this process of artificial selection led to some populations of flies that completed the larval stage in six days (the S, or short, populations) and others that completed

Melon fly (*Bactrocera cucurbitae*). Researchers use these flies as a model system for studying reproductive isolation. These flies have a very fast generation time.

the larval stage in 12 days (the L, or long, populations). The scientists were amazed to discover that the individuals from the S and L populations also differed in the time of day they liked to mate. When they put individuals from the different populations back together, individuals from the S populations mated only with other S individuals. The same was true of the individuals in the L populations. In other words, scientists were able to demonstrate the evolution of reproductive isolation in only 40–50 generations.

Other researchers study reproductive isolation in plants. Figure 6.25 shows two closely related species of plants called monkeyflowers. These species mostly live at different elevations, but they sometimes overlap. The two species do not mate with each other in nature. Researchers suspected that one of the main reasons the two species are isolated is because they have different pollinators. The pink species is pollinated mostly by bees, and the red one is pollinated mostly by hummingbirds. But why are hummingbirds attracted to one species and bees to the other? Researchers suspected that flower color played an important role. In other words, flower color might be causing the reproductive isolation. In a fascinating experiment, scientists were able to use genetic techniques to change flower color (figure 6.25). Then they planted the newly colored species in the wild. They found that the hummingbird-pollinated species that had been changed from a red to a pink color received 74 times more bee visits. The bee-pollinated species that had been changed from pink to red received 68 times more hummingbird visits. These results show that flower color can play a major role in reproductive isolation.

Overall, scientists continue to gather more and more evidence about speciation. Although most evolutionary change is too slow to see, there are examples in which it has been observed. Using models like the melon fly and the monkeyflower, scientists can demonstrate how the processes of evolution can lead to speciation.

Figure 6.25 Reproductive isolation in monkeyflowers. (*a*) Lewis's monkeyflower (named for Meriwether Lewis of the Lewis and Clark expedition) is pollinated by bees. (*c*) Red monkeyflower is pollinated by hummingbirds. Scientists were able to move a hereditary unit for color between the two species (*b* and *d*).

Analysis

1. Revisit your concept map from step 1 and complete the following tasks.
   a. Add the terms "speciation," "reproductive isolation," and "extinction" to your concept map, linking them to other concepts.
   b. Revise your concept map based on your new understandings. Use a different-colored pen or pencil for your revisions.
2. Propose a general explanation of how descent with modification and extinction are involved in shaping the diversity of life on Earth. Be sure to consider all the information you have learned.

Evidence for Common Ancestry

ELABORATE

In chapter 5, you gathered evidence that supports the idea that biological changes are the result of evolutionary processes. In this chapter, you have been learning about how species have all descended from ancestors in the past. In this Elaborate activity, *Evidence for Common Ancestry*, you will learn about additional evidence for evolution—evidence that indicates that organisms share common ancestors in one great tree of life.

PROCESS AND PROCEDURES

1. Work with your team of 4 and decide how to split into 2 pairs. One pair will take on the role of an evolutionary biologist, and the other pair will take on the role of a scientist who studies evolution and development:
   - Evolutionary biologist: A scientist who studies the relatedness of living organisms and how they change across time.
   - Evolutionary developmental biologist: A scientist who studies how the processes of growth and development occur and change over time.
2. With your partner, read and study the information assigned to your role. Complete the procedures outlined for your role. As you work with your partner, make certain you both contribute ideas.

   The information and procedures for both the evolutionary biologists and the evolutionary developmental biologists can be found on page 178 and page 183, respectively. After everyone has had time to study his or her special field of science, you will meet again with your team of 4 so that you can teach each other what you have learned.
3. Return to your team of 4 and prepare to share what you learned with your team and to learn from your other team members.
   a. Open your science notebook to the *Evidence for Evolution* table you created in chapter 5. When your teammates describe their procedures and evidence, take notes in your *Evidence for Evolution* table. Be sure to ask them to clarify any evidence you do not understand.
   b. Work with your partner and describe your specialist role to the other members of the team. Make sure your description includes the following:
      - Briefly describe the procedures you followed to gather your evidence.
      - Explain how your poster shows evidence to support the idea that organisms change across time.

Procedures for the Specialists

Evolutionary Biologists

Evolutionary biologists are scientists who generally study the relationships among organisms and how lineages of organisms change across time. The field of evolutionary biology is broad and includes scientists who work mostly in the laboratory, those who work outside in the field, and some who work primarily on computers. Evolutionary biologists may study ecosystems, whole organisms, fossils, anatomy, molecules, and/or genetic information. In this activity, you will act as an evolutionary biologist who studies anatomy in different organisms. You'll use evidence and knowledge of the tree of life to figure out the origin and history of certain body parts.

Materials (per team of 2)

1 envelope containing examples of vestigial structures
colored pencils
1 piece of poster board
felt-tipped markers
2 pairs of disposable gloves (optional)
2 laboratory aprons (optional)
1 dissecting tray and 1 teasing needle (optional)
1 chicken wing (optional)

1. Read the sidebar, *Interview with an Evolutionary Biologist.* Think about how you will explain the job of an evolutionary biologist to your teammates.
2. Complete the following activities with your evolutionary biologist partner.
   a. Locate the bones listed in the chicken wing anatomy diagram provided by your teacher and color them in using the colored pencils. Color-code the diagram so that each of the underlined bones is a different color.
   b. Obtain 1 bat wing diagram and 1 human hand diagram and their keys.
   c. With your partner, color the bones of the bat wing and human hand with the same color-coding that you used on the chicken wing diagram.
   d. In your science notebook, describe the similarities that you found between a human hand, a bat wing, and a chicken wing.
   e. Mount your diagrams on a poster board to share with your classmates.
   f. (optional) Work with your partner to carefully dissect a chicken wing. As you complete the dissection, compare the bones that you find with the chicken wing anatomy.
   g. **Wash your hands with soap.**

SIDEBAR

ENGLISH SPANISH

Interview with an Evolutionary Biologist

Interviewer: Why did you go into the field of evolutionary biology?

Dr. Edwards: I started first as an ornithologist [bird biologist] and then became interested in using genetic tools to study evolution. With my interest in birds from an early age, I knew that organisms and natural history and working in the outdoors were important to me. As an undergraduate, I realized that to study birds well one has to know about a wide range of topics, from genetics to population studies to behavior. Evolutionary biology is a field that is broad enough to encompass all these interests.

Interviewer: You're a professor with your own research lab. What does that mean, and how did you get to this place in your career?

Dr. Edwards: I am in charge of a group of researchers—some postdoctoral fellows who have their PhDs, some graduate students who are working on their PhDs, and some undergraduates, who may just be dreaming about a PhD. I have even had some high school students and teachers work with me. It's very easy to do research, if one has the curiosity and a little bit of training. I got that sort of training as an undergraduate, and that got me fired up to go on and do more research. As a researcher, your goal is to ask questions and find the answers through science. If you do this long enough, and well enough, you'll eventually get a PhD, and this will be your way of telling society that you've mastered a particular body of knowledge. Running a lab then just becomes an extension of your curiosity—except that now you've got extra hands to help you find the answers!

Interviewer: What do you find most interesting or exciting about your work?

Dr. Edwards: I will never forget my first glimpse of a bird of paradise on the island of

Scott Edwards, PhD. Scott Edwards is Professor of Organismic and Evolutionary Biology and Curator of Ornithology at the Museum of Comparative Zoology at Harvard University in Cambridge, Massachusetts.

New Guinea. The pictures in my field guide were too bizarre and flamboyant to actually convince me that these things really existed. I happened to meet a native who insisted on my visiting his village in the mountains to see the birds of paradise in his area. I didn't believe we'd see any, but still rode for a day and half in the back of an old pickup truck, and finally arrived at his village. A short walk up in the hills brought us to a fruiting tree, and after a few minutes a remarkable iridescent crow-sized bird with a pair of three-foot long, pure white tail feathers flew in to feed on the fruit. It was a ribbon-tailed astrapia, and it was real! It still boggles my mind that such species exist and that they can make their way through the forest with such long, cumbersome tail feathers (which probably hinder rather than help their flight).

I enjoy the mix of fieldwork and lab work that I do, and the challenges of bringing those two areas together. Both are a type of exploration. I would not be in this business if it weren't for the opportunity to travel to some really cool places and see some really cool animals and their habitats. It's an opportunity that I feel extremely fortunate to have had, and it's what often gets people asking how things came to be and how such different habitats and biological communities were produced. But the fieldwork is not everything. I can go only so long without taking a shower or having the control over my surroundings that a laboratory provides. In the lab, I'm able to weave together the wonderful patterns and behaviors observed in the field with the 21st-century tools of genetics. The genetic work is like trawling through yet another jungle—except here it is a morass of DNA letters. Unraveling the evolutionary meaning of those letters and linking them back to the lives of the birds and other animals we study is very exciting.

Interviewer: What do you find most challenging or frustrating about your work?

Dr. Edwards: The level of funding for evolutionary biology in the United States is frustratingly low. There is much more funding, for example, for biomedical science, perhaps because there is a perception that biomedicine is more directly related to human health and happiness. Yet even biomedical scientists are learning that evolutionary biology provides a very compelling context in which to understand the interaction of humans, their behavior, and their environments and, indeed, sometimes to predict the future of the human species. My hope is that society will recognize the important contributions that evolution can make to improving human well-being—and the well-being of the other 12 million species on the planet.

Interviewer: Do you think it is important for all students to study evolution? If so, why?

Dr. Edwards: Yes, I believe it is important for everyone to study evolution. Of course, it is a very sensitive issue, since everyone has their own personal perspective on the world, and has their own story as to how it came to be. These stories are an important part of the diversity of human cultures, and in fact are sacred in many societies, and so I don't believe the teaching of evolution should supplant these other points of view. But I do think that evolution holds a special place, not only in biology, but in science in general. The bottom line is, evolution is one key to answering many questions that we desperately need answers to—from cures for AIDS to the control of invasive species and the response of dwindling habitats to climate change—and for this reason evolution should not just be a luxury but a necessity in our schooling.

3. Read about homologous, analogous, and vestigial structures in the essay, *Modern Life: Evidence for Descent with Modification.*

   Using the *Mark It Up!* annotation strategy may help you understand the ideas in this essay.

   a. On the diagrams that you colored, are the parts of the animal limbs homologous, analogous, or vestigial? Why do you think so?

      How would an evolutionary tree showing common ancestry help you make a decision?

Modern Life: Evidence for Descent with Modification

ENGLISH SPANISH

The fossil skeleton of Lucy has many of the same features as skeletons of modern humans. But modern human skeletons display some significant differences. Comparisons of fossils from long ago with modern organisms provide one line of evidence that living systems share common ancestors, and that they also change. It is important to remember that the process of change occurs not only in the past, such as when Lucy lived. It is going on in all of the living systems alive on Earth today.

Evidence from Comparative Anatomy

The process and pattern of change are explained by the phrase "descent with modification." Major pieces of evidence for descent with modification come from studies of comparative anatomy. This branch of biology concentrates on the similarities and differences in anatomical features of different species. When you take a close look at living organisms, you will see that many species share structural similarities. A pattern of similar characteristics may suggest that organisms share common ancestors. For example, compare the structure of the animal forelimbs shown in figure 6.26. Although they have different functions, these limbs show consistent similarities among the organisms. In addition, the limbs have the same relationship to the body (they all are forelimbs), and they develop in the same way in the young. Because of these consistent similarities, biologists infer that the forelimb structure is **homologous**. Homologous characteristics are similar among different organisms because they evolved from a common ancestor. The forelimb structure seen in modern species is also found in fossils. In contrast, structures that function or appear similar between species, but are not evolved from a common ancestor, are called **analogous** structures. An example of this is a bird's wing and an insect's wing, since flight evolved independently in these two groups.

In making anatomical comparisons, scientists also discover structures that are functional in some species, but seemingly useless in others. We refer to these apparently useless structures as **vestigial**. This term comes from the word "vestige," meaning "a remnant of." The goose bumps that you get when you are cold are an example of vestigial structures. Small erector muscles in the skin produce goose bumps. In other mammals and birds, the contraction of erector muscles causes the fur or feathers to fluff up. This mechanism helps the organism warm up. Modern humans do not have very much hair, but we have retained the erector muscles. Figure 6.27 shows more examples of vestigial structures.

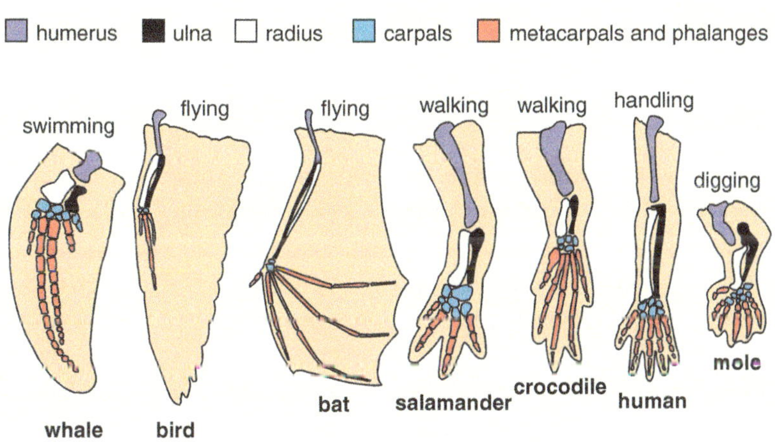

Figure 6.26 Homologies are characteristics that suggest common ancestry. Although adapted for different functions, the bones of these forelimbs (not drawn to scale) show comparable structures. What similarities can you find?

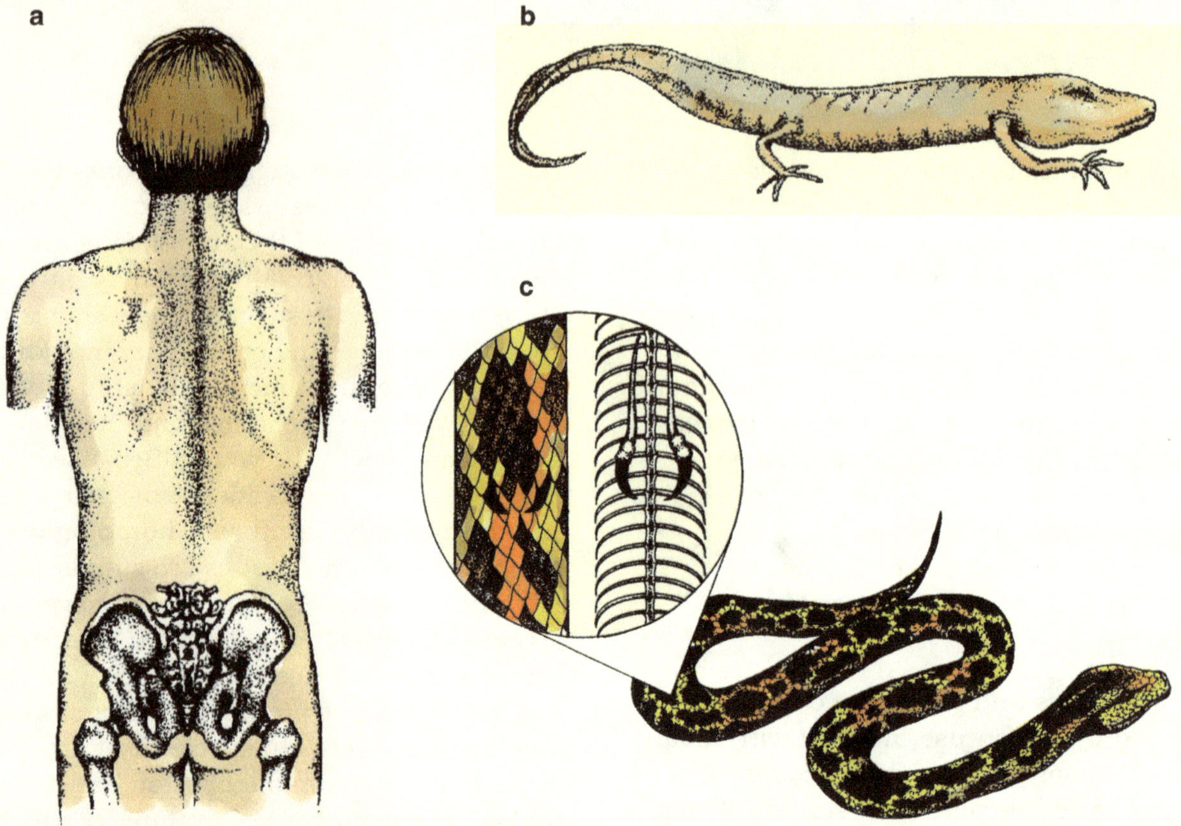

Figure 6.27 **Examples of vestigial structures. (*a*) Human tail structures.** Most mammals have a well-developed tail, but this is lacking in apes and humans. Still, the tail is represented by the last three to five bones in the backbone of humans. Even though an external tail usually is not present, the muscles that move the tail in other mammals also are present in humans. On rare occasions, a fleshy tail in a human extends a few inches. **(*b*) Blind salamander.** These salamanders from Arkansas and Missouri adapted to life in deep caves with no sunlight. They possess eyes but they do not function. The salamanders are blind, but they inherited the ability to make eyes from a sighted ancestor. **(*c*) Snakes with legs.** Notice the bony parts of the python's pelvic girdle, to which tiny limb bones attach. The legs serve no function.

Why do organisms have vestigial structures? Because they inherited the structure from a common ancestor. It is important to remember, however, that a claim about a structure being vestigial is a hypothesis. Future data showing that a vestigial structure has an important function may cause the hypothesis to be revised.

Evidence from DNA

Homologies and vestigial structures are examples of similar features shared by organisms due to their common ancestry. Is there additional evidence to suggest that the tremendous diversity of life arose from a common ancestor? The answer is yes. This evidence is found by comparing modern species. You can study single-celled bacteria such as *Staphylococcus*, protists such as *Amoeba*, fungi such as mushrooms, or the more familiar plants and animals. In each case, you will find that each stores its genetic information in the same complex molecule, DNA (deoxyribonucleic acid). DNA stores important information for making many of the characteristics of each species. Yet the method of storing that information, including the code that is used to interpret the information, is essentially the same in any living organism. For all of its incredible power, DNA is a code based on just four types of molecules, summarized as A, C, G, and T.

Scientists use DNA to make detailed claims about the relationships among organisms. Biologists have compared the DNA among modern species and even among modern and extinct species in rare cases. In fact, new laboratory techniques enable molecular biologists to transfer DNA from one species to another, where the DNA functions successfully. Even cells from a human can receive and decode the DNA from a bacterium. Figure 6.28 shows the similarity of DNA among some primates.

The probability that the primates listed in figure 6.28 developed an identical DNA code without being related to one another is extremely small. A much more likely explanation for this close similarity is that they inherited the genetic code from the same ancestors. The similarity in genetic code across living organisms strongly suggests a common origin for all modern life.

Animal	Percentage of DNA that is the same as human DNA
Monkey	93%
Gibbon	95%
Chimpanzee	98+%

Figure 6.28 Comparing DNA across primates.

b. Title your poster with the correct term (homologous, analogous, or vestigial structures). Write a definition for the term, in your own words, beneath the title.

4. Obtain an envelope containing examples of vestigial structures. Study the structures inside.
   a. On the *Human Vestigial Structures* handout, write a brief, logical explanation about why each structure is considered vestigial.
   b. On the *Whale and Snake Vestigial Structures* handout, write a brief, logical explanation for why these hind limb bones are considered vestigial.

      Make sure to include how the idea of common ancestry helps you explain vestigial structures.
5. Make a poster about vestigial structures, using the space on the back of your poster about homologous structures.
6. Check over your posters. You will use your posters to teach your original team about the evidence contributed by evolutionary biologists to the theory of evolution. Return to *Process and Procedures* step 3.

Evolutionary Developmental Biologists

Scientists in the field of evolutionary development ("evo-devo") biology study the stages of growth and development that living organisms pass through as they grow from fertilized egg to mature adult and beyond. These scientists use information from the tree of life to inform their studies of development. Often, they use evolutionary trees in their studies; the trees are made with data from other sources, like DNA. The work of **evolutionary developmental biologists** helps refine hypotheses about the relationships among organisms and provides clues about how adaptations evolve.

Materials (per team of 2)

online resource (watch as a team)
1 embryo puzzle
1 piece of poster board

1. Read the sidebar, *Interview with an Evolutionary Developmental Biologist.* Think about how you will explain the job to your teammates.
2. Complete these activities with the other evolutionary developmental biologist on your team.
   a. Read and discuss the information in the need to know, *Embryos.* Have your teacher listen to your explanation of the information in the box and confirm your understanding before you proceed.
   b. Study the individual drawings of embryos that your teacher provides. Try to arrange all of the embryonic stages in a developmental order for each animal. When you have finished, your arrangement should show 3 stages of embryonic development for a fish, a frog, a chicken, a calf, and a human.

      These drawings depict various embryonic stages of five different vertebrates. These stages are relative. They do not represent the same point in time, but rather the same relative amount of development.
   c. Compare your work with your teacher's version of the embryo puzzle.

SIDEBAR

Interview with an Evolutionary Developmental Biologist

ENGLISH SPANISH

Interviewer: What inspired you to go into the field of evolutionary developmental biology?
Mr. Pang: I am fascinated by the diversity of animal life and curious as to how it came to be. Yet with all the differences, almost all animals start off as a fertilized egg and go through a developmental program to become an adult. The field I study, evolutionary developmental biology, integrates cell biology, molecular biology, and the study of living organisms.
Interviewer: What is evolutionary developmental biology, and what do you study?

Mr. Pang: Evolutionary developmental biology, or "evo-devo," as it is commonly referred to, is the study of how things evolved by comparing differences or similarities in development.

My work focuses on two groups of animals that diverged from the rest of the animals early in the tree of life. One group includes jellyfish and sea anemones. The second group includes animals called comb jellies. The bodies of these animals are somewhat jellylike and are relatively simple compared to other animals'.

Though their bodies seem simple, the genes that guide their development are similar to those of all other animals. I can find similar genes in the starlet sea anemone that I study, *Nematostella vectensis*, as you would find in vertebrates. This is amazing because these animals shared a common ancestor over 500 million years ago! So I try to see how these different organisms use all these genes to make their bodies. My main area of focus is exploring how the developing embryo "decides" what part of the body goes up, and what goes down.

Interviewer: What do you find most interesting or exciting about your work?

Mr. Pang: The coolest part of my work is the animals that I work on, the warty comb jelly—*Mnemiopsis leidyi*—and the sea anemone. The warty comb jelly can be found off the Atlantic coast of the United States, but it can't be grown in the lab. So for the first year of my project, I was only working with its DNA. When I finally saw a living warty comb jelly, it made me wonder why more people don't study these exquisitely beautiful animals. The way they move by beating their plates of cilia, with light refracting off of them making them appear iridescent, is almost indescribable. Seeing them give off light, or bioluminesce, is also quite a fascinating experience.

I also genuinely enjoy my work. While most of the year is spent in the laboratory, working with embryonic tissue and DNA, during the summer I work in the field. I go to the Atlantic coast to collect my animals. I like the balance of doing molecular work but also being able to see the animals that I work on.

Interviewer: What do you find most challenging or frustrating about your work?

Mr. Pang: Working on an organism that few people study is both challenging and exciting. It is exciting because anything we find out sheds new light on the topic. Sometimes it would be helpful if more people studied these animals. We often apply ideas and techniques from more well-studied animals and see if they work in ours. While some methods carry over quite easily, others require a lot of troubleshooting.

Kevin Pang. Kevin Pang is a graduate student at the University of Hawaii.

NEED TO KNOW

ENGLISH SPANISH

Embryos

Developmental biology focuses on the early development of organisms before they are born or hatched. Scientists call developing organisms "embryos." Evolutionary developmental biologists compare the processes that affect development in the embryos of organisms as diverse as fishes, amphibians, reptiles, birds, and mammals. They find that the embryos of these vertebrate animals (animals that have backbones) resemble one another and use very similar processes.

Make certain that everyone understands these concepts before continuing to step 2b.

d. Discuss the following questions with your partner and record your answers in your science notebook.
   i. In general, which organisms have embryonic stages that are the most similar? The least similar? Explain your answers.
   ii. What do you think these similarities and differences tell scientists about how these organisms are related?

      Consider whether you expect related organisms to look similar or not. Would you also expect related organisms to go through similar stages of development? Consider the later stages of development. Do the more closely related organisms look more or less similar?

   iii. Frogs, chickens, cows, and humans all have four limbs. These structures develop in similar ways, and similar processes control this development. What does this evidence suggest about common ancestry among these organisms?

e. Work with the other evolutionary developmental biologist on your team to make a poster that you will use to teach your original team about embryology.

   Your poster should show the embryo stages that you studied. It also should include a summary of how comparing the development of embryos contributes to scientists' understanding of how different species share common ancestors.

3. Review the essay, *Modern Life: Evidence for Descent with Modification.* Use the information in the essay to help you accomplish the following tasks.

   Using the *Mark It Up!* annotation strategy may help you understand the ideas in this essay.

	Chimp	Gorilla	Orangutan
# DNA letters different from human			
# DNA letters similar to human			
% similarity with human			

Figure 6.29
Comparison Table of β Hemoglobin DNA (between humans and other primates).

a. In your science notebook, explain where DNA is located and what its purpose is.
b. Obtain the *DNA Comparison* handout from your teacher. This handout shows a comparison of the DNA that codes for the β hemoglobin protein, which is found in blood. β hemoglobin is a protein found in hemoglobin. Hemoglobin in blood carries oxygen throughout the cells in the body.
c. Copy figure 6.29 into your science notebook to prepare to use data from the *DNA Comparison* handout. Follow the procedures described on the handout.
d. With your partner, discuss the data you entered in your *Comparison Table of β Hemoglobin DNA*. In your science notebook, write a paragraph summarizing your understanding of what the data indicate about the relatedness of the 4 primates. Draw an evolutionary tree based on the data to accompany your summary.
e. To your poster, add the DNA data from the handout and a copy of your *Comparison Table of β Hemoglobin DNA*. Then add a brief summary about how DNA analysis contributes evidence to the theory of evolution.

4. Review your poster and your notes. You will use this information to teach your original team how evolutionary developmental biologists contribute to our understanding of evolution. Return to *Process and Procedures* step 3.

Analysis

Examine your *Evidence for Evolution* table. Discuss with your partner how having many different types of evidence makes the argument for biological evolution stronger than any of the separate pieces. In your science notebook, record the main ideas from your discussion.

Further Challenges

In the chapter 1 Elaborate activity, *How Similar Are We?*, you discovered that all organisms display the unifying principle of life. In this section of chapter 6, you learned how scientists use evidence to show that all living organisms are related. You can now appreciate that life displays unifying principles *because* living things share common ancestors. Biologists use this understanding of evolutionary relationships to classify the diversity of life on Earth. Read the Need to Know, *Levels of Classification*, and the two Sidebars, *Organizing Diversity* and *Changing Views of the Tree of Life* to learn about the classification system that biologists use. Then answer the following questions to help you integrate your knowledge of evolutionary relationships with your understanding of the unifying principles of life.

1. How does evolution help explain why the major groups of organisms share so many fundamental characteristics (the unifying principles)?
2. Why is the 3-domain system of classification a better representation of life's unity and diversity than the earlier 5-kingdom system?
3. Which of the 6 unifying principles seem to have been most useful to biologists in grouping organisms into the 4 different kingdoms within domain Eukarya (animals, plants, fungi, and protists)? Explain your answer.

NEED TO KNOW

Levels of Classification

Domain, kingdom, phylum, class, order, family, genus, species. What do these terms mean to you? They are all names of groups that biologists use when classifying organisms. Why do biologists have so many different names for groups of organisms? The names that biologists give to different levels of classification can be thought of as corresponding to larger or smaller groups in a nested group diagram. "Domain" is the name of the level of classification that includes the largest diversity of organisms. Members of the same domain form a natural group, but the common ancestor for all of them existed long, long ago. "Species" is the name for organisms that are very closely related.

One level of classification that contains several species is called a family. The next level is called a class, and it may contain many families. The next level is called a phylum, and it may contain many classes.

Biologists use more than three levels in the organizational system for life. You can find out more about all the levels of classification in the essay, *Organizing Diversity*. Consider the following information about a few specific phyla (plural of phylum), classes, and families of organisms.

The kingdom Animalia is organized into 38 living phyla. For example, the phylum Chordata includes organisms with backbones. The phylum Arthropoda includes organisms that have external skeletons and jointed legs. In addition to the 38 living phyla, there are many extinct phyla as well.

The phylum Chordata is organized into 15 living classes. These include the class Aves, which is made up of feathered organisms with other adaptations for flight, like lightweight bones. Another example is the class Amphibia (figure 6.30), which includes all organisms with backbones that have porous skin and lay eggs without shells.

The phylum Arthropoda is organized into 17 living classes. These include the class Insecta, which is made up of organisms with six jointed legs. Another class is Malacostraca, which includes organisms with 10 jointed legs. Most malacostracans spend some part of their lives in the ocean.

Most classes of animals are made up of many orders, and within these orders there are dozens to hundreds of families. In the class Aves, some of these families are the accipters (hawks), the phasianids (pheasants), the cathartids (vultures), the columbids (pigeons), and the tytonids (barn owls). In the class Mammalia, families include the ursids (bears), the delphinids (oceanic dolphins), and the dasypodids (armadillos). In the class Insecta, families include blattids (just one of six families of cockroaches), phasmatids (stick insects), hesperiids (skipper butterflies), and the papillonids (swallowtail butterflies).

Figure 6.30 Which of these chordates is a member of the class Amphibia—the painted turtle, the cane toad, or the reef shark?

SIDEBAR

Organizing Diversity

ENGLISH SPANISH

Whenever people collect information, they develop systems for organizing it. Think of the ways people organize the following: notes for research papers, computer files, recipes, or music collections. Imagine—if you had 2 million songs stored on your computer, how would you possibly organize them? Many people use a nested series of folders on their computers to organize songs, documents, and pictures. For example, you might store all songs in your "music" folder, and within this folder you might store separate folders for each singer or band. Then, within your favorite band's folder, you would likely have several folders, each containing a CD of songs. This organization system has two requirements:

1. A way of forming groups that can be divided into smaller groups; in other words, the groups can be nested
2. A naming system

Biologists have described over 2 million species. That is a tremendous amount of diversity. The process of organizing and describing this diversity is called **biological classification**. Biologists use an organizational system just like the one described for music. Classification plays an important role in biologists' attempts to understand life. First, it helps subdivide the numerous types of living (and extinct) organisms into groups that can be studied productively. Second, classification gives each organism a unique name, which helps biologists communicate.

Though many different types of groups could be used for classification, biologists use groups defined by evolutionary history. Evolution produces organisms that are nested into groups (clades) based on common ancestry. It is a natural form of organization. Biologists use this system of nested groups to organize all of life's species.

Phylum:	Chordata	
Classes:	tunicates	cartilaginous fishes
	lancelets	acanthodians †
	hagfish	ray-finned bony fishes
	conodonts †	lobe-finned bony fishes
	lampreys	amphibians
	cephalaspids †	sauropsid reptiles
	pteraspids †	birds
	placoderms †	mammals

Figure 6.31 The phylum Chordata contains many classes, including all the classes of vertebrates. Some of the classes of Chordata are extinct (†).

Modern biological classification schemes generally contain a number of categories. Each category represents a natural group of organisms with a particular degree, or level, of relatedness to each other. The following groups are frequently used in biological classification, going from the most specific to the most general groups. Organisms that have the greatest number of shared characteristics are grouped together in the lowest level, the category of species. Think of all the dogs you have seen. Great Danes, German shepherds, and toy poodles are all members of the species known scientifically as *Canis familiaris*. The second name, *familiaris*, describes the dog species. Can you see why scientists picked *familiaris* to describe dogs? The first name, *Canis*, is the name of a group of species that form a larger category, known as a genus. Other species in the genus *Canis* include the gray wolf (*Canis lupus*) and the coyote (*Canis latrans*).

Groups of similar genera (plural of genus) form a family. Related families are organized in a large group known as an order. Related orders are grouped together in a class. And a group of related classes form still larger categories known as phyla, sometimes called divisions in plants. See figure 6.31 for an example of the classes that might be contained in one phylum. But classification doesn't stop there. Related phyla are grouped into kingdoms, and related kingdoms are grouped into domains. The domain level of the biological classification system includes the greatest diversity of related organisms. The biological classification assigns each organism to these series of categories. The classification of both dogs and your species is shown in figure 6.32.

Evolutionary trees summarize the nested and hierarchical relationships among species. So evolutionary trees are also used for classification. Let's examine how biologists use evolutionary trees for classification. In the tree in figure 6.33, clades have been assigned different levels of classification. For example, all the odd-toed hoofed mammals are in the order Perissodactyla. Within this order is the family Equidae, and within that family is the genus *Equus*, which contains horses and zebras.

It is important to remember that scientists *will never be able to completely prove* evolutionary relationships. Scientists are always seeking new information. As new information becomes available, understandings of relationships may change. What does this mean?

1. Evolutionary trees are claims based on data. Some parts of the tree of life are supported by large amounts of evidence. Other parts have less evidence.
2. As scientists collect more data and adjust their thinking about these relationships, the ways in which they organize classifications may change. Thus, the ways we classify organisms today are different than they were in the past.

The second aspect of an organizational system involves assigning names. How do biologists handle naming?

Classification level	Dogs	Humans
Domain	Eukarya	Eukarya
Kingdom	Animalia	Animalia
Phylum	Chordata	Chordata
Class	Mammalia	Mammalia
Order	Carnivora	Primates
Family	Canidae	Hominidae
Genus	*Canis*	*Homo*
Species	*familiaris*	*sapiens*

Figure 6.32 **The classification of two familiar species.**

People in every culture throughout history have given names to the living organisms they encountered. But the names typically differed from one culture to the next, making communication difficult. As a result, people recognized hundreds of years ago that a standard naming system had many advantages. Scholars in the Middle Ages started to develop a standard naming system for organisms, beginning with the names used by the ancient Greeks. One major problem, though, was how to handle new organisms. Scholars started adding numerous adjectives to the names of species to help differentiate new species from previous species. This system quickly got complicated. Some species names included nine or more words.

A more effective naming system was proposed by Carolus Linnaeus in 1735. He suggested that the official name for each species should consist of just two words (**binomial nomenclature**). In this system, the name for a species is simply the name of an organism's genus and its specific species name. Scientific names are written in italics. Also, the name for a genus is capitalized. There are two rules for assigning names using binomial nomenclature.

1. The name of each genus needs to be unique.
2. Each species within a genus has a unique species name.

For instance, humans are in the genus *Homo* (Latin for "man") and have the species name *sapiens* (Latin for "wise" or "knowing"). So the scientific name for humans is *Homo sapiens*. Sometimes a genus is only represented by one living species. This is the case for *Homo*. However, there are several extinct *Homo*

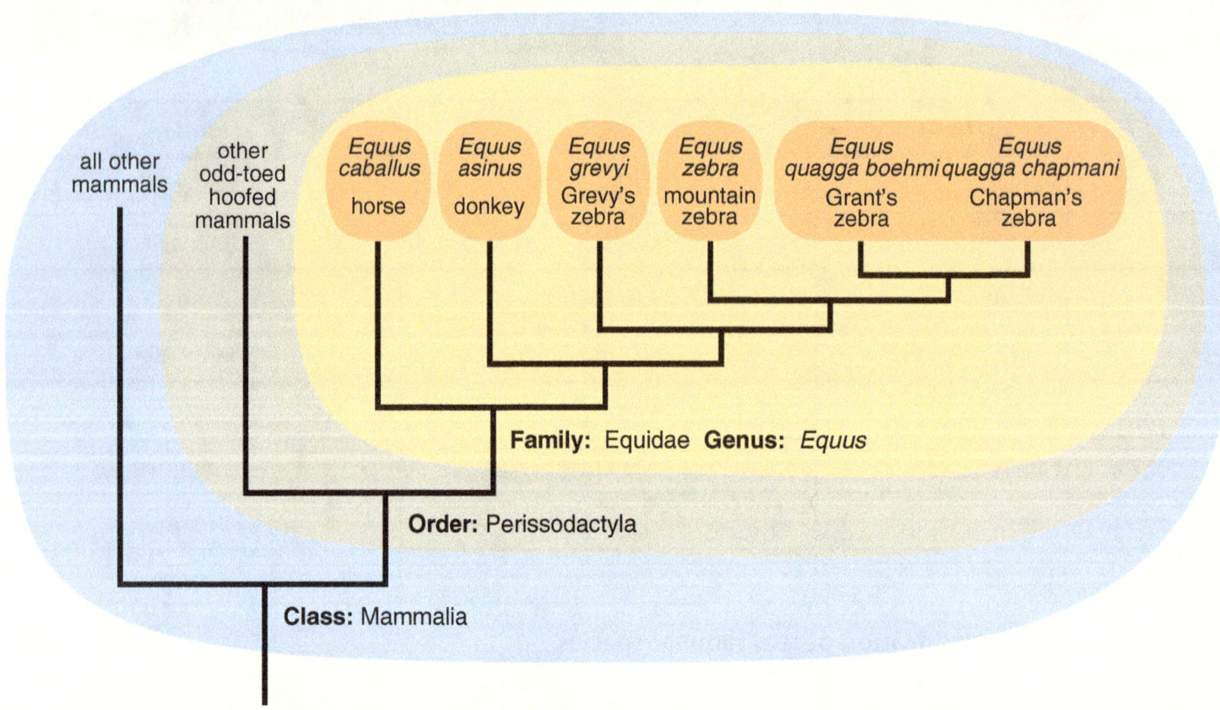

Figure 6.33 **Relationship between classification levels and an evolutionary tree of some mammals, featuring zebras.**

species known from fossils. These include *Homo habilis*, *Homo neanderthalensis*, and *Homo floresiensis*. Also, sometimes the genus name is abbreviated, for example, *H. sapiens*.

Although common names for a species may vary in different regions or different countries, scientific names do not vary (figure 6.34). The standard use of scientific names is very important to accurate communication and efficient research. For instance, there are many common names for Chapman's zebra, including plains zebra, common zebra, and painted quagga. Different African languages include at least 16 more names, including *eutulege*, *lagwar*, *itiko*, *injiga*, *hares*, and *mangwa*. Clearly, if scientists referred to zebras by their common names, it could cause a lot of confusion. Using scientific names reduces the problem.

Does classification matter? Consider the case of Preble's meadow jumping mouse. There is disagreement over whether this mouse (*Zapus hudsonius preblei*) is different enough from other meadow jumping mice to be protected by the Endangered Species Act. Different sets of data indicate different conclusions. The disagreement over this tiny mouse made news headlines in the states of Wyoming and Colorado, because protecting its habitat is at odds with people who would like to develop the land for human use. Millions of dollars are involved. The mouse will continue to be protected until its classification is resolved.

Overall, the usefulness of the biological classification system is illustrated by its enormous lasting power. This system developed gradually over about 100 years from Linnaeus's system of naming species. It has accommodated a tremendous volume of new knowledge that has been added in the past 250 years. And, if biologists are correct in their estimates of the number of species that remain to be discovered and described, we can expect this model to accommodate the data yet to be examined and added.

Figure 6.34 The common bird *Sturnus vulgaris* is called a "starling" in the United States and Britain, but it is called an *étourneau* in France.

SIDEBAR

ENGLISH SPANISH

Changing Views of the Tree of Life

Which organisms shared a common ancestor more recently, the organisms in figure 6.35 or in figure 6.36? Figure 6.35 shows two animals that seem to bear little resemblance to each other. In contrast, figure 6.36 shows three single-celled organisms, each looking very similar.

Surprisingly, from an evolutionary point of view, the two animals in figure 6.35 shared a common ancestor much more recently than the three single-celled organisms. The animals are an African elephant and a close relative, a small mammal known as a hyrax. What you cannot see in figure 6.35 is all the ways in which these organisms are similar, from the basic structures of their cells to the structures of their feet and teeth. These two species inherited these similar structures from their common ancestor.

On the other hand, the organisms in figure 6.36 are very distant in their evolutionary connection, despite the fact that each is a single cell. This means that they share a common ancestor in the very distant past—in this case, probably over a few billion years ago. If you look closely, you can find one of the characteristics that makes *Peranema* different from the other two. Notice that *Peranema* has an interior compartment that is missing in the other cell. That compartment is a nucleus, so *Peranema* is a eukaryotic organism. The second cell is a bacterium called *Escherichia coli*. Like other types of bacteria, its DNA is not separated from the rest of the cell contents by a surrounding membrane. It lacks a nucleus. The third cell is an archaea called *Sulfolobus acidcaldarius*. It, too, lacks a nucleus, but it has features that make it different from both *Peranema* and *E. coli*. Like some other archaea, it lives in an extreme environment, the hot springs of Yellowstone National Park in Wyoming, Montana, and Idaho.

The three single-celled organisms in figure 6.36 illustrate the three largest groups of life that biologists recognize. All the species on Earth are a member of one of these three groups: the Bacteria, the Archaea, and the Eukarya, as shown in figure 6.37. How do you fit into these three big groups, and how are they related to one another?

The organisms in domain Bacteria and domain Archaea look very similar—so similar that the two groups were originally

Figure 6.35 **(*a*) African elephant (*Loxodonta africana*).** The average male African elephant is 350 cm high and weighs 5,000 kg. (*b*) **Rock hyrax (*Procavia capensis*).** A rock hyrax may be 30 cm high and weigh 4 kg.

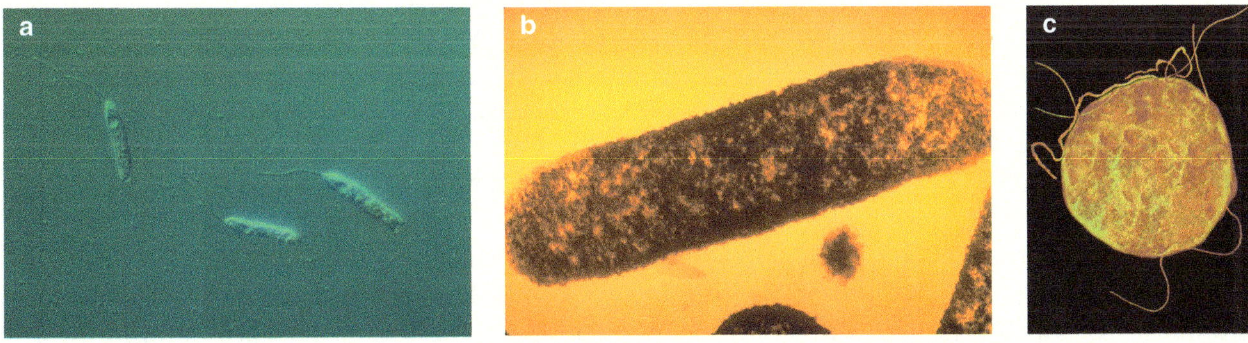

Figure 6.36 (*a*) This *Peranema* is about 40 μm. (*b*) This *Escherichia coli* is 3.5 μm long (photographed at 35,000×). (*c*) This *Sulfolobus acidcaldarius* is photographed at 85,000× magnification.

Figure 6.37 The three-domain system. The domain Bacteria includes the organisms that do not have membrane-bound organelles. Those in domain Archaea are also single celled, but are genetically distinct. Plants, animals, fungi, and protists are all in domain Eukarya. What differences do you think separate the organisms in each domain?

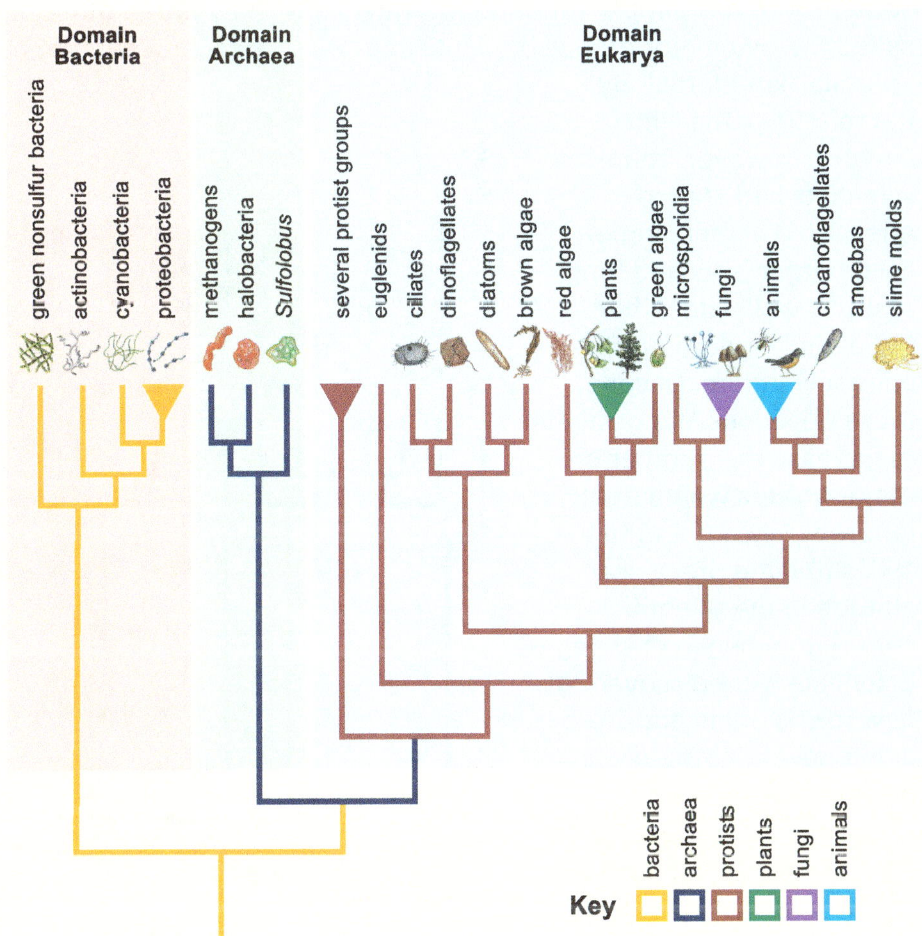

considered to be the same. But analysis of the DNA of archaea has shown that they form a clade that is distinct from the bacteria. Archaea differ from bacteria in the ways that they use their DNA, build their membranes, and construct their flagella (structures that they use for movement). Eukaryotes, including humans and *Peranema*, have cells with nuclei. Eukaryotic cells also may have additional membrane-enclosed compartments that perform a variety of functions, such as energy transformation and protein storage and packaging. Although many similar processes go on in the cells of archaea and bacteria, these cells do not contain compartments.

How has our growing understanding of the relationships of life on Earth affected classification schemes? Consider how views of the major groups of life have changed. Does the modern classification scheme shown in figure 6.37 and used in this course seem a bit foreign to you? After all, as children you learned to focus on the two groups of organisms you can easily see: plants and animals.

Children are not alone in this way of thinking. From the days of Aristotle to the mid-1800s, almost everyone was content with this simple subdivision. They called the two major groups of plants and animals "kingdoms." By the middle of the 19th century, however, some scientists had started to question whether single-celled organisms and fungi really fit well with either the plant kingdom or the animal kingdom. Despite these questions, suggestions to increase the number of major groups were largely ignored. It was not until the 1960s that the prevailing attitude in the scientific community began to change. Scientists were discovering new forms of life and were using new microscopic and biochemical techniques to examine cell structure and function in even well-known organisms. This led to an increasing amount of evidence that supported proposals, beginning in 1968, to increase the number of major groups to five: bacteria (also called kingdom Monera), protists, animals, plants, and fungi.

Maybe you learned about the five-kingdom system in the past. Take a look at figure 6.38. Which of the two groupings (three domain or five kingdom) best identifies natural clades within the giant tree of life?

Near 1980, new data convinced many biologists that archaea were very distinct from bacteria. The prokaryote "kingdom" did not form a natural group. The archaea are more closely related to eukaryotes than they are to other bacteria. Genetic data also showed that the protists, animals,

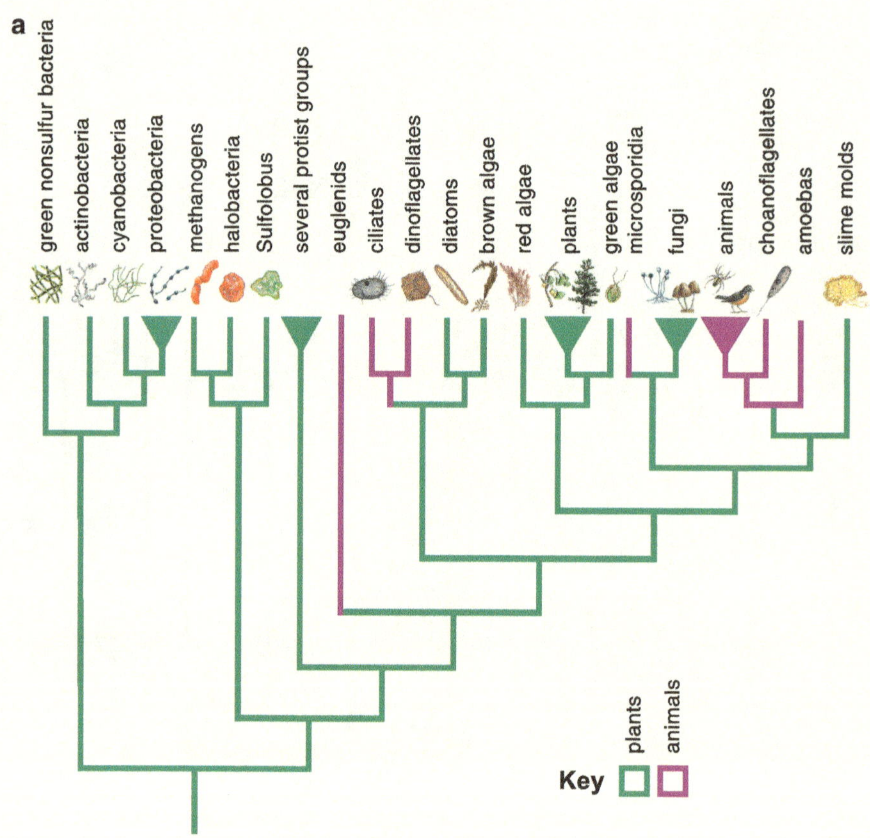

Figure 6.38 Scientific ideas change across time. (*a*) The first attempts to categorize life resulted in a two-kingdom division between plants and animals. (*b*) Scientists developed a four-kingdom scheme when they realized the great differences between eukaryotes and prokaryotes. (*c*) The five-kingdom system is currently the most familiar to students and teachers today, although it was recently replaced by the three-domain system. Are all the kingdoms natural groups or clades?

b

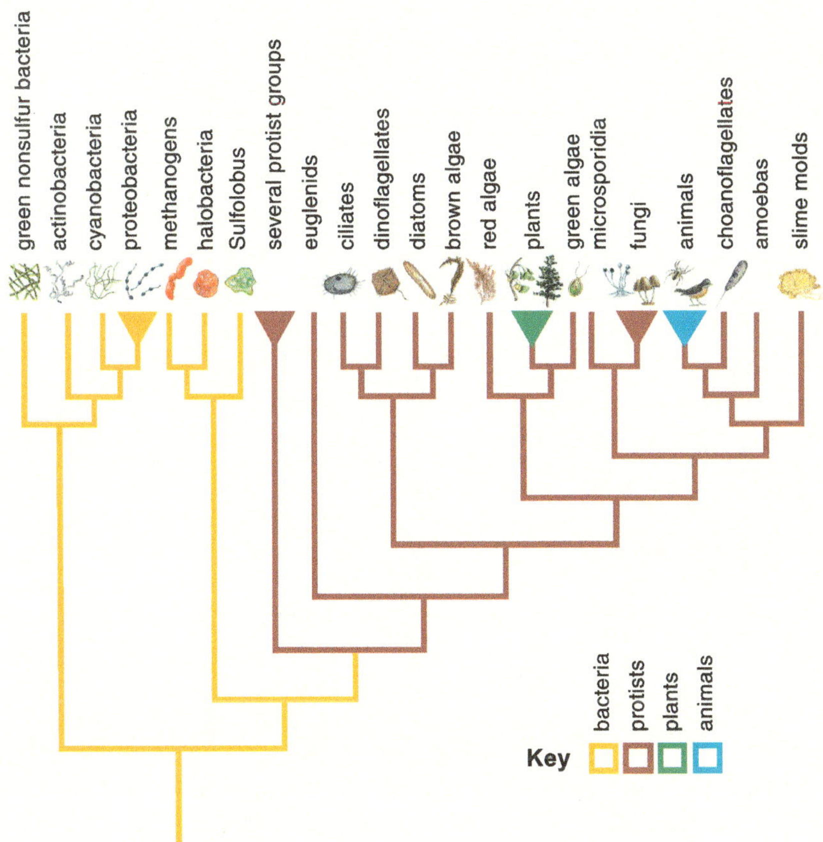

c

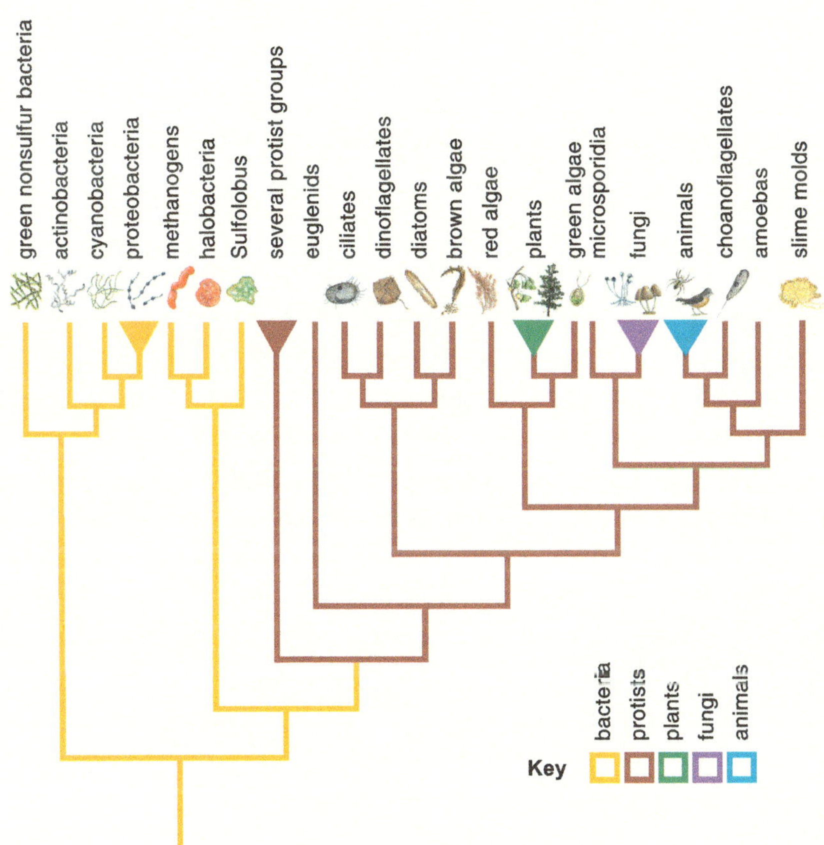

plants, and fungi weren't as distinct from one another as they would appear. This spurred the development of the three-domain system. Figure 6.38 illustrates the earlier classification schemes. These ideas will help you trace the changes that have occurred in scientists' thinking to bring us to the three-domain system that is the standard today.

Scientists are continually rethinking the classification system as they continue to learn more about the organisms that inhabit Earth. For example, birds are traditionally classified in the class Aves, as you learned in the need to know, *Levels of Classification*. Reptiles are traditionally classified as a separate class, Reptilia. However, evolutionary studies clearly show birds are a group *within* the reptiles. In fact, dinosaurs are more closely related to birds than they are to other reptiles. Because biologists strive to describe natural groups, or clades, of organisms, there are proposals to change the classification level of birds. This is a great example of the nature of science—claims of relationships change as more evidence is gathered.

EVALUATE

Your Critter's Adaptations and Ancestry

Throughout the activities and essays in this chapter, you have focused on how evolution has produced the tremendous diversity of living systems that exist (and have existed) on Earth. You also have seen that, in addition to explaining the diversity of life, evolution accounts for its unity. Evolution has produced organisms that are very different from one another, adapted to life in different habitats, and even adapted differently to the same habitat. Evolution *from a common ancestor*, however, also has resulted in organisms that show important similarities to one another.

This Evaluate activity invites you to revisit your critter as you evaluate your understanding of evolution. In *Your Critter's Adaptations and Ancestry*, you will make hypotheses about some of your critter's adaptations and illustrate how it is related to other organisms.

Materials (per person)

descriptions and diagrams of your critter from chapter 4
1 environmental change card

PROCESS AND PROCEDURES

1. Look at the diagrams and descriptions of your critter from chapter 4. Identify the adaptations that help it survive in its habitat, and label them as adaptations on your drawings.

   Remember, your description of the critter in this activity is the evidence that your teacher will use to evaluate what you have learned about evolution, biological diversity, and the unity of life.

2. Write a paragraph about the adaptations you labeled on your critter drawing. Select 1 adaptation as an example and describe your understanding of how scientists would explain how adaptations arise in a population.

   It is important to include all the necessary steps and requirements for the evolution of an adaptation.

3. Write a second paragraph describing a scientifically reasonable scenario by which your critter may have diverged from an ancestral species to form a new species.

4. Write a third paragraph that describes at least 3 characteristics that your critter shares with *all* other forms of life. Identify the unifying principle(s) of life that relate to those characteristics.

   As you complete this task, you may want to review the notes and table you made about the unifying principles of life in the chapter 1 Elaborate activity, *How Similar Are We?*

5. Obtain an environmental change card from your teacher. Write down the environmental change in your science notebook. Then write a fourth paragraph predicting what might happen to your critter's species as a result of this change. If you predict that it will go extinct, explain how this may affect the evolution of other species. You may want to use the following terms in your explanation:
   - Natural selection
   - Extinction
   - Adaptation
   - Variation
   - Heritable

6. Obtain a *Critter's Adaptations and Ancestry Rubric* from your teacher that explains the grading criteria for your critter. Read the rubric carefully. Use it as a checklist to be certain that you have described your critter effectively.

   You will know that you have generated a good description if it meets the following criteria:
   - It describes how your organism resembles all known forms of life.
   - It describes how your organism is adapted to the habitat in which you found it.
   - It identifies how your species is related to other organisms, and it describes the evidence on which you base your answer.

Analysis

Which of the following 2 statements is most consistent with scientists' understanding of the process of evolution, and why? What is wrong with the other statement? Record your responses in your science notebook.

- Statement 1: "The habitat that I worked in is very wet and salty. My critter has a common ancestor that evolved a rubberlike skin because it wanted to keep from shriveling up."
- Statement 2: "The habitat that I worked in is very wet and salty. My critter evolved a rubberlike skin because this trait provides protection from the harmful effects of all that salt.

Further Challenges

Put your critter into a broader perspective by constructing an evolutionary tree that includes examples of organisms from across the tree of life.

SEP

1. Create an evolutionary tree that illustrates the relationships between your critter and the following organisms that have appeared in unit 1.
   - **a.** *Sulfolobus acidcaldarius*
   - **b.** Fly agaric mushroom
   - **c.** Painted turtle
   - **d.** *Homo sapiens*
   - **e.** *Lycopodium* moss
   - **f.** Hooded oriole
   - **g.** Click beetle
   - **h.** *Escherichia coli*
   - **i.** Tube sponge
   - **j.** Reef shark
   - **k.** *Fuligo septica*
   - **l.** Cane toad
   - **m.** Volvox

   You may find example evolutionary trees to help you with this task. See, for example, figure 6.37 in the essay, *Changing Views of the Tree of Life,* and figure 6.33 in the essay, *Organizing Diversity.*

2. Add the following characters in the appropriate places on your evolutionary tree.
   - **a.** Multicellularity
   - **b.** Flowers
   - **c.** External digestion
   - **d.** Cells with membrane-enclosed compartments
   - **e.** Backbone
   - **f.** Amniotic egg
   - **g.** Four legs
   - **h.** External skeleton

APPENDIX A

SFUSD Science Laboratory Safety

Science in San Francisco Unified School District is a hands-on laboratory class. You will be doing many activities that require the use of chemicals and laboratory equipment. While engaging in science class, safety must be the #1 priority for all. This SFUSD Student Science Safety Contract has been developed to ensure a safe science classroom. These rules must be followed at all times.

Your science teacher will provide you with a copy of the SFUSD Student Science Safety Contract that must be reviewed and signed by both you and a parent/guardian prior to participating in the laboratory.

General Science Rules

1. Conduct yourself in a responsible manner at all times in the laboratory.
2. No student may work in the laboratory without a science teacher present. Students are not permitted in the science storage rooms or preparation areas.
3. Do not touch any equipment, chemicals, or other materials in the laboratory area until you are instructed to do so by your science teacher.
4. Do not eat food, drink beverages or chew gum in the laboratory.
5. Be prepared for your work in the laboratory. This includes reading all procedures thoroughly before doing a lab or activity. Perform only those experiments authorized by your science teacher. Carefully follow all instructions, both written and oral. Unauthorized experiments are prohibited.
6. Follow all written and verbal instructions carefully. If you do not understand a direction of part of a procedure, ask your science teacher for clarification before proceeding.
7. Work areas should be kept clean and tidy at all times. Bring only your laboratory instructions, worksheets, and/or reports to the work area. Please keep aisles clear and push your chair or lab stool in when not in use.
8. Labels and equipment instructions must be read carefully before use. Set up and use equipment as directed in laboratory instructions or by your science teacher.
9. Always work in a well-ventilated area. Use the fume hood as instructed in chemistry or biology class when working with volatile substances.
10. Dispose of all chemical waste properly—Never down the sink! Sinks are to be used only for water. Solid chemicals, metals, matches, filter paper, and all other insoluble materials are to be disposed of in the proper waste containers as directed by your science teacher.
11. Keep hands away from face, eyes, mouth and body while using chemicals.

12. Handle all living and preserved organisms used in a laboratory activity in a humane manner.
13. When using sharp instruments, always carry with tips and points pointing down and away. Always cut in a direction away from your body.
14. If you have a medical condition (allergies, etc.), check with your physician prior to working in lab.
15. Be alert and proceed with caution at all times in the laboratory. Notify your science teacher immediately of any unsafe conditions you observe.
16. Know the locations and operating procedures of all safety equipment including the first aid kit, eyewash station, safety shower, fire extinguisher, and fire blanket. Know where the fire alarm and the exits are located.
17. Know what to do if there is a fire drill during a science lab; containers must be closed, gas valves turned off, fume hoods turned off, and any electrical equipment turned off.
18. Wash your hands with soap and water after performing all experiments. Clean all work surfaces and equipment at the end of each experiment.

Science Safety Attire

19. Dress properly during a laboratory activity. Long hair must be tied back and dangling jewelry and loose or baggy clothing must be secured. Shoes must completely cover the foot. Sandals are not allowed in the science laboratory. Lab aprons or coats should be worn when provided.
20. Students will wear appropriate laboratory goggles any time chemicals, heat or glassware is used.
21. Contact lenses should not be worn in the laboratory unless you have permission from your science teacher.

Accidents and Injuries in the Science Laboratory

22. Report any accident (spill, breakage) or injury (cut, burn) to your science teacher immediately.
23. If a chemical splashes in your eye(s) or on your skin, immediately flush the area with running water from the eyewash station or safety shower for at least 20 minutes. Have someone notify your science teacher immediately.

Chemical Handling in the Science Laboratory

24. All chemicals in the laboratory are to be considered dangerous. Do not touch, taste, or smell any chemicals unless specifically instructed to do so. Never remove chemicals or other materials from the laboratory area.

25. Check the label on chemical bottles twice before removing any of the contents. Only take as much chemical as you need. Never return unused chemical to their original containers.
26. Hold a container away from your body when transferring reagents between containers.
27. Your science teacher will demonstrate the proper method for handling and diluting acids prior to using them in the lab.
28. Handle flammable hazardous liquids over a pan to contain spills. Never dispense flammable liquids anywhere near an open flame or source of heat.

Handling Glassware and Equipment in the Science Laboratory

29. Examine all glassware before each use. Never use chipped, cracked or dirty glassware.
30. Never handle broken glass with your bare hands. Immediately inform your science teacher of any broken glass. Use a brush and dustpan to clean up broken glass and place it in the designated glass disposal container.
31. Do not immerse hot glassware in cold water; it may shatter.
32. Use care when inserting and removing glass tubing from rubber stoppers. Always lubricate glassware (thermometers & tubing) before inserting it in a stopper.
33. Grasp the plug, not the electrical cord when removing an electrical plug from its socket. Hands must be completely dry before touching an electrical switch, plug, or outlet.
34. Report any damaged equipment to your teacher immediately.

Heating Substances in the Science Laboratory

35. Exercise extreme caution when using a gas burner. Ensure that your hair, clothing, and hands are a safe distance from the flame at all times. Do not put any substance into the flame unless specifically instructed to do so by your science teacher.
36. Never leave a lit burner unattended. Never leave anything that is being heated or is visibly reacting unattended. Always turn the burner or hot plate off when not in use.
37. You will be instructed by your science teacher in the proper method of heating liquids in test tubes. Do not point the open end of a test tube being heated at yourself or anyone else.
38. Never look into a container that is being heated.

You will see the signal words **CAUTION** and **WARNING** used in activities throughout this curriculum. These words tell you about the potential for harm and / or injury. They remind you to observe specific safety practices. Always read and follow these statements. They are meant to help keep you and your fellow students safe.

CAUTION statements advise you that the material or procedures has some potential risk of harm or injury if directions are not followed.

WARNING statements advise you that the material or procedure has a moderate risk of harm or injury if directions are not followed.

It is important that you pay particular attention to the safety information within this curriculum so that you learn how to recognize and avoid potentially hazardous situations. Become familiar with the safety symbols identified in the NEED TO KNOW.

NEED TO KNOW

ENGLISH SPANISH

Safety Symbols Used in This Program

Caution

The caution symbol alerts you to procedures or materials that may be harmful if directions are not followed properly. You may encounter the following common hazards during this course:

Sharp object

Sharp objects can cause injury, either a cut or a puncture. Handle all sharp objects with caution. Use them only as your teacher instructs you. Do not use them for any purpose other than the one intended. If you do get a cut or puncture wound, call your teacher and get first aid.

Irritant

An irritant is any substance that, on contact, can cause reddening of living tissue. Wear safety goggles, lab apron, and protective gloves when handling any irritating chemical. In case of contact, flush the affected area with soap and water for at least 15 minutes. Call your teacher. Remove contaminated clothing.

Reactive

Reactive chemicals are capable of reacting with another substance, including water. They can cause a violent reaction. ***Do not*** mix a reactive chemical with any other substance, including water, unless directed to do so by your teacher. Wear your safety goggles, lab apron, and protective gloves.

Corrosive

A corrosive substance injures or destroys body tissue on contact by direct chemical action. When handling any corrosive substance, wear safety goggles, lab apron, and protective gloves. In case of contact with a corrosive material, immediately flush the affected area with water. Call your teacher immediately.

Biohazard

Any biological substance that can cause infection through exposure is a biohazard. Before handling any material so labeled, review your teacher's specific instructions. ***Do not*** handle in any manner other than as instructed. Wear safety goggles, lab apron, and protective gloves. Any contact with a biohazard should be reported to your teacher immediately.

Safety goggles

Safety goggles are for eye protection. Wear goggles every time you see this symbol. If you wear glasses, be sure the goggles fit comfortably over them. If a material from the laboratory splashes into your eyes, flush your eyes (including under the lid) at an eyewash station for 15–20 minutes. If you wear contact lenses, remove them *immediately* and flush your eyes as directed. Call your teacher.

Lab apron

A lab apron is intended to protect your clothing. Whenever you see this symbol, put on your apron and tie it securely behind you. If you spill any substance on your clothing, call your teacher.

Gloves

Wear gloves every time you see this symbol or whenever your teacher directs you to do so. Wear them when using ***any*** chemical or reagent solution. Do not wear your gloves for an extended period of time. Wash your hands after wearing gloves.

Flammable

A flammable substance is any material capable of igniting under certain conditions. Do not bring flammable materials into contact with open flames or near heat sources unless instructed to do so by your teacher. Remember that flammable liquids give off vapors that can be ignited by a nearby heat source. Should a fire occur, ***do not*** attempt to extinguish it yourself. Call your teacher. Wear safety goggles, lab apron, and protective gloves whenever you handle a flammable substance.

Poison

Poisons can cause injury by direct action within a body system through direct contact with skin, inhalation, ingestion, or penetration. Always wear safety goggles, lab apron, and protective gloves when handling any material with this label. If you have any preexisting injuries to your skin, inform your teacher before you handle any poison. In case of contact with a poison, call your teacher immediately.

APPENDIX B1

How to Use the Science Notebook

In the SFUSD Biology Core Curriculum, you will use a science notebook on a regular basis. In SFUSD, science notebooks can look very different and range from a composition notebook to a binder with loose-leaf paper. All science notebooks have many purposes: They provide a place to record data, take notes, reflect on your progress, and respond to questions. This science notebook will become your permanent record of your work, and you will refer to it often during discussions and assessments. The more complete your science notebook is, the more valuable it will be for you. The following sections describe the major ways in which you will use your science notebook in this program.

Recording Data

Science depends on accurate data. No one—not even the original observer—can trust the accuracy of confusing, vague, or incomplete data. Scientific record keeping is the process by which you maintain neat, organized, and accurate records of your observations and data. Use a pen to record data. Although your interpretation of data may change, *the original data are a permanent record.* If you learn new or additional things and your thinking changes, make changes in your science notebook in a different-colored pen or pencil. That way, you and your teacher have a record of your ongoing learning.

Keep records in a diary form and record your name and the date at the beginning of each entry. Keep the records of each activity separate. Be brief but to the point when recording data in words. It may not be necessary to use complete sentences, but single words seldom are descriptive enough to represent accurately what you have observed or done.

Sometimes the easiest way to record data is in the form of a drawing or sketch. Such drawings need not be works of art, but they should be accurate representations of what you have observed. Place your sketches or drawings in the middle of the page, leaving room for captions, revisions, and highlights. Keep the drawings simple, use a hard pencil, and include clearly written labels. Often, the easiest way to record numerical data is in the form of a table. When you record data for counts or measurements with numbers, include the units of the measurements you used, for example, degrees Celsius or centimeters.

Do not record your data on other papers and then copy them into your science notebook. Doing so may increase neatness, but it will decrease accuracy. Your science notebook is your book, and cross outs and stains are normal circumstances of field and laboratory work.

You will do much of your laboratory work as a member of a team. Your science notebook, therefore, will contain data that other team members have contributed. Keep track of the source of observations by circling (or recording in a different color) the data that others reported.

Responding to Questions

When you answer discussion or activity questions from your science notebook, record the date and the activity title. Then number each response. You also may find it useful to record the questions. Sometimes you will respond to questions individually and sometimes with your team; indicate whether your responses are your own or your team's. As you are writing your responses, practice writing in complete sentences; this will help you when you synthesize and present ideas. After each answer that you write, leave a blank space where you can add questions or comments that arise as your understanding grows.

Taking Notes

When you need to record notes about information you wish to remember, always begin with the date. Then record the source of information. Often, this is a person or a book, but it could be a video, a website, or a computer program. When you are recording notes, start each new idea on a new line. Try to group related ideas under broad headings that will help you remember the important ideas and how they are connected. Write down more than you think you will need; it is hard to make sense of a few words when you look back at them later. Include diagrams and charts to clarify ideas.

It is often valuable to take notes during team and class discussions as well as when your teacher is presenting ideas or instructions. In addition, taking notes in your science notebook as you read essays or sidebars helps you better absorb the written information.

You can use the information in your science notebook to prepare for discussions or to review what you have learned. At times, you also will use the information that you have recorded to complete assessment activities.

Keeping Track of Your Questions

Often, as you read or work through an activity, a question will come to mind or you will find that you are confused about something. If you are not able to talk with your teammates or your teacher right away, jot down your question or what you find confusing in your science notebook so that you will remember to ask about it when you have the opportunity. You also may use this technique to record questions that you want to answer yourself.

Keeping Track of Your Responsibilities

Because you will use your science notebook every day in science class, it is a good place to record your class assignments and responsibilities. Each day you may want to record these in red in the upper corner of your science notebook page.

Using Your Science Notebook During Assessment

At times throughout this program, you will use your science notebook during assessments—both ongoing assessments, such as class discussions and team presentations, and more formal, end-of-unit assessments. Your teacher will collect your science notebook periodically to assess your progress. Using a science notebook for assessment will be a rewarding experience if your entries are complete, detailed, and well organized. Remember to make it easy for someone else reading your science notebook to understand what you have recorded. Using blank space to separate activities, notes, and data will make your science notebook easier to assess, and it will provide space for you to add new information if needed. Keep this in mind as you make entries in your science notebook.

APPENDIX B2

How to Use Chapter Organizers

Some of the skills you will improve upon this year will be your skills of organization. You will work this year at organizing your thoughts and knowledge in a science notebook. In addition, you will learn to use the organizational tools that the student book provides. You will see that we have included a chapter organizer at the beginning of each chapter. This organizer will help you see the big picture. Your understanding will deepen and strengthen as you see that what you have learned today connects to what you will learn tomorrow.

To familiarize yourself with the chapter organizer tool, work with a partner to complete the following tasks. Record your answers and thoughts in your science notebook under the title "Using the Chapter Organizers."

1. Look through the table of contents of your student book and find a chapter title that most interests you. Do not turn to your chapter yet; just look at the title. Make sure that you and your partner choose *different* chapters. If you both like the same one, work out a plan to examine 2 different chapters. Learning to compromise in a group is another skill you will develop this year. Complete the following tasks based on the chapter you selected.
   a. Describe to your partner why you think this chapter will be interesting to you. Listen as your partner explains his or her thinking to you. Record the title of your chapter and at least 1 statement explaining why you think this chapter will be interesting.
   b. Think of as many concepts and ideas as you can that might be included in your chapter. List these concepts in your science notebook.
   c. Share your list with your partner and listen as your partner shares with you. Think about your partner's chapter selection. Can you add to his or her list of topics? Add new topics to your list that emerge during this discussion.
2. Find the chapter organizer for your chapter. It is found at the beginning of the chapter. Look at it carefully. All the chapters in this book follow the BSCS 5E instructional model. The 5Es provide a structure for active learning that will have you *doing* and *understanding* science, not just reading about it. Taken together, the 5Es will help you build a strong understanding of science. Can you find each of the 5Es included in your chapter organizer? List them in your science notebook.
3. What do you think each *E* represents in the learning sequence? Record your ideas in your science notebook.

   Include in your answer what you think you should be doing in each activity. For example, in the Explore activity, what will you be doing? How will you be interacting with your teacher and your teammates?

4. Every *E* is an activity that builds on the previous one and helps prepare you for the next one—the next *E*. Do you notice that between each activity there is a linking question? Discuss with your partner what you think the purpose of the linking question might be. Record your best ideas in your science notebook.
5. Look back at your original ideas about concepts you thought would be included in your chapter.
   a. Circle the ones that appear to be covered in this chapter.
   b. Look at other chapters in this same unit. Highlight topics that will be covered in other chapters.
   c. What feature of the chapter organizer helped you determine the topics covered in your chapter?
6. Discuss with your partner how this chapter organizer can help you with your learning. Record at least 3 ways that you can use the chapter organizers to enhance your learning.
7. Using the chapter organizer you chose as a guide, record what you think the main idea of that chapter is. Try to sum it up in 1 sentence.
8. What part or parts of the chapter organizer did you use to write your main idea sentence? What part of the organizer helped you the most?

APPENDIX B3

How to Use the Think-Share-Advise-Revise (TSAR) Strategy

Does learning stop when your paper comes back with a grade on it? It shouldn't. The same is true for experiences *during* class. That is, you get the most out of school when you get ongoing feedback on your thinking, then revise your original ideas to reflect what you've learned. This cycle of thinking on your own, sharing your ideas, getting advice from others, and revising what you think is essential in the workplace as well as in school. Work with a partner to learn about the think-share-advise-revise (TSAR) strategy.

1. Chapter 1, *The Human Animal*, has an example of using the TSAR strategy. Find it in the first Explore activity, *Primates Exploring Primates*, step 11.
2. Match each of the following steps, tailored to carrying out step 11 of *Primates Exploring Primates,* to the descriptions listed in the table in figure B3.1. You'll see generalized tasks in the table and specific examples in the steps below. The combination of the tasks and the examples provides you with why, what, and how to use the TSAR process. Use this strategy for any problem, especially in team situations.
    a. Review silently and *think* about whether each of the questions you and your partner wrote in step 8 of *Primates Exploring Primates* is testable. Think about how you can use the criteria from the need to know, *Testable Questions* to back up your ideas. Write down your reasoning *in complete sentences.*
    b. *Share* your opinion about the testability of the first question with your partner. To do this, read your reasoning aloud exactly as you wrote it. Then listen as your partner reads his or her reasoning.
    c. Ask for *advice* on whether your partner thinks your reasoning is sound, and if there are additional reasonings he or she thought of. Then switch roles and discuss your partner's reasoning.

        You might ask your partner questions about whether there were places that she or he was confused or where an example might help.

    d. *Revise* your work if you think your partner's advice makes good sense and improves your understanding. Use a different-colored pen or pencil for your revisions.

        Try to revise your original ideas rather than rewriting your whole answer. This will help you keep track of your learning and will ensure that you have a complete answer in one place in your science notebook. Sometimes, you will not revise an answer until later activities and will want to have a complete answer with all the information you have learned in one place in your science notebook.

    e. Repeat this cycle for each of the questions you listed.

Figure B3.1
TSAR table.

Step	What you do	What others do
Think	• access what you already know and understand and the skills you already have • work individually • pinpoint what you do and don't know • generate questions • document your thoughts in your science notebook using complete sentences	• respect your private thinking time by working quietly
Share	• read aloud your thinking to a teammate, exactly as you wrote it • explain any diagrams, charts, or sketches • respond to requests for clarification	• listen attentively • ask questions respectfully
Advise	• ask questions about your own answers • offer suggestions, elaborations, or alternative explanations to what your teammate read • respond to questions about your advice	• listen to your advice without interruption • ask for clarification if needed
Revise	• work individually to revise your ideas using a different-colored pen or pencil • record why you changed your original answer in response to advice (remember, not all advice leads to changes)	• respect your private time to revise your first thoughts

APPENDIX B4

How to Use and Create Organizing Tables

Organizing information helps you see patterns and better understand text materials. There are many different kinds of organizing tables. For example, you can use tables to organize data in an investigation, to make comparisons and analogies, and to show relationships between information in reading passages. Here are three common organizing tables you might use.

1. *Information tables* can show relationships between information listed in the horizontal rows (figure B4.1). These tables can have 2, 3, or even 4 columns. You can use tables to show similarities or differences, or to organize what you know before or after you read. In this program, sample information tables are often given to help you start developing your own. But you can use this table format anytime you have information to organize, for example, while reading essays or reviewing for tests.

Reading about Genetics	
Fact or idea I read	**Questions I have about the fact or idea**

Figure B4.1 **T-table example.** This is an example of a T-table you could use as you read about genetics. As you read a passage, record your ideas in a table to help you organize your thoughts.

Feature of a road trip	is like . . .	aspect of the process of science	They are alike because . . .
A detour on the road	is like . . .	getting unexpected results from an investigation.	when you encounter something you do not expect, you change the way you approach your plan.
Circling back on a portion of the road to look for a turn	is like . . .	adjusting the design of an investigation.	if you miss what you are looking for the first time, you adjust your plan to try again.
Trying different routes on a road trip	is like . . .		
Encountering car trouble and returning home	is like . . .		
Abandoning your car on the road	is like . . .		
Starting your trip and changing the destination	is like . . .		

Figure B4.2 Analogy map example. This analogy map is one you could use to compare a road trip you might take with the process of science.

2. *Analogy maps* are a special type of table that allows you to connect new ideas with ideas you are familiar with (figure B4.2).
3. *Data tables* provide a place to record observations or data from an investigation. You can create graphs from the information in these tables or interpret your data directly from the tables themselves (figure B4.3).

Material	Volume of liquid sample (mL)	Mass of cylinder with liquid sample (g)	Mass of cylinder alone (g)
Sample A	100	142.54	2.54
Sample B	100	93.21	2.54
Sample C	100	83.44	2.54

Figure B4.3 Data table example. Data tables are a place to record both qualitative and quantitative observations and data from an investigation. This data table shows data recorded as students conduct an investigation about density. The data can be used to make a graph or do calculations.

APPENDIX B5

How to Use and Create Venn Diagrams

Venn diagrams are a powerful strategy for comparing topics or concepts. You can use them to visually show similarities and differences. A Venn diagram is made up of two or three overlapping circles. Each circle represents one topic or concept. The region inside each circle lists characteristics of that topic or concept. The part of the circle that overlaps contains characteristics common to both concepts. See the example in figure B5.1. Then try creating your own Venn diagrams using steps 1–5 to help you.

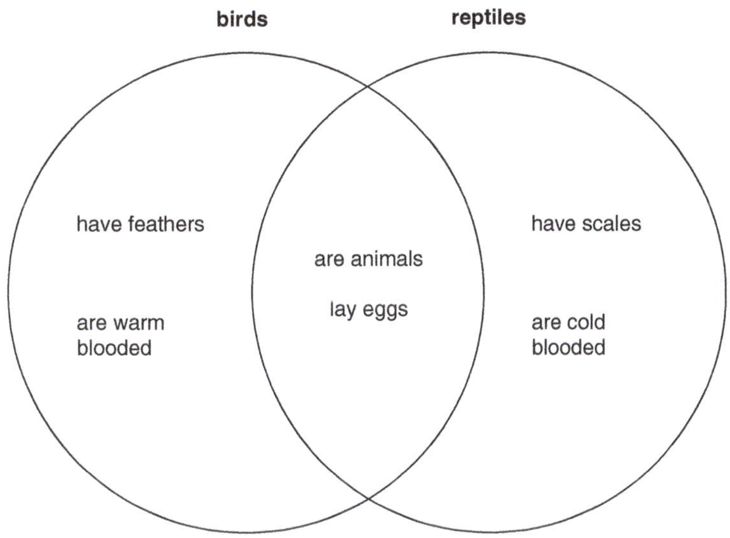

Figure B5.1 Venn diagram comparing birds and reptiles.

Venn Diagram Guidelines

1. Draw 2 overlapping circles like the ones shown in figure B5.1. Use at least a half sheet of paper for the circles to give you enough room to write inside the circles. You might also want to turn your notebook to a landscape orientation.
2. Label each circle with the topic or concept you are comparing.
3. Identify the important characteristics of each topic or concept.
4. Write the characteristics that are specific to only 1 of the 2 topics or concepts inside the circle, but outside the overlapping area.
5. Write the characteristics that are common to both topics or concepts inside the area where the circles overlap.

APPENDIX B6

How to Write Highlight Comments and Captions

How do you make sense of charts, diagrams, graphs, and sketches? You do what scientists have been doing for centuries. You note what you see, then you try to say what it means. This process helps you connect evidence to interpretations, a hallmark of the process of science.

Highlight comments help you link observations from graphs, charts, and other spatial forms of representation to possible interpretations. Captions assemble highlight comments into sentences that form a coherent paragraph. This paragraph tells the story of the graph, chart, or sketch, and communicates a summary of the essential understandings displayed. You may have noticed captions for pictures in magazines, newspapers, or textbooks. The combination of highlight comments and captions helps you communicate scientific information with increasing effectiveness, improving your performance and deepening your understanding of the natural world.

Suppose you investigated the uptake of a nutrient by a tree over 24 hours. How would you make sense of the data? Follow the steps in figure B6.1 and use them as a general guide for any graph, chart, diagram, or sketch you make.

Figure B6.1 Steps for writing highlight comments and captions.

1. Look for changes, trends, or differences. Draw an arrow to each of these you notice in the graph.

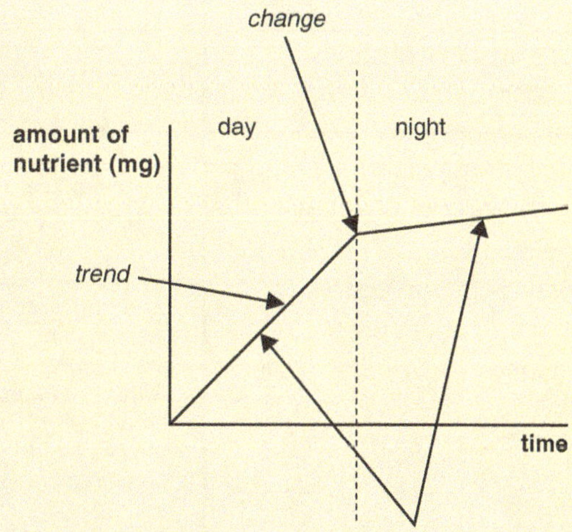

2. Write the phrase "What I see" and then describe what you see. Each arrow should have a different description. Be concise. Write only the essence, or *highlights*, of what you see. "What I see" statements should be limited to observations, not interpretations of what the observations mean.

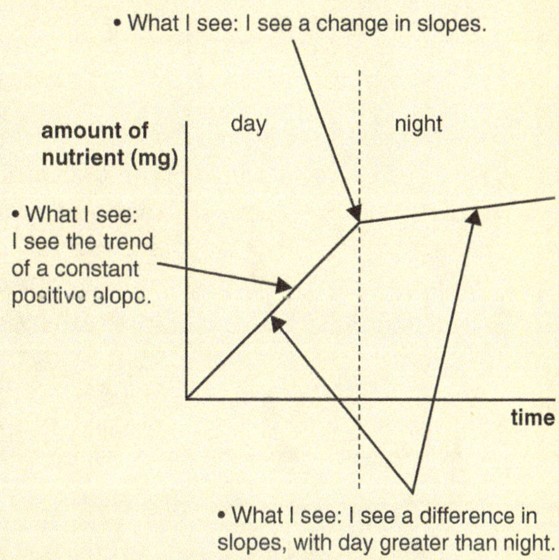

3. Interpret what you see. Write what each observation means. Do not interpret the entire figure at once, just one observation at a time.

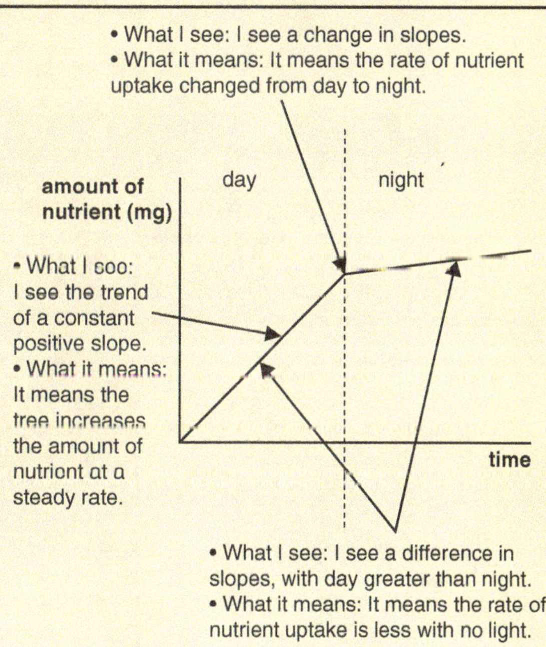

Figure B6.1 **Steps for writing highlight comments and captions.** *(continued)*

4. Write a caption. Think of the caption as an executive summary. Start by writing a topic sentence to describe what is in the graph or figure. Then join each "What I see" to its "What it means" to form a sentence. Build a coherent paragraph out of the sentences.

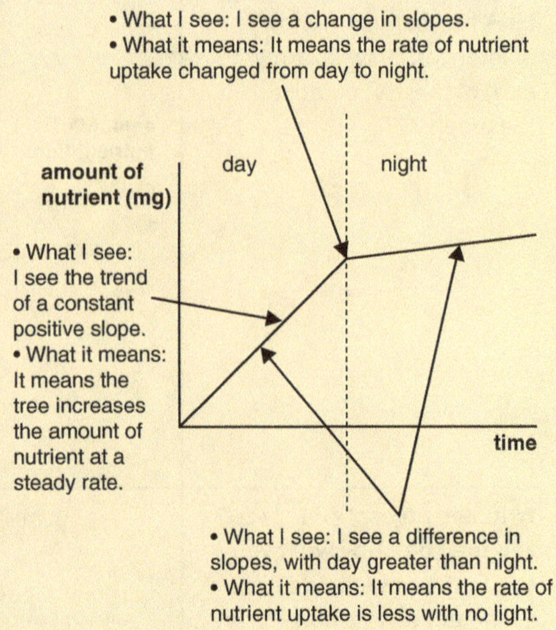

This graph shows the uptake of nutrients in a tree over a 24-hour period. During the day, the graph shows a constant, positive slope, meaning there is a steady rate of uptake. At night, the rate changes as shown by a change in slopes. This suggests that light changes the rate of uptake. Finally, the night slope is less than the day slope, meaning the uptake of nutrients slows at night.

APPENDIX B7

How to Construct a Concept Map

Concept maps are tools that help you organize ideas in a way that shows the relationships among them. There is no single correct concept map for a body of information. But together, the concept words, connecting lines, and linking words should be an accurate representation of the content. To create a concept map, follow these steps.

Concept Map Guidelines

1. Identify the major concept that you will map. Then list several words or phrases that are important to understanding this concept. These should be words or phrases that identify parts of your major concept, such as parts of a system, a key idea, or an important process.
2. On a new page in your science notebook, write the major concept that you will map at the top of the page and draw a box around it. Choose related words or phrases from your original list below this box. Leaving plenty of space around your original concept, arrange these related words so that the bigger ideas are near the top and the more specific ideas are near the bottom. Draw boxes around these words, too.
3. Draw lines between the boxes to show relationships between the concept words. Lines can crisscross to show complex relationships.
4. Label the lines with linking words that describe the relationships.

Study the sample concept map of AIDS concept words in figure B7.1. Identify the connecting lines and linking words on the map. Can you see how this concept map was generated by following the four steps in these guidelines? Although the concept map in figure B7.1 starts at the top and can be read down the page, not all concept maps need to be constructed in this way. Concept maps may branch in different directions. Also consider how to link words between branches. For example, can you think of a link between "immune system" and "virus" in figure B7.1?

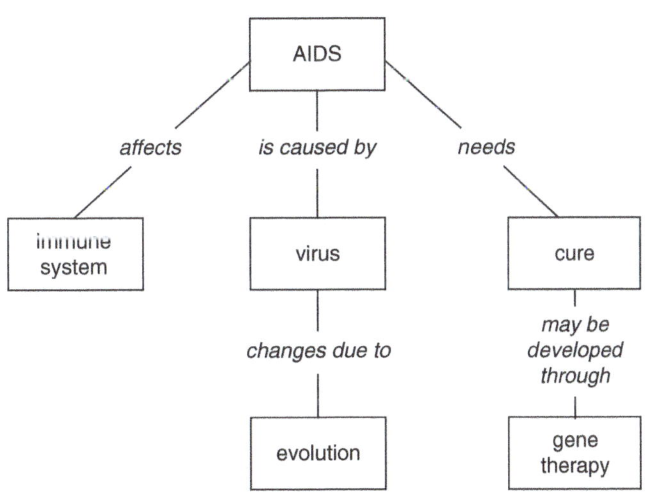

Figure B7.1 **AIDS concept map.**

APPENDIX B8

How to Develop a Personal Glossary

It is often said that biology students learn more new vocabulary than students studying a new foreign language. Sometimes the amount of terminology can feel overwhelming or hard to remember. A personal glossary can help you build your understanding and memory for terms that will be used throughout this curriculum. The use of your personal glossary will be determined by your science teacher and is designed to serve as a resource to you. It is a reference bank that you will build yourself, in your own words. Once you start building this glossary, you can refer to it as you re-encounter terms in your reading across the year.

Some activities in this curriculum suggest that you add terms to your personal glossary as you read, or give a sample list of terms that you can choose to enter. Keep an eye out for additional words to add to your personal glossary. Words that are unfamiliar or bold are good candidates. Any time you come across a word that you do not know or that seems important to understanding biology, you should add that word to your glossary. The more you make it a habit to use your glossary, the more it will help you understand and use the language of biology.

Personal Glossary Guidelines

Complete your personal glossary in a way that best supports your understanding of the terms. We recommend using a chart similar to the one shown in figure B8.1 for each activity or reading that includes new terms. You may only need to fill in one or two of the columns of the chart for terms you find easy to understand. We suggest that you reserve the last 25 pages of your science notebook to allow you to add to your personal glossary over time, and start a new page for each new chapter you begin in the program.

1. **Title** your chart and include a page reference to your textbook.
2. List the **term** you wish to understand or remember in the left-most column.
3. Add the **pronunciation** if you like. For some terms, this may not be necessary.
4. Write **your definition** of the term based on information in the essay, the definition in the text, and other resources. Use your own words.
5. Draw a **picture** that will either help you define or remember the term if you can.
6. Add a personal connection to help you.

Term	Pronunciation	Definition	Picture	Personal Connection
Limiting factor		Anything that can slow or limit the growth of a population		Time I spend doing homework is like a limiting factor—for my grades!
Carrying capacity		The maximum population of one species that a habitat can support		
Biotic	bī ät'ik	Living		Example: my brother
Abiotic	ā bī ät'ik	Nonliving		Example: my pet rock
Population density		The number of individuals per unit of habitat	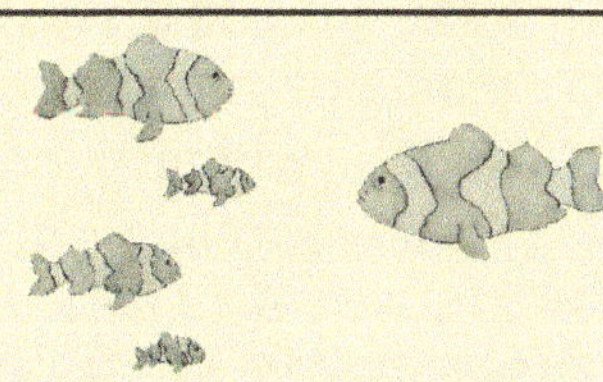	Mosquitoes by the school pond have high population density in spring.

Figure B8.1 **Sample student personal glossary for terms included in the chapter 15 essay, *Interdependence Involves Limiting Factors and Carrying Capacity.***

APPENDIX B9

How to Conduct an Effective Internet Search

Searching for information on the Internet can be rewarding as well as frustrating. It may take hours to sift through the thousands of sites that pop up from a poorly designed search. It can also be difficult to know which information on the Internet is credible, especially when conflicting information is found on different websites. *How to Conduct an Effective Internet Search* gives you a few pointers for using any search engine to look for credible information on the Internet. There are times when you want to broaden the search to include more documents, and there are times when you will want to narrow the search to return fewer documents. The following steps will give you 1) ways to conduct a balanced search that returns documents that pertain to your topic and 2) a method to analyze whether information comes from credible sources.

Internet Searching Guidelines

1. ***Choose your keywords carefully.*** You will type keywords that relate to your topic into a search engine. Choose nouns and objects as your keywords. For example, if you were searching for information about new planets discovered outside our solar system, using the keyword *planet* or *planets* would be a good start. Verbs, adjectives, adverbs, and similar will either be thrown out by the search engine or will be too variable to be useful.
2. ***Use several keywords in your search.*** Using 6 to 8 appropriate keywords can greatly reduce the number of documents that are returned with your search. Using the same example as before, the keywords *new, planet, solar, system,* and *discovery* would return useful documents.
3. ***Use appropriate variations in your words connected by* OR.** For example, use *planet OR planets* to make sure the search engine picks up both variations of the word "planet."
4. ***Use synonyms connected by* OR *where possible.*** *Discovery OR find* is an example of using 2 synonyms connected by *OR* that will cover the different ways a concept can be described.
5. ***Combine words into phrases where possible and place phrases in quotation marks (" ").*** For example, *"solar system"* is a phrase in our example that should be combined and put in quotation marks. This will restrict the search to exact matches of the phrase.
6. ***Combining 2 or 3 concepts in 1 search, distinguished by parentheses, will narrow your results and possibly give you just what you want.*** For our current example, using *("solar system")("new planet")(discover OR find)* would be the best selection.

7. ***Order your concepts with the main subject first.*** Search engines tend to rank documents that match the first keywords in the search higher than those that match the later keywords. For our example search, you would order the concepts as *("new planet")(discover OR find)("solar system")*.

Identifying Credible Websites

1. ***Know what* credible *means.*** The word credible means you can reasonably believe the information that is presented. Credible information tends to be written by respected authorities or experts with a long history of studying a particular topic. It also tends to be consistent across different sources of information or patterns of reasoning. It is based in a presentation of evidence followed by reasoning that explains that evidence. Where opinion is an important part of the information, it is clearly identified as opinion rather than evidence. Historically, opinion-oriented articles in newspapers have been identified as "Op-Ed" columns. Today, entire websites may be based on opinion. In particular, blogs tend to be used to express opinion, frequently intermixed with facts used to support the opinion expressed.
2. ***Identify the author of the information.*** Most websites authored by organizations include the organization logo and a link to their home page at the top of every web page. By clicking on the logo, you can find out more about the organization that sponsors the web page. Most websites authored by individuals include the author name at the top or bottom of the web page. They may also include whether the author holds an advanced degree, like an MA, MS, or PhD in the field that he or she has written about. Many websites have additional links to biographies of the staff that wrote the content of the site, or to the organization(s) that funded their work. Websites that lack identifying or funding information are less credible than those that contain this information.
3. ***Choose information based on credibility.*** Most often, you will find several websites that offer information on the topic of your search. Review websites that you consider to be credible first. Credible sources tend to come from scientific, governmental, and major nonprofit organizations. Government websites always end with ".gov." Examples include the National Institutes of Health and the National Science Foundation. Major nonprofit organizations frequently operate nationally or internationally and their websites always end with ".org." Examples include the American Heart Association and the National Center for Science Education. Scientific websites are usually associated with the publication of a scientific journal, a university, a museum, or a national conference of scientists. An example includes the American Association for the Advancement of Science, which publishes *Science* and *Science News*.

Another example is the website called Understanding Evolution, which is associated with the University of California Museum of Paleontology. Credible information is also contained on the websites of smaller organizations, individuals, and sites presenting blogs or community-editable wikis. However, the credibility of this information is harder to determine. One way to determine whether such sources are credible is to check to see if the information contains citations. Citations are marked by parentheses or placed in superscript next to individual pieces of information, and they link to a set of references that can be traced back to credible sources. Websites that contain reference information are more credible than those that lack this information.

APPENDIX B10

How to Use the Mark It Up! Annotating Text Strategy

Description

Annotating text is a way to identify key information and/or parts of a text that you do not understand. This can include main idea(s), claims, relevant data/evidence, significant phrases, and/or vocabulary.

Materials Needed

- A highlighter, pen or pencil (different colors can be used for different elements)
- A piece of text
- Post-Its (optional)

Sample Annotation Marks

Highlight key information: (use sparingly)

- Main idea
- Claims
- Relevant data/evidence
- Significant phrases

Underline words, phrases, or sentences that are unclear or stand out to you.

Circle known words used in unfamiliar ways.

Marginal Notes

Left Margin:	Right Margin:
• What is the author saying? o Summarize a pre-determined chunk of text.	• Questions • Comments • Connections • Drawings

Additional shorthand notes that you can use. Be sure you remember what you meant when it's time to share.

? = I have a question about this.
* = This idea is new to me.
✓ = I knew this.

APPENDIX B11

How to Use the Scientific Argument Writing CER Strategy

The use of scientific argument is the means by which scientists advance hypotheses that may later become scientific theory. The *Claim-Evidence-Reasoning (C-E-R) Graphic Organizer* is meant to assist students in preparing for an oral or written presentation of their conclusion(s)/scientific argument(s). Scientific argument is a means for students to demonstrate their understanding of scientific principles by investigating a claim or claims and directly connecting evidence that they generate or learn from other sources to a claim using scientific reasoning.

Helpful Sentence Starters and Examples for CER responses

Claim: an explanation or conclusion that provides a sufficient answer to a question or problem.

- Example: Video games are detrimental (harmful) to children.

Evidence: provides observation or data to support the claim. Evidence can come from an investigation that you conducted. It can also come from other reliable sources such as text, animations, video, investigations that were conducted by others.

- According to ______________________________, ...
- As stated in ______________________________, ...
- The text directly states "______________________________."
- In our ______________________________ class, we learned that...
- Based on my experience with...
- In the ______________________________ lab, I saw ....

Reasoning: provides justification (why or how) that the evidence and claim are linked. The reasoning should involve a scientific principle or idea to describe why or how the evidence supports the claim

- In this example...
- According to ______________________________
- This means ________________________ because________________________
- This is similar to ______________________________, which shows ______________________________
- This result can be explained by ______________________________

CER Graphic Organizer

Question:

Claim:

Evidence:

Source:

Reasoning:

Transition Words:

Evidence:

Source:

Reasoning:

Transition Words:

Evidence:

Source:

Reasoning:

Transition Words:

Conclusion:

Write each separate piece of detailed evidence in the boxes above.

Write a scientific explanation that connects the evidence to the claim.

Write a conclusion about the question.

APPENDIX B12

How to Use the Think-Pair-Share Strategy

Description

The Think-Pair-Share strategy is designed to differentiate instruction by providing a structure to individually think and respond to a scenario/problem/question before sharing and discussing with a partner and then share with a whole group or cross groups. Although this strategy can be used for a variety of classroom activities, it has been found to be most effective when students are provided with a prompt that requires initial individual thinking. Often times, the "Engage" portion of a lesson provides the right place for this. The strategy also serves as a tool for the teacher to uncover student misconceptions or flaws in student reasoning—information that can be used to help plan and/or adjust instruction.

Pair-Share Sentence Stems

To ask for more ideas:

- How could we think about this in a different way?
- Did anyone approach this differently?
- Does anyone have anything new to add?
- How do we know that?
- How could we design an investigation to ______________________?
- What would happen if ______________________?
- What kind of model could we use to ______________________?

To clarify an idea that was shared:

- Can you tell me more about that?
- Can you help me understand...?
- Could you explain what you meant when you said...?
- How can I explain this to my friend?
- Can you explain that to me in another way?
- Does ________________ always result in ________________?

To summarize an idea that was shared:

- I heard you say ______________________. Did I summarize that correctly?
- [Name] suggested ______________________. What other thoughts were shared?

To add on to others' ideas:

- I value what you're saying, and I'd like to add...
- What [Name] said made me realize...
- Your idea makes me think about...
- What type of data would support what you are saying?
- I first thought ... but after hearing what ______________________ said I realize...

To disagree respectfully:

- I value what you're saying, but what about…
- I'm not sure about ___________________________ because….
- When I look at the data, I'm not sure it supports what you are saying because...

To give recognition to an idea:

- That's a unique approach to…
- Very impressive idea, [Name].

APPENDIX C1

How to Use a Compound Microscope

The human eye cannot distinguish objects much smaller than 0.1 mm in diameter. The compound microscope is a tool often used in biology to extend vision. It allows observation of much smaller objects. The most commonly used compound microscope is monocular (that is, it has one eyepiece). Figure C1.1 shows a binocular microscope. Light reaches the eye after it has passed through the objects being examined. In this appendix, you will learn how to use and care for a microscope.

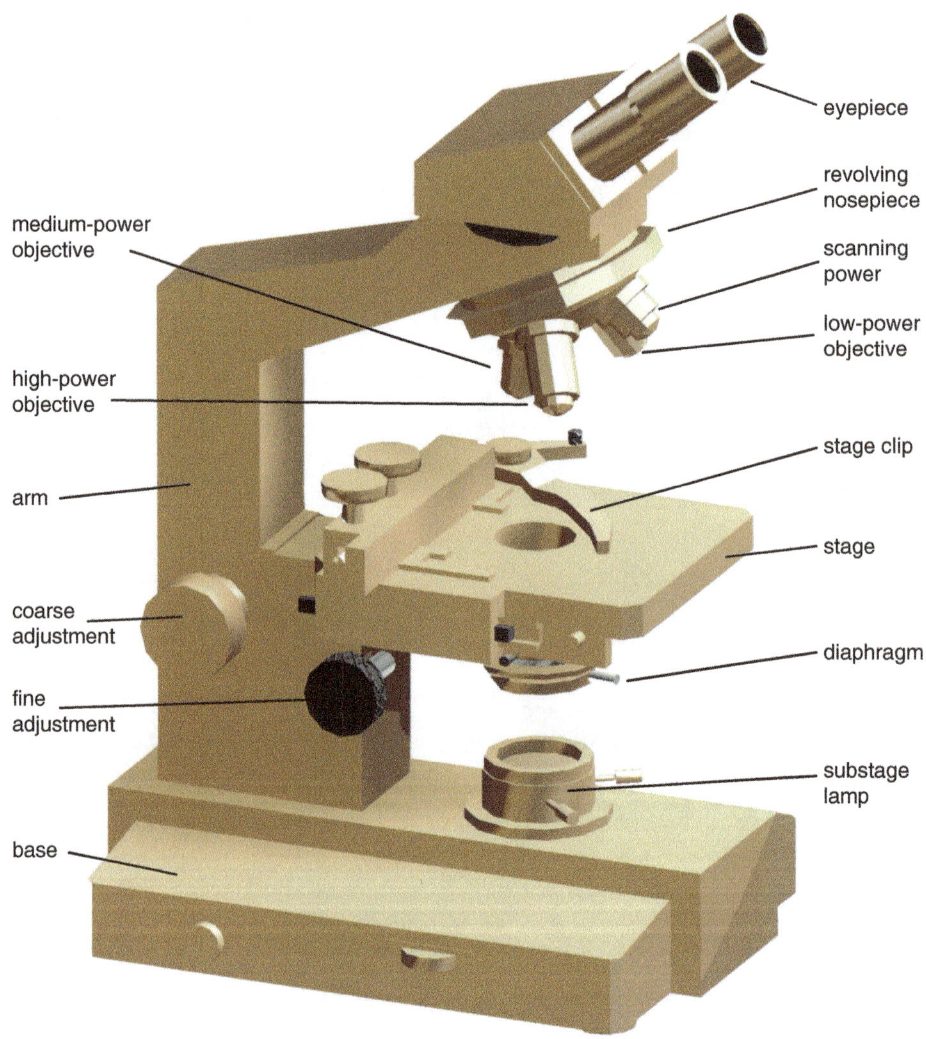

Figure C1.1 **Compound microscope.** Use this figure to help locate the parts of a compound microscope.

Part I Setting Up the Microscope

Materials (per team of 2)

3 coverslips
3 microscope slides
1 100-mL beaker or small jar
1 dropping pipet
1 compound microscope
1 pair of scissors
lens paper
newspaper
water

PROCESS AND PROCEDURES

1. Read *Care of the Microscope* to learn how to properly care for a microscope.

Care of the Microscope

- The microscope is a precision instrument that requires proper care. Always carry the microscope with both hands. Put one hand under its base, the other on its arm (see figure C1.2).
- Keep the microscope away from the edge of the table. If a lamp is attached to the microscope, keep its cord out of the way. Move everything not needed for microscope studies off your lab table.
- Avoid tilting the microscope when using slides made with water.
- The lenses of the microscope cost almost as much as all the other parts put together. Never clean lenses with anything other than the lens paper designed for this task.
- Always return the microscope to the low-power setting before putting it away. The high-power objective extends too close to the stage to be left in place safely.

Figure C1.2 How to carry a microscope. Always place one hand under the base and the other hand on the arm.

2. Rotate the low-power objective into place if it is not already there. When you change from one objective to another, you will hear the objective click into position.

3. Move the mirror so that you obtain even illumination through the opening in the stage. Or turn on the substage lamp. Most microscopes are equipped with a diaphragm for regulating light intensity. Some materials are best viewed in dim light, others in bright light.

CAUTION: Never use a microscope mirror to capture direct sunlight when illuminating objects under a microscope. The mirror concentrates light rays, which can permanently damage the retina of the eye. Always use indirect light.

4. Make sure the lenses are dry and free of fingerprints and debris. Wipe lenses with lens paper only.

Part II Using the Microscope

Materials (per team of 2)

supplies from Part I

PROCESS AND PROCEDURES

1. In your science notebook, draw a data table similar to the one in figure C1.3.

Object being viewed	Observations and comments
Letter *o*	
Letter *c*	
Letter *e* or *r*	

Figure C1.3 **Microscope observations.**

2. Cut a lowercase letter *o* from a piece of newspaper. Place it right side up on a clean slide. With a dropping pipet, place 1 drop of water on the letter. This type of slide is called a wet mount.
3. Wait until the paper is soaked before adding a coverslip. Hold the coverslip at about a 45° angle, with the bottom edge of the slide touching both the slide and the drop of water. Then slowly lower the coverslip. Figure C1.4 shows these first steps.

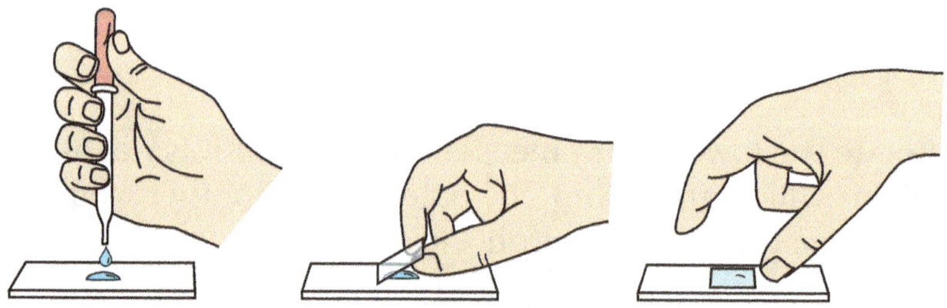

Figure C1.4 **Preparing a wet mount.** This figure shows the steps to prepare a wet mount with a microscope slide and coverslip.

4. Place the slide on the microscope stage. Clamp it down with the stage clips. Move the slide so that the letter is in the middle of the hole in the stage. Use the coarse-adjustment knob to lower the low-power objective to the lowest position.
5. Look through the eyepiece. Use the coarse-adjustment knob to *raise* the objective slowly, until the letter *o* is in view.
6. If you cannot find the *o* on the first try, start the process again by repeating steps 4 and 5.
7. Once you have the *o* in view, use the fine-adjustment knob to sharpen the focus. Position the diaphragm for the best light. Compare the way the letter looks through the microscope with the way it looks to the naked eye. Record your observations in your data table.
8. To determine how magnified the view is, multiply the number inscribed on the eyepiece by the number of the objective lens being used. For example:

   eyepiece (10×) × objective lens (10×) = total (100×).

9. Follow the same procedure with a lowercase *c*. Describe in your data table how the letter appears when viewed through a microscope.
10. Make a wet mount of the letter *e* or the letter *r*. Describe how the letter appears when viewed through the microscope. What new information (not revealed by the letter *c*) is revealed by the *e* or *r*?
11. Look through the eyepiece at the letter as you use your thumbs and forefingers to move the slide slowly *away* from you. Which way does your view of the letter move? Move the slide to the right. Which way does the image move?
12. Make a sketch of the letter as you see it under the microscope. Label the changes in image and in movement that take place under the microscope.

Part III Using High Power

Materials (per team of 2)

supplies from Part I
1 light-colored hair
1 dark-colored hair

PROCESS AND PROCEDURES

1. Make a wet mount of 2 different-colored hairs, 1 light and 1 dark. Cross 1 hair over the other. Sketch the hairs as they appear under low power.
2. With the crossed hairs centered under low power, adjust the diaphragm for the best light.

3. Turn the high-power objective into viewing position. Do *not* change the focus.
4. Sharpen the focus with the *fine-adjustment knob only. Do not focus under high power with the coarse-adjustment knob.* The high-power objective will touch the slide if the objective is in its lowest position. So you must not make large adjustments toward the slide. *Doing so can damage the objective and the slide by driving the objective into the slide.*
5. Readjust the diaphragm to get the best light. If you are not successful in finding the object under high power the first time, return to step 2. Repeat the entire procedure carefully.
6. Using the fine-adjustment knob, focus on the hairs at the point where they cross. Can you see both hairs sharply at the same focus level? How can you use the fine-adjustment knob to determine which hair is crossed over the other? Sketch the hairs as they appear under high power.

Analysis

Work with your partner to answer the following questions in your science notebook.

1. Summarize the differences between an image viewed through a microscope and the same image viewed with the unaided eye.
2. When you view an object through the high-power objective, not all of the object may be in focus. Explain why.

APPENDIX C2

How to Plan and Carry Out an Investigation

An important and exciting part of science is setting up your own investigation to answer your own questions. But some people are hesitant to set up their own experiments. Maybe they don't know where to start. Maybe they don't know what types of data to collect. Maybe they don't consider themselves capable of *doing* science.

One of the major goals of this curriculum is to help you develop the skills and abilities to *do* science. As a student of the process of science, you often find yourself considering some problem or aspect of a problem that requires scientific thinking. You often develop a question and then set out to answer your question by carefully designing an investigation. You then conduct your investigation and record and analyze your results. Your results might help you answer your question and they might present you with new questions to investigate. Your experience with activities in this program should convince you that the careful design of your investigation is critical to obtaining information that is useful for developing an explanation.

Scientists construct explanations based on the evidence they collect during their research. This dynamic practice of wondering and asking questions, of designing investigations, of making detailed observations, and of developing explanations based on those observations and evidence is the *process of science*. It is not always a straightforward process. Look at figure C2.1, *How Science Works*, to see the different paths the process of science can take. These are some of the possible ways scientists study the natural world.

Designing Your Own Investigation

Though the whole process of science is not prescribed and sequential, careful consideration and attention to detail must accompany planning and carrying out investigations. You want to make sure that your design will give you the information you are seeking. When scientists focus on the plan of an investigation, they have to make sure that the design itself guides them to the answer to their question and controls for error. If the design is poor, the results are not valid.

When you find yourself pondering a question in science, you may find it useful to go through the following checklist of questions. If you can answer these questions well, you will know you have a useful experimental design:

- ❑ What is your question or the problem you are trying to solve?

  Plan an investigation that answers the question in the most appropriate and direct manner. Make sure you understand how the question ties into the major concept you are studying.

- ❑ Are you sure that your question is testable?
- ❑ What variables can change to affect the result?

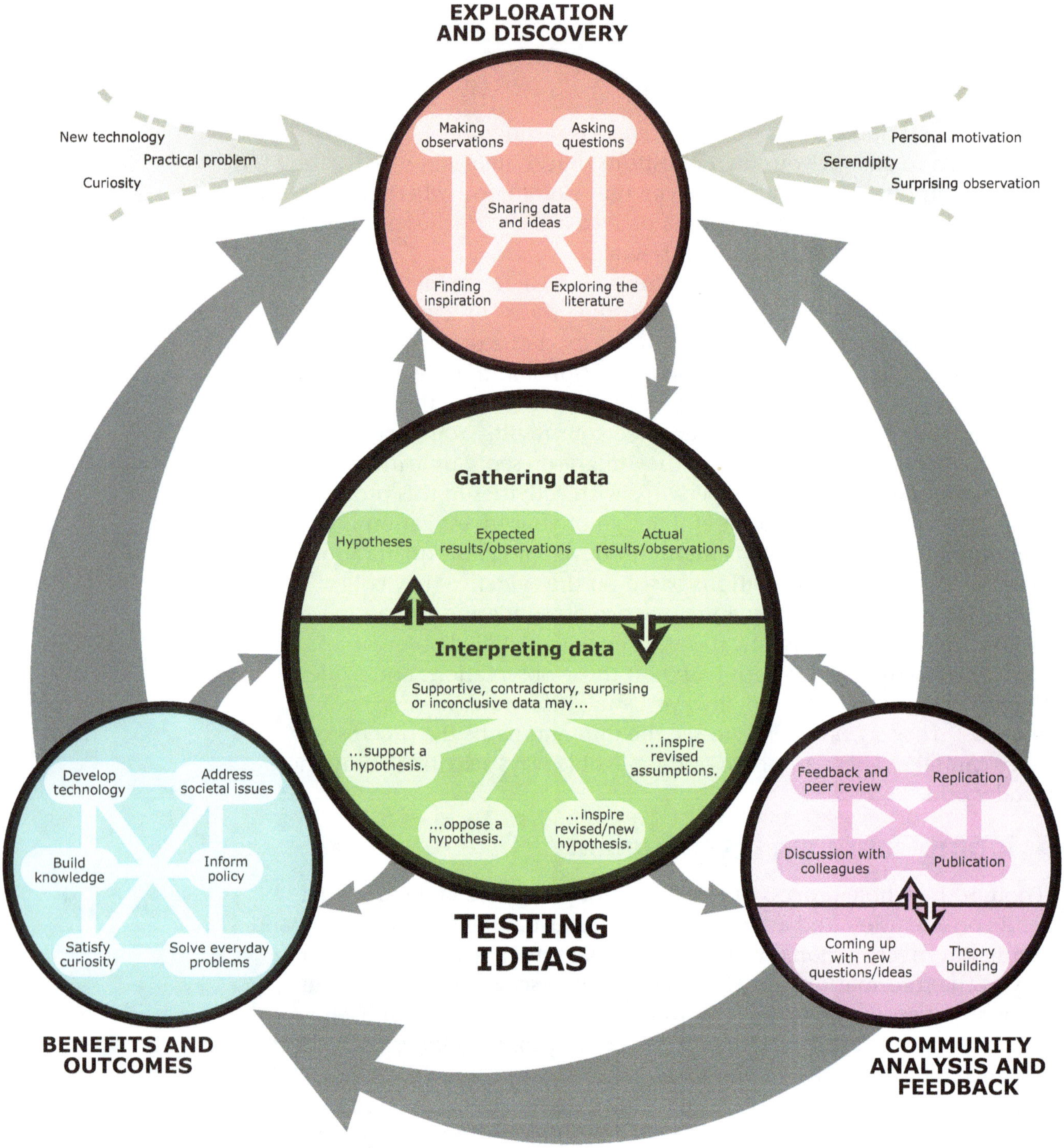

Image courtesy of the University of California Museum of Paleontology—Understanding Science www.understandingscience.org.

Figure C2.1 How Science Works. The process of science is a dynamic and exciting practice used to study the world around you. Where would controlled experiments fit onto this diagram?

- ❑ What are the dependent variables?
- ❑ What are the independent variables?
- ❑ What is one variable that you can change using the materials available?
- ❑ Do you need a control for the experiment? If so, describe it.

  Some experiments have independent and dependent variables. Some also have control conditions and experimental conditions. However, studies based only on observations have neither of these. All of these variations on the process of science are valid ways of collecting evidence.

  In a controlled experiment, scientists try to control as many variables as possible while changing only one independent variable. One group is not exposed to the variable you are testing, but is similar with respect to other variables. This "control group" serves as a comparison group to determine that the results were due to the tested variable and not to some other factor. In other words, changes in a dependent variable caused by an independent variable can be understood. For example, if scientists are studying the effectiveness of a new arthritis medicine, one group of individuals does not take the medicine. Scientists compare this group with the group taking the medicine. Not all experimental designs involve a control, but many do.
- ❑ What materials do you need for the investigation?
- ❑ What procedures will you follow to conduct your investigation?

  You should write the steps down and describe them in sufficient detail that someone else in the class can replicate your experiment using only these steps. It is a good idea to read over your procedure and have a classmate read it over as well, to see if you missed any important steps. It is very helpful to provide diagrams for setting up the equipment and carrying out the investigation. You should have your teacher review your procedure before you begin. Think about whether or not you have enough time to complete the experiment as designed. If you start the experiment and notice that you need to modify the procedures, make sure you document the changes and justify why you made them.
- ❑ What data will you collect?

  You should ask yourself, "How will the data help me answer the question?" and "Will the data I collect help me understand how the variable I am testing affects the outcome?"
- ❑ What result do I predict?
- ❑ How will I record and organize the data I collect? How will I represent the data when the investigation is complete?

  Data tables often make the work of collecting data more efficient and better organized. Graphing data often helps reveal patterns and trends.
- ❑ How many trials are needed so that I can be confident in the results?

APPENDIX C3

How to Interpret Patterns and Trends

A graph tells a story about data. The story evolves as you discover patterns and trends in the graph. The story will have more meaning when you can describe relationships among the variables plotted on the graph. To tell an accurate story, you must learn to recognize patterns and trends. You must also be able to describe both what patterns you see in the data and what you think the patterns mean.

Noticing patterns and trends is something you do naturally. You noticed that over time your friends got taller in elementary school—a trend in the growth rate of your friends. You also noticed that every school day started with the same kinds of activities and that pattern continued each school day. Being able to recognize patterns and trends is a skill that is very useful in using evidence to support explanations in science. Often, patterns or trends among variables are easier to see on a graph. And if these patterns or trends occur, then some underlying relationship among the variables that are graphed may exist.

Trends are changes that move in a particular direction. Begin identifying trends in graphically displayed data by looking for

- what kinds of data the graph is showing;
- an overall change (increase or decrease) in the data across the entire data set;
- changes (increases or decreases) over smaller sections of the data; and
- no change.

These trends tell part of the story. It is important to describe the trend in the graph with precise wording. Avoid saying simply "it increases" or "it decreases" in your description because "it" is vague and could be interpreted to refer to different things. Remember that you want your story to be clear. So for the graph in figure C3.1, you might say that as the mass of the material increases, the volume also increases.

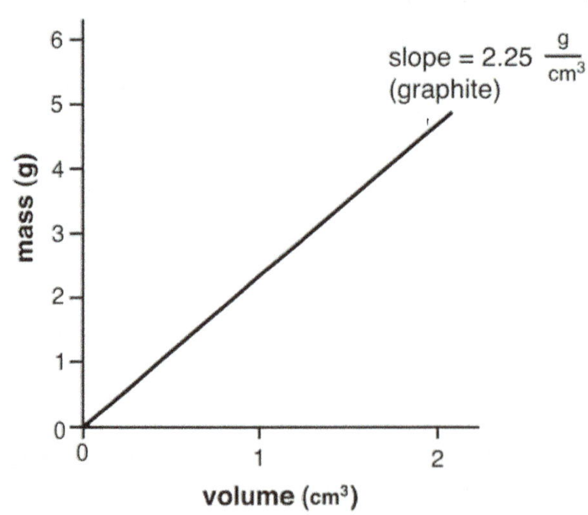

Figure C3.1 Mass versus volume. As the mass of the material increases, the volume also increases. This graph shows a direct relationship between variables.

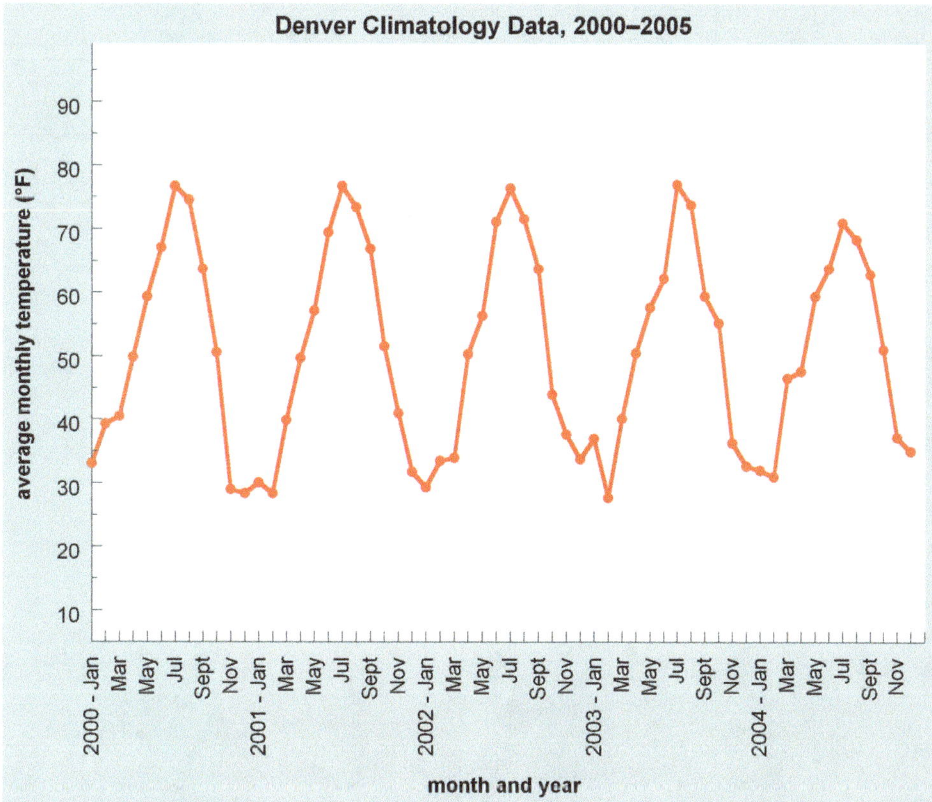

Figure C3.2 **Temperature data for Denver, Colorado.** You can see a pattern of increasing and decreasing temperatures in this graph. The temperature in Denver, Colorado, increases during the summer months and decreases during the winter months.

Source: NOAA's National Weather Service Forecast Office. Retrieved from: http://www.crh.noaa.gov/den/cli/climo.php

Patterns are changes that repeat periodically. You can recognize patterns in graphs by looking for increases or decreases that occur regularly. Look at the graph in figure C3.2. What repeating pattern do you see? Patterns tell another part of the story in a graph. Again, you want to be precise in your description of patterns. Avoid describing the pattern as "it goes up and down." A clearer description would be that the temperature in Denver, Colorado, increases during the summer months and decreases during the winter months.

You may identify no pattern or trend on a graph. This information is important, too. One variable may not be affected by the other variable if you see no pattern or trend. This would indicate that there is no dependency among the data plotted in the graph. An example of a graph that shows no relationship is figure C3.3. Can you see how the lack of a relationship between these variables can be informative?

In this curriculum, you are asked many times to interpret what you see on a graph. Sometimes you may use the highlight comments and captions strategy (see appendix B6, *How to Write Highlight Comments and Captions*). Use the strategies you have learned here to help you write clearer and more descriptive comments and captions for your graphs or to answer questions related to the data. Work through the following exercises to develop your skills at interpreting patterns and trends in data.

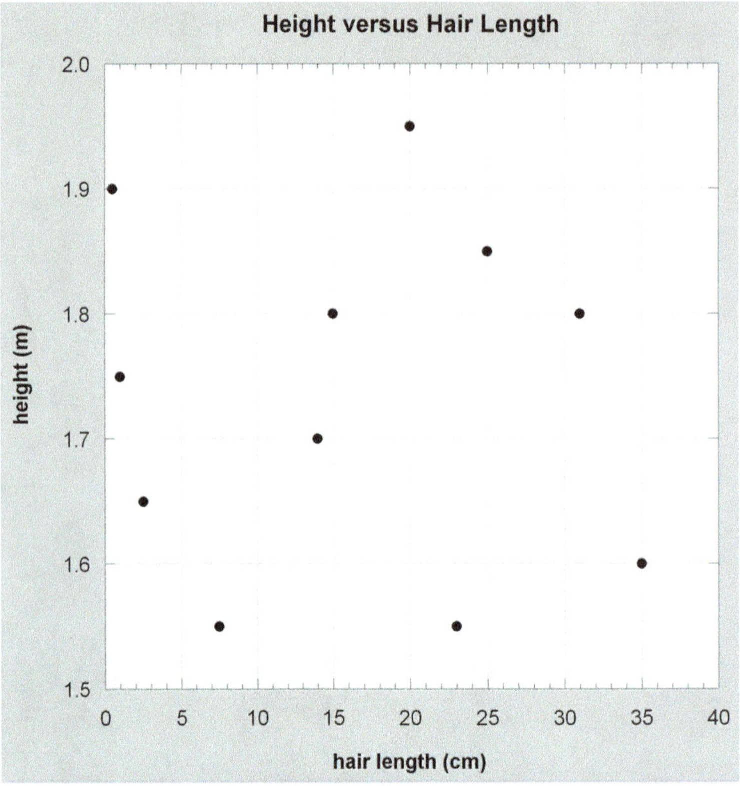

Figure C3.3 **Height versus hair length.** This graph shows no pattern or trend—the data points are scattered randomly across the graph. You can infer from this graph that there is no relationship between a person's hair length and his or her height.

Analysis

1. Analyze the graph in figure C3.4. This graph shows the number of paramecia, a microscopic protist, in a population over time. Complete the following steps, using the graph.
   a. Sketch this graph in your science notebook.
   b. What trends do you see?

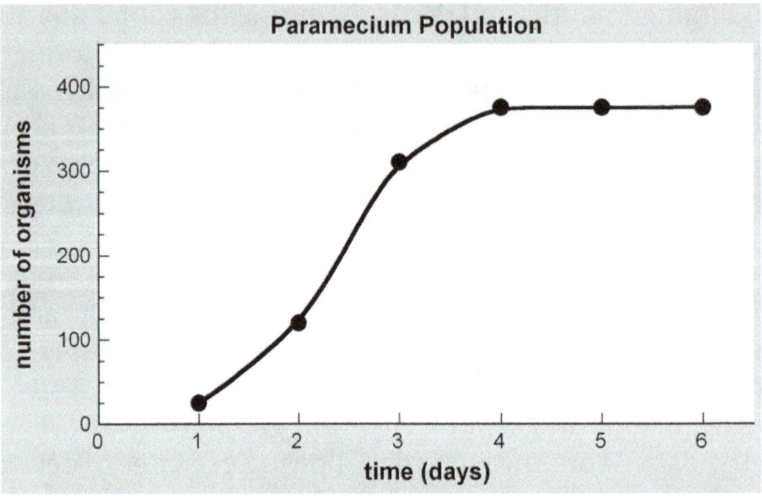

Figure C3.4 **Paramecium population.** Use this graph to complete steps 1a–f.

c. What patterns do you see?
d. What do you notice about the shape of the graph? Record your observations as the "What I see" portion of the highlight comments.
    Describe any trends and patterns on this graph that you notice.
e. Interpret what you see by recording the "What it means" portion of the highlight comments.
f. Summarize your analysis by writing a caption below your graph.

2. Use figure C3.5 to analyze both patterns and trends.
    a. Sketch the graph in figure C3.5 into your science notebook.
    b. Describe patterns and trends you see on the graph by using highlight comments and captions.
    c. Summarize your ideas by combining your comments into a caption for the graph.

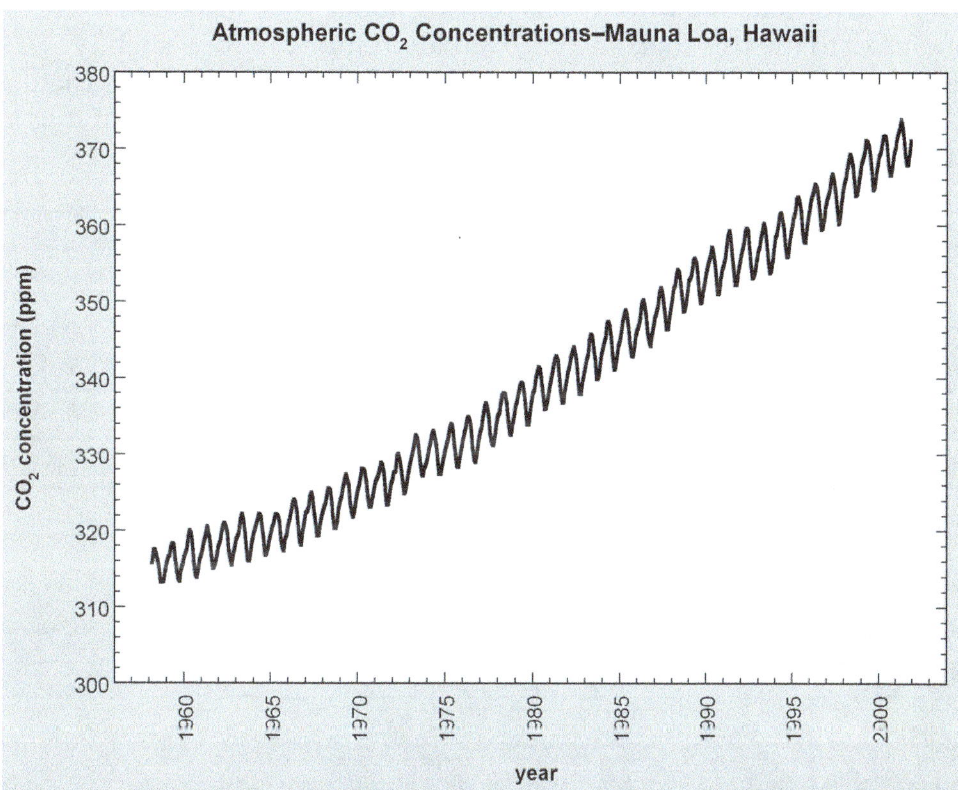

Figure C3.5
Carbon dioxide (CO_2) concentrations at Mauna Loa, Hawaii. Use this graph to complete steps 2a–c.

Source: Keeling, C.D. and T.P. Whorf. 2005. Atmospheric CO_2 records from sites in the SIO air sampling network. In *Trends: A Compendium of Data on Global Change.* Carbon Dioxide Information Analysis Center, Oak Ridge National Laboratory, U.S. Department of Energy, Oak Ridge, TN, USA.

APPENDIX C4

How to Construct an Explanation

Scientists support their claims and conclusions about how the natural world works by using data and evidence about the topic they are studying. As you learn about the practices that are part of the process of science, you will learn to support your claims and conclusions with evidence. This evidence may come in the form of data that you have collected or that someone else has collected. The evidence may be from work that you have done. You may also use evidence from reports and summaries from scientists or even from other students. These data provide you with the evidence you need to support your ideas, conclusions, and claims in science.

The explanation tool (figure C4.1) provides you with a framework to structure your claims, evidence, and reasoning in a way that makes the most sense to someone who will be reading about your work. There are many ways to use this tool, and each situation may call for a slightly different approach to using it. Do not think of this tool as a rigid structure that must be followed precisely for every situation. Rather, think of it as a graphic organizer for your ideas.

Question	
Evidence from data and observations	**Reasoning, including science ideas**
Claim ***(Your claim should answer the question.)***	
Explanation ***(Make sure to link your claim, evidence, and science ideas.)***	

Figure C4.1 **Explanation Tool.** This explanation tool can help you formulate an explanation from a claim, evidence, and reasoning.

Notice that the explanation tool in figure C4.1 has five basic parts. It can also have several additional features that you and your teacher may decide to use for different circumstances. Following is a summary of the basic parts of the tool:

1. **The question that you are trying to answer.** Science centers on answering questions about the natural world. You will conduct investigations and research to answer questions. Testable questions in science are those that can be answered by investigations. The questions that you ask guide the forms of data you will collect. Ultimately, you will use data to try to explain the answers to the questions.
2. **The evidence that you gather.** This part of the tool includes a list of all of the evidence you have collected. Do not list individual data points but rather write a summary of the data. This evidence may come from a number of sources. The data (evidence) should relate to the given claim.
3. **The reason or rationale.** In this section, explain why each piece of evidence supports your claim. Your reasoning is a justification that logically links the claim and evidence. These statements show why the data count as evidence that supports the claim according to the appropriate scientific principles.
4. **Your claim or claims.** What does the evidence say about the question you are trying to answer? You will write your claims, or your answer to the question, in this section of the tool. Your claim should make an assertion or conclusion that addresses the original question or problem. This may be in the form of a statement of trend, a behavior, or a generality that is based on the evidence presented.
5. **The explanation.** An area for your summary statement is included at the bottom of the tool. In this area connect each piece of evidence and reasoning to your claim. This explanation will likely be more than one complete sentence long. The goal is to write a coherent explanation or conclusion that is supported with your evidence and reasoning.

Study the example in figure C4.2 to develop your understanding of how the explanation tool can be applied to a set of data, like the mass and volume data graphed here.

Figure C4.2
Sample explanation tool. The graph (*a*) was reviewed to complete the sections of the (*b*) explanation tool. You should strive for complete explanations to connect evidence and reasoning to support your claim.

a

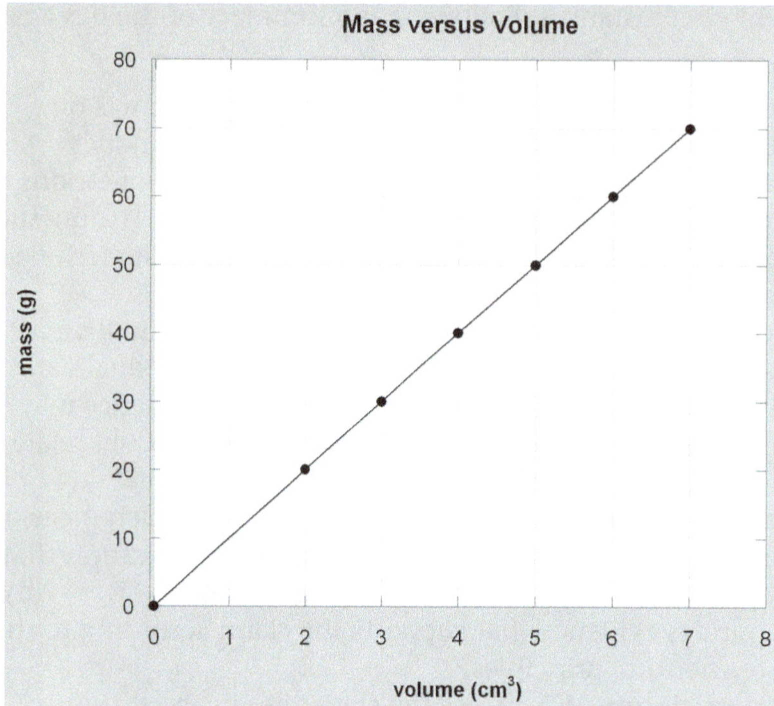

b

Question	
What is the relationship between the mass and volume of a substance?	
Evidence from data and observations	**Reasoning, including science ideas**
Data points for mass and volume make a straight line with a constant positive slope.	A constant positive slope means the variables plotted on each axis are directly proportional.
Claim ***(Your claim should answer the question.)***	
Mass and volume are directly proportional.	
Explanation ***(Make sure to link your claim, evidence, and science ideas.)***	
______________ results in ______________ because ______________. (evidence) (my claim) (reason/rationale) A constant positive slope for my mass versus volume graph *results in* my claim that mass and volume are directly proportional. This is *because* linear graphs represent variables that are directly proportional.	

APPENDIX D1

How to Use Very Large and Very Small Numbers

Astronomers have to work with numbers in the billions—and bigger—all the time. For example, how many stars are in the sky? Are there more stars that you cannot see? All stars that you see are part of the galaxy in which we live, the Milky Way Galaxy. A galaxy is an enormous group of stars in a massive cluster. The Milky Way has more than 100 billion stars. The universe contains hundreds of billions of different galaxies, many of which are made up of hundreds of billions of stars. Such large numbers can be difficult to manage. It would be awkward for astronomers to use terms such as a million billion billion or to write out numbers like 1,000,000,000,000,000,000,000,000. Rather, astronomers (and other scientists) use a special way of expressing numbers called scientific notation. In this appendix, you will learn about and practice using scientific notation.

Part I Big Numbers

Materials

calculators

Scientific notation is a way to abbreviate numbers to make them easier to work with. To show numbers with scientific notation, you must first be comfortable with exponents. Exponents are shorthand for the number of times a number, called the base, is multiplied by itself. A base with an exponent is said to be "raised to the power" of that exponent. For example, the number 2^4 means $2 \times 2 \times 2 \times 2$, or 16. Here, 2 is the base and 4 is the exponent. In scientific notation, the base number is always 10. Having 10 as the base works well because the exponent shows how many zeros you would need to write out in the long form of the number. So, 10,000 is expressed as 10^4, because $10 \times 10 \times 10 \times 10 = 10{,}000$.

In the previous example, you may have noticed that with scientific notation, you simply move the decimal point of a number to obtain a more manageable number. Then you write the number of places you moved the decimal as an exponent of 10. For example, you would write the number 4,600,000,000 as 4.6×10^9 with scientific notation. The second number is a lot simpler, and it says the same thing as the first. You write it like this because you moved the decimal nine places to the left to get to the numeral 4.6. This is shown in figure D1.1.

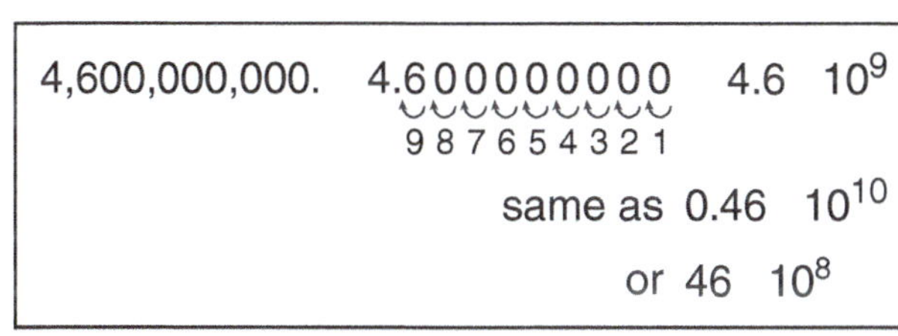

Figure D1.1
Example of moving decimals for positive exponents. This diagram shows the conversion of the number 4,600,000,000 to scientific notation, 4.6×10^9. The illustration in the center shows the decimal place moving left nine times. The table in figure D1.2 explains why 4.6×10^9 is the same as 4.6 billion.

Pronounced	Number	Powers of ten (scientific notation)	Unit prefix in the metric system (SI)
Trillion	1,000,000,000,000.	10^{12} or 1×10^{12}	tera-
Billion	1,000,000,000.	10^{9} or 1×10^{9}	giga-
Million	1,000,000.	10^{6} or 1×10^{6}	mega-
Thousand	1,000.	10^{3} or 1×10^{3}	kilo-
Hundred	100.	10^{2} or 1×10^{2}	hecto-
Ten	10.	10^{1} or 1×10^{1}	deka-
One	1.	1×10^{0} or 10^{0}	
Three hundred twenty-seven thousand	327,000.	3.27×10^{5}	

Figure D1.2 **Table for large numbers.**

Other large numbers are also easy to write using scientific notation. You would write the number 34,000 as 3.4×10^4. You would write the number 286,000,000 as 2.86×10^8. You might remember that the metric system is based on multiples of 10. The table in figure D1.2 shows how large numbers convert to powers of 10. It also shows prefixes for these numbers in the metric system. With the example above, 286,000,000 is also the same as 286×10^6, or 286 million.

Analysis

Work through these problems individually and write your answers in your science notebook. When you finish, join with another student and compare your answers. Discuss and resolve any differences you have in your answers.

1. One kilometer (km) is the same as 1,000 meters (m). How would you write 1,000 m using scientific notation?
2. A googol is one of the biggest named numbers. It is written as the number 1 followed by 100 zeros. Write this number using scientific notation.
3. The speed of light is 3.0×10^8 meters per second (m/sec). What is this value written without using scientific notation?
4. Use scientific notation to write $87 billion and 248 million stars.

Part II Small Numbers

Materials

calculators

Biologists work with large numbers, especially when studying fossil life-forms. They also work with extremely small numbers, especially when working with microscopes. A key property of working with microscopes is the wavelength of light, which is a key feature of colors. Wavelengths of light are commonly about 1 billion times shorter than a meter.

You write numbers less than 1.0 in scientific notation in the same general way that you write large numbers. The key difference is that the power of 10 is a negative exponent. The exponent still tells you how many places the decimal is from the number 1.0, but the decimal is moved in the other direction (to the right). When writing small numbers using powers of 10, you imagine moving the decimal to the right. The number of places you move the decimal is the power of 10 expressed as a negative number. We would write the number 0.0000001 in scientific notation as 1×10^{-7} because the decimal moves seven places to the right to get to the number 1.0. Another example is shown in figure D1.3 for 0.00000035, which is the same as 3.5×10^{-7}.

0.00000035   0.0000003.5 3.5×10^{-7}

1 2 3 4 5 6 7

same as 0.35×10^{-6}

or $350. \times 10^{-9}$

Figure D1.3
Example of moving decimals for negative exponents. This diagram shows the conversion of the number 0.00000035 to scientific notation, 3.5×10^{-7}. The illustration in the top center shows the decimal place moving to the right seven times.

Analysis

Work through these problems individually and write your answers in your science notebook. When you finish, join with another student and compare your answers. Discuss and resolve any differences you have in your answers.

1. A micron is an abbreviation for the term micrometer (µm). How would you express 1 µm using scientific notation? Look at the prefixes in the table in figure D1.4.
2. How many meters are in 1 millimeter (mm)?
3. Scientists often measure wavelengths of light in units called nanometers (nm). A nanometer is 0.000000001 m. Write this number using scientific notation.

Pronounced	Number	Powers of ten (scientific notation)	Unit prefix in the metric system (SI)
Tenth	0.1	10^{-1} or 1×10^{-1}	deci-
Hundredth	0.01	10^{-2} or 1×10^{-2}	centi-
Thousandth	0.001	10^{-3} or 1×10^{-3}	milli-
Millionth	0.000001	10^{-6} or 1×10^{-6}	micro-
Billionth	0.000000001	10^{-9} or 1×10^{-9}	nano-

Figure D1.4
Prefixes for small numbers.

APPENDIX D2

How to Convert Measurements

Units of Measurement

When you measure something, it is important to record what units you used for your measurement. For example, suppose someone told you that her cat had a weight of "20"; that number doesn't mean much without units. Does the cat weigh 20 pounds, 20 newtons, 20 ounces, or 20 tons?! Distance measurements (length) also need units such as feet, inches, meters, kilometers, and miles. If you are measuring time, you use units such as seconds, minutes, hours, and years.

Scientists measure things according to the *Système international d'unités* (International System of Units), more commonly referred to as "SI." SI is a modification of the older metric system. It was used first in France, and now is the common system of measurement throughout the world.

Among the basic units of SI measurement are the meter (length), the kilogram (mass), the Kelvin (temperature), and the second (time). All other SI units are derived from these four. Some of these units are described in the following sections. You will use some of these units in your laboratory work.

Length

1 kilometer (km)	=	1,000 meters	=	10^3 m
1 hectometer (hm)	=	100 meters	=	10^2 m
1 dekameter (dkm)	=	10 meters	=	10^1 m
1 meter (m)				
1 decimeter (dm)	=	0.1 meter	=	10^{-1} m
1 centimeter (cm)	=	0.01 meter	=	10^{-2} m
1 millimeter (mm)	=	0.001 meter	=	10^{-3} m
1 micrometer (μm)	=	0.000001 meter	=	10^{-6} m
1 nanometer (nm)	=	0.000000001 meter	=	10^{-9} m

The meter is the basic unit of length. Its increments increase or decrease by the power of 10. For example, measurements under microscopes often are made in micrometers, which is one-millionth of a meter. Still smaller measurements are made in nanometers. You might use nanometers to measure wavelengths of light that plants use in photosynthesis. The units of length you will use most frequently in the laboratory are centimeters (cm) and millimeters (mm).

Units of area are derived from units of length by multiplying two lengths. One hectometer squared is one measure that often is used in ecological studies. It commonly is called a hectare and equals 10,000 m^2. Measurements of area made in the laboratory most frequently are in centimeters squared (cm^2).

Mass

1 kilogram (kg)	=	1,000 grams	=	10^3 g
1 hectogram (hg)	=	100 grams	=	10^2 g
1 dekagram (dkg)	=	10 grams	=	10^1 g
1 gram (g)				
1 decigram (dg)	=	0.1 gram	=	10^{-1} g
1 centigram (cg)	=	0.01 gram	=	10^{-2} g
1 milligram (mg)	=	0.001 gram	=	10^{-3} g
1 microgram (µg)	=	0.000001 gram	=	10^{-6} g
1 nanogram (ng)	=	0.000000001 gram	=	10^{-9} g

Measurements of mass are based on the gram. Like the meter, the basic units of mass also increase or decrease by the power of 10. In the biology laboratory, measurements usually are made in kilograms, grams, centigrams, and milligrams.

Volume

1 kiloliter (kL)	=	1,000 liters	=	10^3 L
1 hectoliter (hL)	=	100 liters	=	10^2 L
1 dekaliter (dkL)	=	10 liters	=	10^1 L
1 liter (L)				
1 deciliter (dL)	=	0.1 liter	=	10^{-1} L
1 centiliter (cL)	=	0.01 liter	=	10^{-2} L
1 milliliter (mL)	=	0.001 liter	=	10^{-3} L

SI units of volume are derived from units of length by multiplying length by width by height. One meter cubed (m^3) is the standard unit. Although not officially part of SI, liters are often used to measure the volume of liquids. There are 1,000 liters in 1 meter cubed (m^3). In other words, it is 1 L = 0.001 m^3. One milliliter equals 1 centimeter cubed (1 mL = 1 cm^3).

Because 1 meter cubed (m^3) is too large for practical use in the laboratory, we use centimeters cubed (cm^3). Volume measurements in the laboratory usually are made in glassware marked for milliliters and liters.

Temperature

Units of temperature that you will use in this course are degrees Celsius. On the Celsius scale, 0°C is the freezing point of water, and 100°C is the boiling point of water. Figure D2.1 illustrates the Celsius scale alongside the Fahrenheit scale, which still is used in the United States. On the Fahrenheit scale, 32°F is the freezing point of water and 212°F is the boiling point of water. Figure D2.1 is useful for converting from one scale to the other.

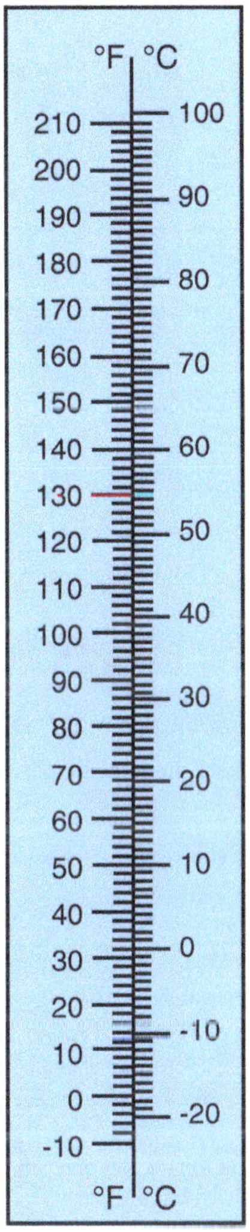

Figure D2.1
A comparison of Fahrenheit and Celsius temperature scales.

Another type of measurement you will encounter is molarity (labeled with the letter *M*). Molarity measures the concentration of a dissolved substance in a solution. A high molarity indicates a high concentration. Some of the solutions that you will use in the activities are identified by their molarity.

If you wish to learn more about SI measurement, check out the Web site of the National Institute of Standards and Technology (NIST).

Calculating Unit Conversions

What if a friend told you that he would phone you in 86,400 seconds? What time would that be on the clock? When faced with a problem like this, you need to be able to convert from one unit to another. After this activity, you will be able to use the technique of unit conversion to show that 86,400 seconds is the same as one day.

Unit conversions are also important for comparing two measurements made with different units. For example, suppose a person who is 5 feet 8 inches tall has a hat on that is 0.30 meters tall. What is the total height of the person, including the hat? Unfortunately, you cannot simply add the lengths. You must convert all of them to the same unit, and then you can add the lengths. You may have to convert again to a more reasonable unit. The total height of the person would be 79.8 inches, 6 feet 7.8 inches, or 2.03 meters tall.

How do you make these conversions? The method is called unit analysis (or dimensional analysis). These terms may sound complicated, but the method is pretty simple. The method uses conversion factors to convert units step-by-step, canceling units at each step. Using the guidelines in the protocol, unit analysis is simple.

PROTOCOL

PROTOCOL

Unit Analysis Guidelines

1. Conversion factors relate different units and are different ways of expressing the number 1. For example, because there are 12 inches (in) in 1 foot (ft), and 12 in = 1 ft, we can say that $\frac{12\text{ in}}{1\text{ ft}} = 1$. This fraction, which contains two different units and equals 1, is the conversion factor for distances measured in inches or feet.
2. Conversion factors can be flipped (inverted) as long as the units stay with the number. For example, you can write

$$\frac{12\text{ in}}{1\text{ ft}} = 1, \text{ or } \frac{1\text{ ft}}{12\text{ in}} = 1.$$

3. Units behave as numbers do in fractions. For example, the units in the numerator of fractions will cancel the same units in the denominator of fractions. For example,

$$\frac{12\ \text{in}}{3\ \text{in}} = 4.$$

4. In unit analysis, your goal is to cancel the same units in the numerators and denominators until you end up with the units you are striving for.
5. When you convert between units, follow these steps.
   a. Identify the units that you have.
   b. See which units you want.
   c. Note the conversion factors that you can use to convert the measurement from the unit in step 5a to the unit in step 5b.
   d. Set up your conversion similar to the examples found in this appendix.

Work through the following three conversions to practice using the guidelines for unit analysis from the protocol box.

1. How many inches are there in 1 mile (mi)?

   To convert miles to inches, start with what you know (1 mi) and use conversion factors (figure D2.2) to cancel units as you go until you get to the units that you want (inches). When the same units are on both the bottom and top, they cancel. Work with your teacher to see how to cancel these units on the top and bottom.

$$(1\ \text{mi}) \times \left(\frac{(5{,}280\ \text{ft})}{1\ \text{mi}}\right) \times \left(\frac{12\ \text{in}}{1\ \text{ft}}\right)$$

Conversion step (from Protocol Step 5)	Answers
What unit do you have now?	1 mi
What unit do you want?	"How many inches"
What are the conversion factors?	1 mi = 5,280 ft, or $\left(\frac{5{,}280\ \text{ft}}{1\ \text{mi}}\right) = 1$ 1 ft = 12 in, or $\left(\frac{12\ \text{in}}{1\ \text{ft}}\right) = 1$

Figure D2.2 Conversion table for miles to inches.

The units cancel, so you are left with units of inches. You can then multiply the numerator numbers together for the answer in inches.

$$(1\,\text{mi}) \times \left(\frac{5{,}280\,\text{ft}}{1\,\text{mi}}\right) \times \left(\frac{12\,\text{in}}{1\,\text{ft}}\right) = \left(\frac{1 \times 5{,}280 \times 12\,\text{in}}{1 \times 1}\right) = 63{,}360 \text{ in}$$

2. Work through a more complicated example. Suppose that you want to convert 75 miles per hour (mph) into feet per second (ft/sec).

   Now what? Take your conversions (figure D2.3) one step at a time, canceling units as you go until you arrive at the units you want.

$$\left(\frac{75\,\text{mi}}{1\,\text{hr}}\right) \times \left(\frac{5{,}280\,\text{ft}}{1\,\text{mi}}\right) \times \left(\frac{1\,\text{hr}}{60\,\text{min}}\right) \times \left(\frac{1\,\text{min}}{60\,\text{sec}}\right) = \left(\frac{75 \times 5{,}280\,\text{ft} \times 1 \times 1}{1 \times 1 \times 60 \times 60\text{ sec}}\right)$$

$$= \left(\frac{396{,}000\,\text{ft}}{3{,}600\,\text{sec}}\right) = \left(\frac{110\text{ ft}}{1\,\text{sec}}\right) = 110\,\frac{\text{ft}}{\text{sec}}$$

Figure D2.3
Conversion table for miles per hour to feet per second.

Conversion step (from Protocol Step 5)	Answers
What unit do you have now?	$75\,\frac{\text{mi}}{\text{hr}}$ (mph)
What unit do you want?	$\frac{\text{ft}}{\text{sec}}$ (ft/sec)
What are the conversion factors?	1 hour (hr) = 60 minutes (min) 1 minute (min) = 60 seconds (sec) 1 mi = 5,280 ft

3. Try the following conversions on your own. Use the following conversion factors for steps 3a–c. Show your calculations for each.

   1 slink = 7 zips
   1 sliff = 5 zips
   4 voles = 3 sliffs
   8 lampos = 7 flies
   12 voles = 1 lampo

   a. How many sliffs are in 1 lampo?
   b. One vole is how many zips?
   c. How many flies are in 1 slink?

APPENDIX D3

How to Create Graphs

Graphs are visual representations of numerical data. They help us see patterns that we might not see if we looked at numbers alone. Graphs make it easy to see at a glance what happened in an investigation. We also can use the patterns we observe in graphs to predict future changes or events.

Different graphs serve different purposes. For example, a line graph is a good way to show the relationship between two sets of numbers. Figure D3.1 shows a line graph that relates the number of days and the number of individuals in a laboratory population of microorganisms. A bar graph is a good way to show the frequency of measurements that fall into different categories. Figure D3.2 shows a bar graph of shoe sizes among the girls in a high school biology class. In this case, the different categories are the shoe sizes. A histogram is similar to a bar graph, but it shows data that occur over particular intervals. Figure D3.3 shows a histogram of student scores on a geology test, where the scores are grouped into 5-point intervals. These groups are known as bins.

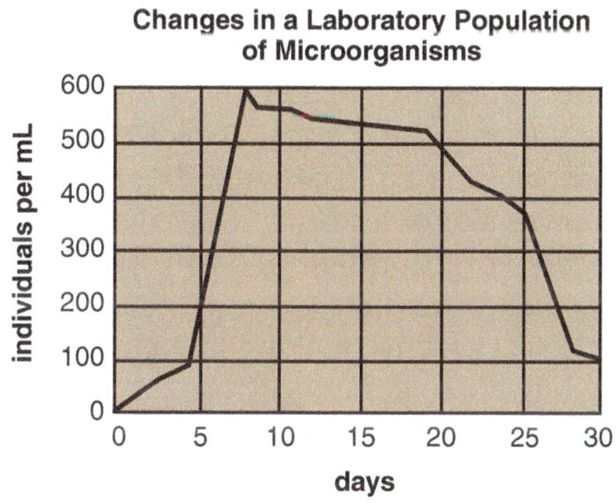

Figure D3.1 Line graph. A line graph shows the relationship between two sets of numbers. Line graphs are frequently used to show how something changes over a period of time. This line graph shows the growth and decline of a population of microorganisms over 30 days.

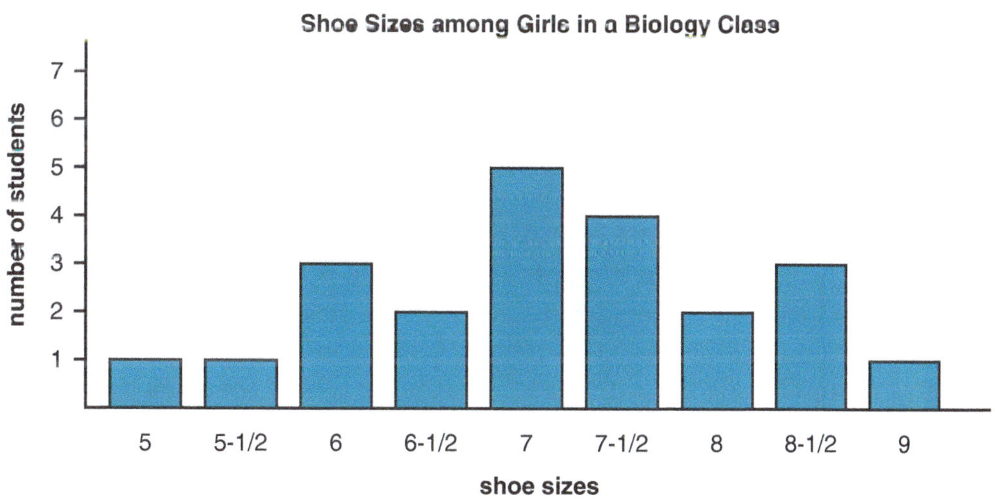

Figure D3.2 Bar graph. A bar graph shows the frequency of a set of observations. This bar graph shows how many girls in a biology class wear shoes of different sizes.

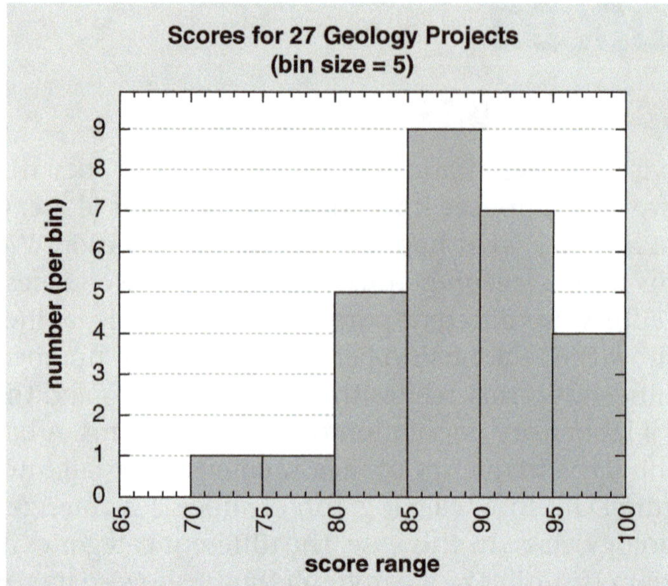

Figure D3.3
Histogram.
A histogram is a bar graph where observations are "binned" into particular intervals. This histogram shows the scores on a geology test by grouping them into 5-point intervals.

There are many types of graphs. In addition to line and bar graphs, you have probably seen pie graphs. These are often used to show percentage data. For example, you could use a pie graph to show the percentage of students in your class with brown, blonde, red, and black hair. You may also have seen a scatter plot. These graphs are often used to visualize a large number of data points. Lines are not used to connect each data point to the others. For example, you might use a scatter plot to show the time of day that students in your class started a particular homework assignment. Next time you look at a newspaper or a science magazine, pay attention to how graphs are used to represent information. Can you find examples of line, bar, and pie graphs, and scatter plots?

In this activity, you will practice making line graphs, bar graphs, and histograms because they are so commonly used in biology. These graphs have two major lines. One runs horizontally, and one runs vertically on the page. These lines are called the *axes* (singular: axis). The horizontal line is the *x*-axis; the vertical line is the *y*-axis. The point at which these two lines meet is the place where the graph begins, or the origin.

Both axes often include a sequence of numbers called a *number scale*. The numbers on the *x*-axis read from left to right; those on the *y*-axis read from bottom to top. The number scale is not necessarily the same on both axes. Look at figure D3.1. The number scale on the *x*-axis reads from 0 to 30; the *y*-axis reads from 0 to 600. What are the number scales on the graph in figure D3.2?

On some graphs, one or both axes show categories, rather than numbers. However, regardless of what a graph's axes represent, graphs are always clearly labeled to help the reader understand them. Look again at figures D3.1, D3.2, and D3.3. Each axis is labeled to explain to the reader what the numbers represent. Furthermore, each graph has a title that describes the relationship that the graph displays.

Making a graph involves several important steps. The steps in part I, *Line Graphs,* will help you draw any line graph. The steps in part II, *Bar Graphs,* will help you draw any bar graph. The steps in part III, *Histograms,* will help you draw any histogram.

Materials (per person)

2 sheets of graph paper
ruler

PROCESS AND PROCEDURES

Part I Line Graphs

1. Review the data in figure D3.4.

   Before you can make a graph using data that you have collected, you need to organize your data into a data table. For this practice in graphing, you can use the data in figure D3.4 or data that you have collected over some interval of time.

Number of mice caught in a field	
Day	**Number of mice caught per 100 traps per night**
0	25
30	45
60	38
90	30
120	20
150	14
180	13
210	8
240	7
270	11
300	4
330	13

Figure D3.4 **Table showing the number of mice caught over three months of time.**

2. Draw the *x*-axis and the *y*-axis for the graph.

   Use a ruler and graph paper so that your lines will be straight and perpendicular (see figure D3.5a).

Figure D3.5 How to draw a line graph.

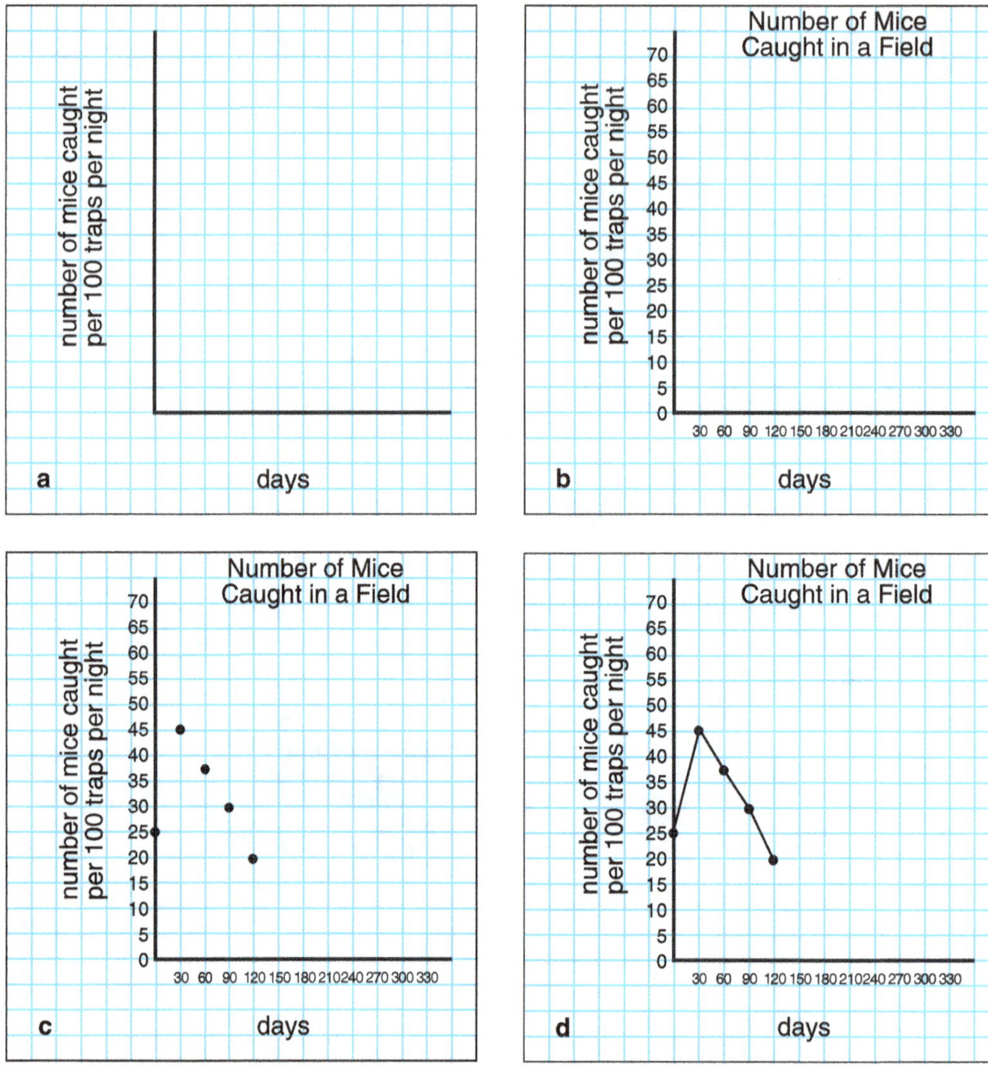

3. Identify what information will go on the *x*-axis and what information will go on the *y*-axis.

   Information on the *x*-axis is usually predictable in nature. For example, this information might be a series of dates, numbers, distances, or sizes. Information on the *y*-axis is usually variable in nature. For example, this could be the number of mice caught, the number of people of a certain height, or the number of butterflies captured.

4. Label each axis of the graph according to what each represents. Remember to include the units for measurements.

   The headings of the data table you are using can be a good place to start when thinking about how to label axes.

5. Set up the number scales on each axis.

   Allow space on each axis for a range of numbers just a bit greater than the numbers that are included in the data table (see figure D3.5b). Note that a number scale does not have to start with the number 1, but the numbers do need to be spaced in equal increments.

6. Give your graph a descriptive title.
7. Plot the data on your graph by doing the following.
   a. Read 1 row of data from the data table, for example, day 0 and 25 mice caught.
   b. Find the number on the *x*-axis where the piece of corresponding data fits (for example, 0).
   c. Move up from the number on the *x*-axis to the place on the *y*-axis where the corresponding piece of data fits (for example, 25).
   d. Draw a dot, called a *data point*, at that place.
   e. Repeat steps a–d for all the pieces of data in the data table.

      Figure D3.5c illustrates five data points. Can you determine where the remaining data points go?
8. Draw a smooth line from left to right that connects the data points. This should help you see the relationship between the data points.

   In this case, the line illustrates how the number of mice caught changed throughout the year (figure D3.5d).

Part II Bar Graphs

1. Review the data in figure D3.6.

   Alternately, use data that you have collected. Remember that bar graphs are used to show data that fall into different categories.
2. Draw the *x*- and *y*-axes for the graph.
3. Decide which information goes on each axis (refer to figure D3.7a).
4. Label each axis according to what each represents. Remember to include the units for measurements.
5. Decide on the number scales or category labels for each axis. Position the categories on the *x*-axis so that you can draw bars in the spaces between the lines. Place the numbers on the vertical axis next to the lines so that you can end a bar between 2 numbers, if necessary. Figure D3.7b will help you with this step.

Population of heath hens, Martha's Vineyard, MA	
Year	**Population size**
1900	90
1905	45
1910	280
1915	2,010
1920	550
1925	40
1930	10

Figure D3.6 Table showing the population of heath hens.

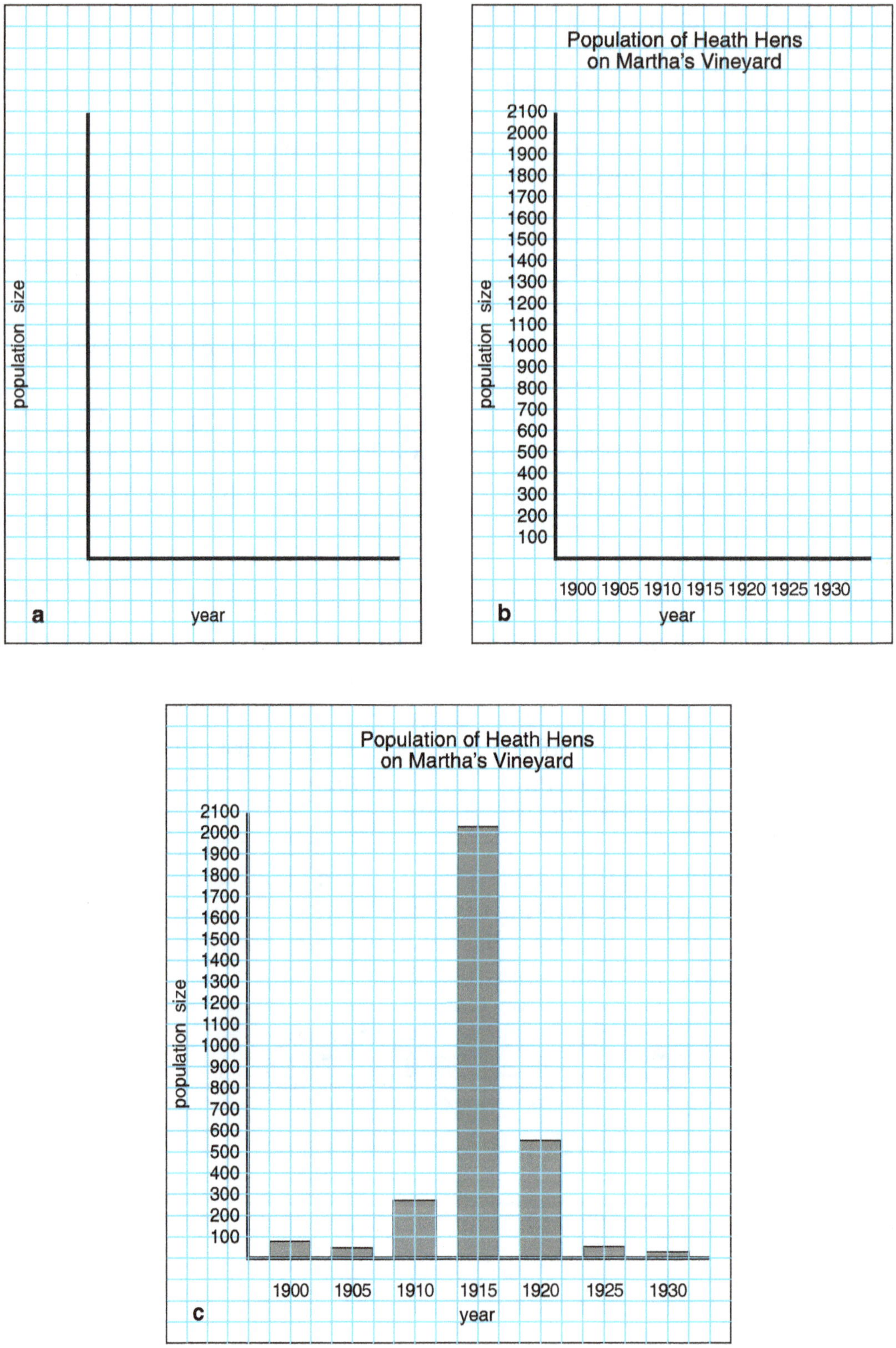

Figure D3.7 How to draw a bar graph.

6. Add a title to your graph.
7. Plot the data by following these steps.
    a. Read 1 row of data from the data table in figure D3.7, for example, year 1900 and population size 90.
    b. Find the corresponding category label on the *x*-axis scale of the graph (for example, 1900).
    c. Move up the column above the *x*-axis number you found, until you get to a point at the level of the appropriate number for that piece of data on the *y*-axis (for example, 90).
    d. Draw a horizontal line at that number to make the top of the bar.
    e. Color in the bar from that line down to the *x*-axis number you identified in step 7b.
    f. Repeat steps a–e for all the pieces of data in the data table (see figure D3.7c).

Part III Histograms

1. Review the data in figure D3.8.
2. Draw the *x*- and *y*-axes for the graph.
3. Decide which information goes on each axis.

    Refer to the example in figure D3.3 if you are not sure.

4. Label each axis.

Height (meters)	Number of aspen trees
0–0.5	15
0.6–1.0	10
1.1–1.5	7
1.6–2.0	3
2.1–2.5	3
2.6–3.0	1

Figure D3.8 **Table showing the heights of a group of aspen trees in 0.5 m intervals.**

5. Decide on the number scales for each axis. Position the numbers on the x-axis so that you can draw bars in the spaces between the lines. Place the numbers on the vertical axis next to the lines so that you can end a bar between 2 numbers, if necessary.

   When writing in the intervals, you can choose to either write the full interval (for example, 0–0.5) or just the end point of the interval (for example, 0.5).

6. Add a title to your graph.
7. Plot the data by following these steps.
   a. Read 1 row of data from the data table, for example, height 0–0.5 m and number of trees 15.
   b. Find the interval for the corresponding piece of data on the x-axis of the graph (for example, 0–0.5 m).
   c. Move up the column above the x-axis interval you found, until you get to a point at the level of the appropriate number for that piece of data on the y-axis (for example, 15).
   d. Draw a horizontal line at that number to make the top of the bar.
   e. Color in the bar from that line down to the x-axis.
   f. Repeat steps a–e for all the pieces of data in the data table.

APPENDIX E

Tables

Periodic Table of Elements

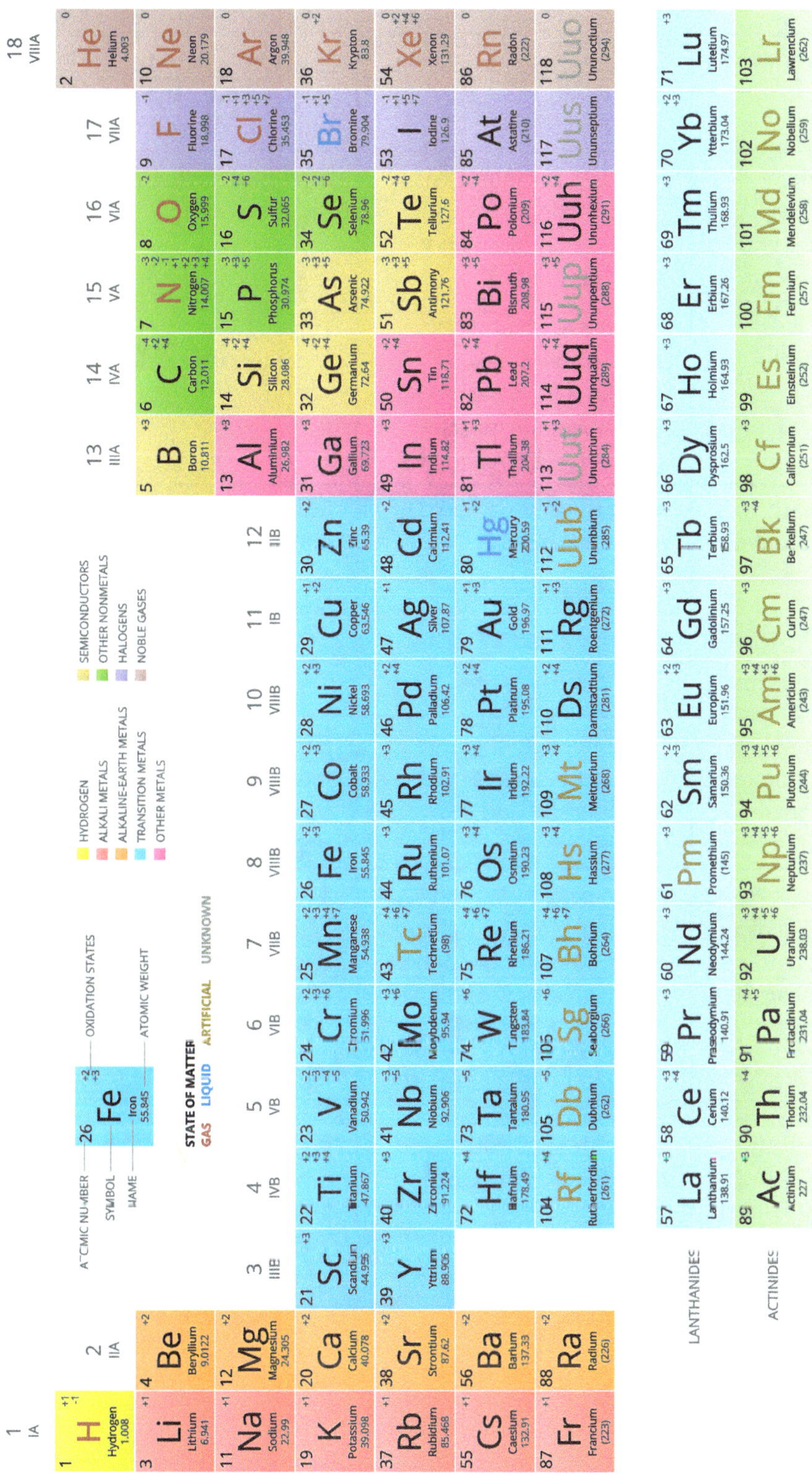

Figure E1.1 Periodic table of elements.

Vitamins

Important vitamins for human health			
Vitamin *fat-soluble*	**Sources**	**Functions**	**Deficiency symptoms**
A (retinol)	Liver, green and yellow vegetables, fruits, egg yolks, butter	Forms eye pigments; helps cell growth, especially of epithelial cells	Night blindness, flaky skin lowered resistance to infection, growth retardation
D (calciferol)	Fish oils, liver, action of sunlight on lipids in skin, fortified milk, butter, eggs	Increases calcium absorption from gut; is important in bone and tooth formation	Rickets (defective bone growth)
E (tocopherol)	Oils, whole grains, liver, mayonnaise, margarine	Protects red blood cells, cell membranes, and vitamin A from destruction; helps maintain muscles; synthesizes DNA and RNA	Fragility of red blood cells, muscle wasting, sterility
K (menadione)	Synthesis by intestinal bacteria; green and yellow vegetables	Assists liver in synthesis of clotting factors	Internal hemorrhaging (deficiency can be caused by oral antibiotics, which kill intestinal bacteria)
Vitamin *water-soluble*	**Sources**	**Functions**	**Deficiency symptoms**
B_1 (thiamine)	Whole grains, legumes, nuts, liver, heart, kidney, pork, macaroni, wheat germ	Facilitates carbohydrate metabolism, nerve transmission, and RNA synthesis	Beriberi, loss of appetite, indigestion, fatigue, nerve irritability, heart failure, depression, poor coordination
B_2 (riboflavin)	Liver, kidney, heart, yeast, milk, eggs, whole grains, broccoli, almonds, cottage cheese, yogurt, macaroni	Forms part of electron carrier in electron transport system; aids production of FAD; activates B_6 and folic acid	Sore mouth and tongue, cracks at corners of mouth, eye irritation, scaly skin, growth retardation
Pantothenic acid	Yeast, liver, eggs, wheat germ, bran, peanuts, peas, fish, whole grain cereals	Facilitates energy release and biosynthesis; stimulates antibodies and intestinal absorption	Fatigue, headaches, sleep disturbances, nausea, muscle cramps, loss of antibody production, irritability, vomiting
B_3 (niacin)	Yeast, liver, kidney, heart, meat, fish, poultry, legumes, nuts, whole grains, eggs, milk	Serves as coenzyme in energy metabolism; is part of NAD+ and NADP	Pellagra, skin lesions, digestive problems, nerve disorders, diarrhea, headaches, fatigue

Table E1.2 **Important vitamins for human health.**

Important vitamins for human health			
Vitamin	**Sources**	**Functions**	**Deficiency symptoms**
B_6 (pyridoxine)	Whole grains, potatoes, fish, poultry, red meats, legumes, seeds	Serves as coenzyme in amino acid and fatty acid metabolism and in the synthesis of brain chemicals, antibodies, red blood cells, and DNA; is essential to glucose tolerance	Skin disorders, sore mouth and tongue, nerve disorders, anemia, weight loss, impaired antibody response, convulsive seizures
Biotin	Cauliflower, liver, kidney, yeast, egg yolks, whole grains, fish, legumes, nuts, meats, dairy products; synthesis by intestinal bacteria	Serves as coenzyme in fatty acid, amino acid, and protein synthesis; promotes energy release from glucose; facilitates insulin activity	Skin disorders, loss of appetite, depression, sleeplessness, muscle pain, elevated blood cholesterol and glucose levels
Folate (folic acid)	Liver, yeast, leafy vegetables, asparagus, salmon	Serves as a coenzyme in nucleic acid synthesis and amino acid metabolism; is essential for new cell growth	Failure of red blood cells to mature, anemia, intestinal disturbances, diarrhea
B_{12} (cobalamin)	Liver, organ meats, meat, fish, eggs, shellfish, milk; synthesis by intestinal bacteria	Serves as coenzyme in nucleic acid synthesis; helps maintain nervous tissues; plays a role in glucose metabolism	Pernicious anemia, fatigue, irritability, loss of appetite, headaches
C (ascorbic acid)	Citrus fruits, tomatoes, green leafy vegetables, peppers, broccoli, cauliflower	Is essential to formation of collagen, an intercellular substance that holds cells together; protects against infection; maintains strength of blood vessels; increases iron absorption from gut; plays important role in muscle maintenance and stress tolerance	Scurvy, failure to form connective tissue, bleeding, anemia, slow wound healing, joint pain, irritability, premature wrinkling and aging

Table E1.2 **Important vitamins for human health.** (*continued*)

Essential Elements

Name	Food	Function	Deficiency symptoms	Excess symptoms
Calcium (Ca)	Dairy products, green vegetables (broccoli, greens), legumes, tofu (bean curd), small fish (with bones)	Helps in bone and tooth development; facilitates muscle contraction, blood clotting, nerve impulse transmission, and enzyme activation	Osteoporosis, stunted growth, poor quality bones and teeth, rickets, convulsions	Excess blood calcium (rare), loss of appetite, muscle weakness, fever
Chlorine (Cl)	Table salt, soy sauce, processed foods	Helps maintain acid/base balance; assists hydrochloric acid formation in stomach; promotes bone and connective tissue growth	Metabolic alkalosis (rare), constipation, failure to gain weight (in infants)	Vomiting
Magnesium (Mg)	Whole grains, liver, kidneys, milk, nuts, dark green leafy vegetables, seafood	Serves as a component of chlorophyll, bones, and teeth, and as a coenzyme in carbohydrate and protein metabolism	Infertility, menstrual disorders	Loss of reflexes, drowsiness, coma, death
Phosphorous (P)	Soybeans, dairy foods, egg yolks, meat, whole grains, shrimp, peas, leafy green vegetables	Serves as component of bones, teeth, nucleic acids, phospholipids, proteins, and ATP	Bone fractures (rare), disorders of red blood cells, metabolic problems, irritability, weakness	Decreased levels of calcium, muscle spasms, jaw erosion
Potassium (K)	Whole grains, meats, fruits, vegetables, milk, peanut butter	Helps maintain body water and pH balance; plays important role in nerve and muscle activity, insulin release, glycogen and protein synthesis	Muscle and nerve weakness, poor digestion	Abnormalities in heartbeat or heart stoppage, muscle weakness, mental confusion, cold and pale skin (all are rare)
Sodium (Na)	Table salt, soy sauce, processed foods, baking soda, baking powder, meat, vegetables	Helps maintain body water and pH balance; plays role in nerve and muscle activity and glucose absorption	Weakness, muscle cramps, diarrhea, dehydration, nausea	High blood pressure, edema, kidney disease
Sulfur (S)	Dairy products, nuts, legumes, garlic, onions, egg yolks	Serves as component of some amino acids; plays role in enzyme activation and in blood clotting	None known; protein deficiency would occur first	Excess sulfur-containing amino acid intake leads to poor growth

Figure E1.3 Important elements for human health.

APPENDIX F1

Microorganisms

Diseases Result from Interrelationships

Disease is a condition that interferes with an organism's ability to perform a vital function. An infectious disease involves a disease-causing agent, or **pathogen**. The disease results from the interaction between an organism, or **host**, and the pathogen. The pathogen infects the host. If the pathogen harms the host, symptoms of disease may appear.

How serious a disease becomes depends on the characteristics of both the host and the pathogen. The ability to cause disease is called **virulence**. The ability of an infected host to cope with a pathogen is called **resistance**. A pathogen with high virulence may cause death in a host with low resistance. A host with high resistance, however, may show only mild symptoms of the same disease. Sometimes, a moderately virulent pathogen may produce serious illness. During a famine, for example, more people die from disease than from starvation. This is because a poorly nourished host may have much less resistance than a well-fed host.

Although genetics does play a role in the ability of an organism to resist disease, much of an individual's resistance is acquired during the lifetime of the individual, rather than having been inherited. Resistance, whether acquired or inherited, is called **immunity**. When a pathogen infects a human host, the host produces proteins called **antibodies** that can help destroy the pathogen. In addition to antibodies, certain cells of the body's defense, or immune system, have a chemical memory. If the host survives the initial infection, these cells retain the ability to produce the same antibodies. If the same pathogen infects the host a second time, the host's body can act against it more rapidly, preventing symptoms from developing or reducing their severity.

Figure F1.1
The English physician Edward Jenner demonstrated in 1796 that by inoculating patients with a vaccine made of cowpox viruses he could protect them from the more serious disease of smallpox.

Each type of antibody is effective against only the particular pathogen that brought about its production, or pathogens that are very similar. For example, a person can have chicken pox only once, because antibodies produced during the first infection usually prevent symptoms from developing after a second infection. These antibodies are not effective against measles, however. Because there are many types of pathogens that cause colds, immunity to one type does not confer immunity to all the others.

Many diseases can be prevented by vaccines. A **vaccine** contains only enough of the killed or weakened disease-causing agent to stimulate production of antibodies by the immune system, usually without producing symptoms of the disease. Memory cells keep later infections, by the same pathogen, from causing disease. Vaccines first were used by Edward Jenner, an English physician (see figure F1.1).

Disease may result from a variety of causes. The pathogens involved in infectious diseases may be viruses, eubacteria, or other organisms. For example, athlete's foot, ringworm, potato blight, and corn smut are caused by fungi. Different protists cause malaria, African sleeping sickness, and amoebic dysentery. Many worms and insects cause diseases of plants and animals, and are important vectors, or carriers, of disease. Other diseases, such as scurvy, are the result of diet deficiencies; still others are a result of advancing age; and some, such as asthma, are brought on by reactions to substances or pollutants in the environment. Finally, some disorders, such as cystic fibrosis or Huntington's disease, are hereditary.

Some Eubacteria Are Pathogens

Although the vast majority of microbes are beneficial or at least harmless to other organisms, some eubacteria are pathogenic. At the present time, however, no archaebacteria are known to cause disease.

Eubacterial diseases may be spread through water, food, or air. Many plant diseases are caused by eubacteria, and almost all types of plants are susceptible to one or more types of eubacterial disease. Fire blight, shown in figure F1.2, is a common eubacterial disease that can destroy fruit trees. In Florida in 1984, an outbreak of a eubacterial disease called citrus canker led farmers to destroy more than 4 million citrus seedlings in an effort to halt the spread of the disease. Citrus canker is caused by one of more than 100 distinct varieties of the gram-negative eubacterium *Xanthomonas campestris*. Other varieties of this eubacterium cause diseases with similar symptoms in beans, cabbages, peaches, and other plants. Citrus canker is a continuing threat to the existence of the Florida citrus industry, an annual $8.5 billion business.

Eubacteria also cause human diseases such as cholera, leprosy, tetanus, eubacterial pneumonia, whooping cough, and diphtheria. Many eubacteria multiply in the human digestive tract and leave the body in the feces. If untreated feces enter the water supply, the eubacteria may enter another organism when it drinks the water. Waterborne diseases such as typhoid kill 6 million children each year worldwide. Most cases of food poisoning are due to eubacteria. In some cases, the eubacteria produce poisonous

Figure F1.2 Fire blight is a disease of apples, pears, and related trees. It is caused by the bacterium *Erwinia amylovora*.

substances, or toxins, that are released in the food. These toxins can cause nausea, vomiting, and diarrhea within a few hours of ingestion. In the case of botulism, even a small amount of the toxin can be fatal. In other cases, eubacteria ingested with food multiply in the intestine, causing symptoms even several days after ingestion.

Figure F1.3
Droplets from a sneeze carry disease-causing agents through the air.

Many eubacteria are airborne. Whenever you open your mouth and exhale, you release droplets of moisture, each of which may contain one or two eubacteria. With a sneeze, you release many thousands of droplets at great speed. Droplets from a sneeze, visible in figure F1.3, have been clocked at 200 miles an hour. Droplets from a cough travel at about half that speed. Each time you sneeze, you can expel from 10,000 to 100,000 individual eubacteria. The eubacteria travel through the air without being killed and may infect another person standing close to you.

Group A streptococcus (GAS) eubacteria commonly infect people and cause a variety of health problems. Scientists believe that there are more than 120 different strains of GAS that each produces its own set of proteins. Some infections caused by GAS are minor, but some can be life-threatening. For example, some people may have been infected with the bacteria but don't have any illness or symptoms. These people can pass the infection to others; GAS infections are spread by direct contact with saliva or nasal mucus.

One of the more common diseases caused by GAS is strep throat; people who have strep throat have a sore throat, white patches on their tonsils, swollen lymph nodes, fever, and headache. GAS can also cause minor skin infections. Medical experts estimate that more than 10 million mild GAS infections occur each year. These minor infections can normally be treated effectively with antibiotics. However, some GAS bacteria cause severe, or invasive, infections. Severe infections include infections of the bloodstream, streptococcal toxic shock syndrome (not the same illness that is associated with tampon use in menstruating females, although that disease is also caused by a bacterium), and necrotizing fasciitis (sometimes referred to as the flesh-eating disease). All of these severe infections can lead to shock, organ failure, and even death. Although illnesses such as the flesh-eating disease are scary because of their severe nature, they are fairly rare. Approximately 10,000 to 15,000 cases of invasive GAS diseases occur each year in the United States, compared with millions of cases of strep throat and minor skin infections caused by GAS.

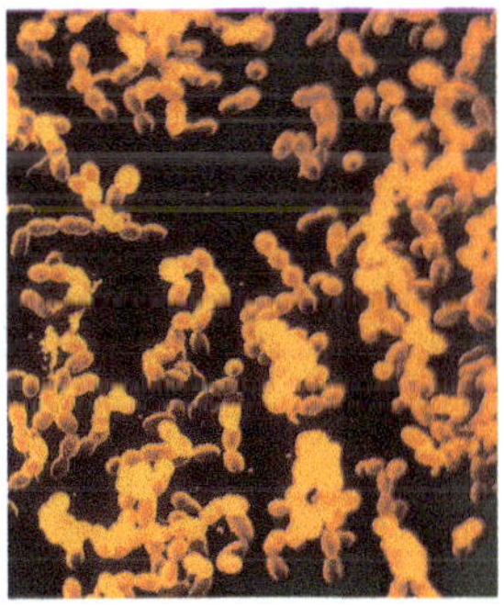

Figure F1.4
Streptococcus mutans is a bacterium associated with tooth decay.

Scientists continue to learn more about how GAS bacteria cause such a wide range of diseases. In recent years, scientists have found that a strain of the GAS bacteria that causes rheumatic fever, the most common infectious cause of childhood heart disease, contains gene sequences that appear to come from a bacteriophage, a virus that infects bacteria. The viral genetic sequences that were inserted into the bacterial DNA cause the bacteria to produce new toxic proteins. A strain of GAS without the viral genes does not cause rheumatic fever. Continuing scientific research will help scientists understand how GAS bacteria can cause such a wide variety of illnesses in humans.

Even dental caries, or tooth decay, is an infection caused by eubacteria. The decay begins on the surfaces of the teeth in a biofilm known as dental plaque. This biofilm consists of large numbers of eubacteria in a complex sugar matrix (shown in figure F1.4), and builds up on unbrushed teeth and in crevices not reached by a toothbrush. Sugars and carbohydrates in the diet serve as food sources for the eubacteria (*Streptococcus mutans*) that are present in the mouth. The eubacteria break the sugars and carbohydrates down into simple sugars and produce lactic acid as a waste product. The acid then causes the tooth enamel to lose minerals, thus causing tooth decay. Fluoride makes the teeth more resistant to decay because it retards the loss of minerals, including calcium and phosphate, from the teeth. It also promotes the replacement of these minerals to help keep the teeth healthy.

Viruses Are Unusual Pathogens

What do AIDS, hepatitis, measles, mumps, influenza, colds, and polio have in common? These diseases all are caused by viruses such as those shown in figure F1.5. A virus is an infectious agent that contains a nucleic acid (either DNA or RNA) and a protein coat. Viruses are so small that they can be seen only with an electron microscope. In fact, viruses can pass through most bacteriological filters. Viruses may play one of two roles when they enter a host cell. As agents of disease, viruses enter the host cells, disrupt the cells'normal functioning, and sometimes kill them. As agents of heredity, viruses can enter cells and cause permanent, heritable changes. Often, the role the virus plays depends on the host cell and environmental conditions.

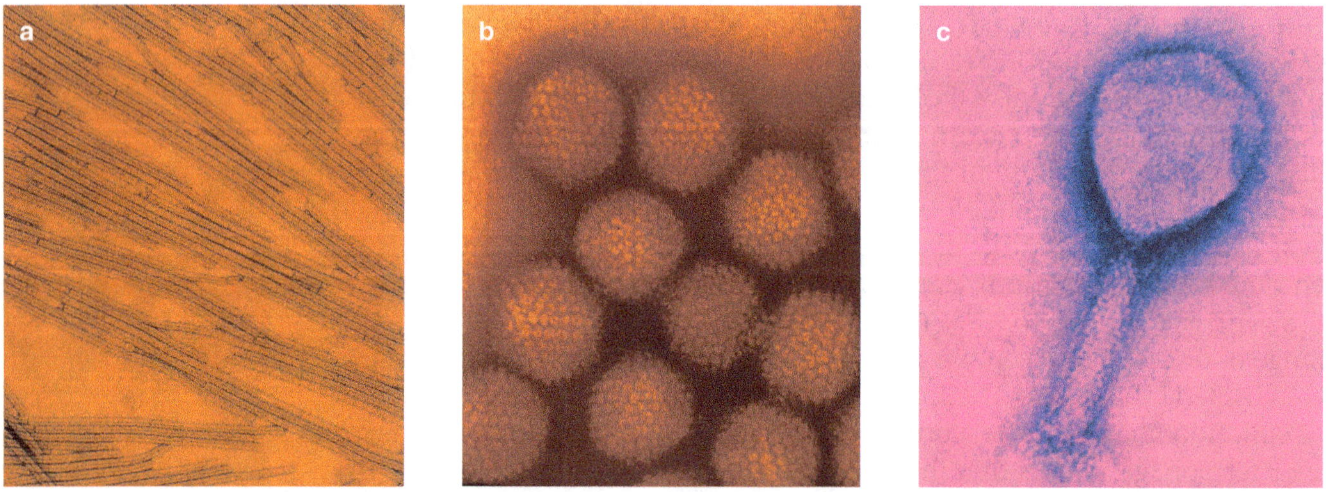

Figure F1.5 All viruses are pathogenic. Tobacco mosaic virus (*a*) causes a disease of plants (185,000×); adenovirus (*b*) causes respiratory illnesses in humans (11,000×); T4 bacteriophage (*c*) (240,000×) infects and kills bacterial cells.

Viruses differ from living things in several important ways. They are able to produce copies of themselves only inside a living organism. Outside the host cell, viruses do not reproduce, feed, or grow. They have no metabolism of their own. They do not take in and use energy. They do not have cell parts. Some can be crystallized and survive for years in that state. Only when they enter the appropriate host cells can they resume reproduction. Viruses infect plants (see figure F1.6). Viruses also infect bacteria, as shown in figures F1.7 and F1.8. To reproduce, a virus first attaches to a host cell and then injects its viral DNA or RNA into the host cell. The viral nucleic acid takes over the machinery of the host cell (the proteins that replicate host DNA and that make host cell proteins). Viral proteins direct the host cell to make more viral protein and viral nucleic acid. The proteins and nucleic acids then are assembled into new virus particles, and the infected cell ruptures, releasing hundreds of newly made viruses. Each new virus has the ability to infect a single new host cell.

Usually, genetic information is stored in DNA, transcribed into mRNA, and then translated into protein. In many viruses, genetic information is stored as DNA. In some viruses, however, the viral genetic information actually is stored as RNA. Certain RNA viruses, known as retroviruses, must go through an additional step before they can reproduce. Retroviruses must make copies of DNA from their RNA, the reverse of the normal flow of stored information in biological systems. Like other viruses, a retrovirus binds to the surface of a host cell and injects its RNA into the cell. Reverse transcriptase, a special enzyme associated with the RNA, allows the retrovirus to make a complementary DNA copy of its RNA (see figure F1.9). The newly formed viral DNA then can be integrated into the host cell DNA, replicated at the same time, and transmitted to offspring cells. When the cell produces RNA from its own DNA, it also produces viral RNA, which becomes the source of new viral particles, thus continuing the infection.

Retroviruses are known to cause cancer in some animals and have been associated with certain types of cancers in humans. Human oncogenes have similarities to genes in these retroviruses; both can result in cancerous growth when they are somehow disturbed. The virus that causes AIDS is a retrovirus that attacks and kills certain cells of the immune system. Once these cells are killed, the immune system is unable to perform its normal function of defending the organism against disease.

Figure F1.6 **TYMV (turnip yellow mosaic virus) in a young cabbage plant.**

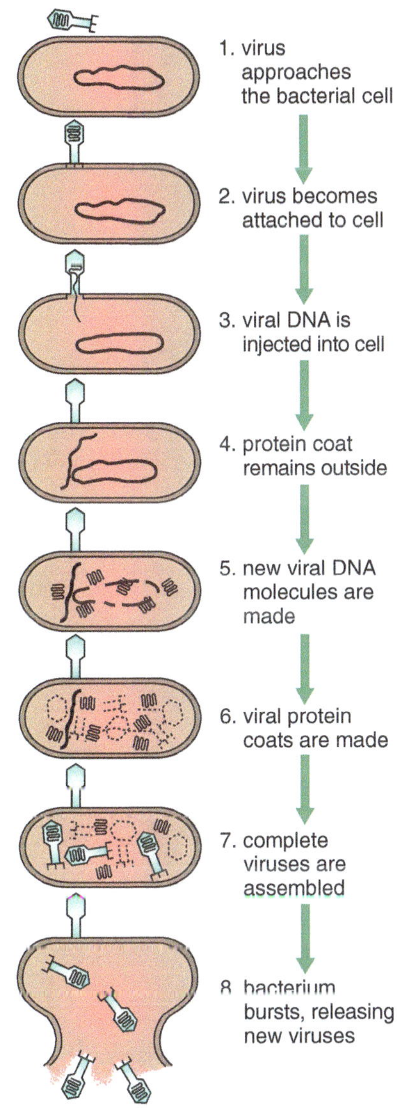

Figure F1.7 **A bacterial virus (called a bacteriophage or simply, a phage) reproducing within a single eubacterium.**

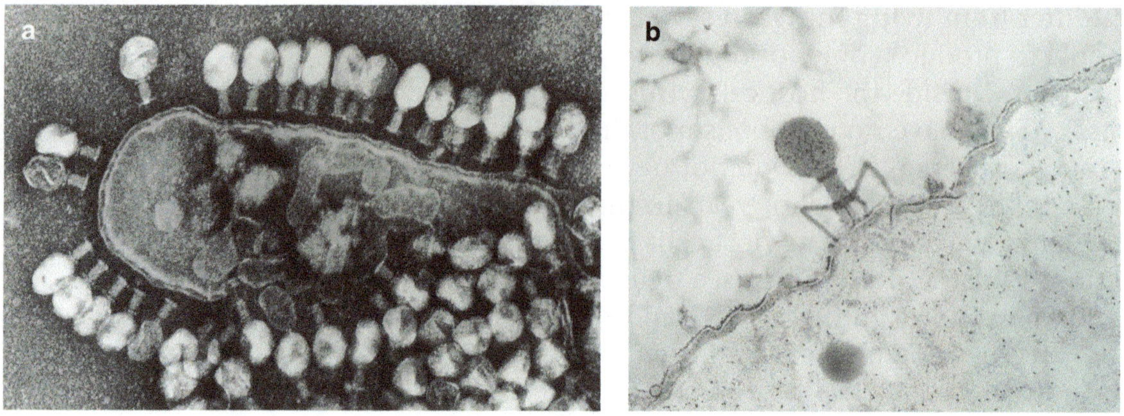

Figure F1.8 The eubacterium *Escherichia coli* (*a*) with T4 phage viruses attached to the surface and inside (50,000×). T4 phage (*b*) attached to the eubacterium after the phage has injected its DNA (200,000×).

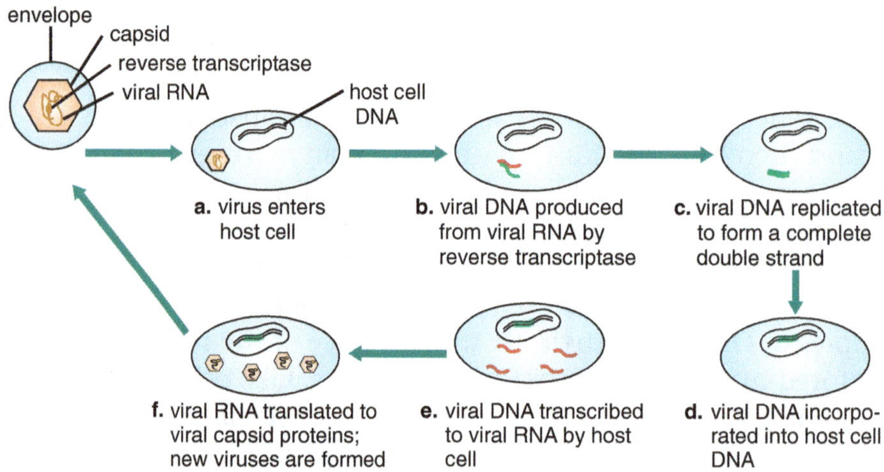

Figure F1.9 Retroviruses are composed of an envelope (from the plasma membrane of a host cell), a protein coat, viral RNA, and the enzyme reverse transcriptase. Retroviruses use the enzyme reverse transcriptase to make a DNA copy of their RNA. The viral DNA then is incorporated into the host cell DNA, and the host cell directs production of new viruses.

Analysis

1. Describe the interaction between host and pathogen that may result in disease.
2. Distinguish between infection and disease.
3. What is immunity? How can a vaccine bring about the development of immunity?
4. Describe, using examples, the major ways by which eubacterial diseases can be spread.
5. What is the role of eubacteria in dental caries?
6. What is the structure of a virus, and how does it infect a host cell?
7. How does a virus direct the formation of other virus particles?
8. How does a retrovirus differ from a virus?

APPENDIX F2

Plants

Several hundred million years ago, the land surface of Earth was barren. Life existed only in the water. One-celled algae colored the surface waters of the oceans shades of green. Some algae joined together into multicellular organisms and evolved a degree of cell specialization. These organisms share common ancestry with today's modern plants.

Figure F2.1 The giant sequoias, *Sequoiadendron giganteum,* are the largest organisms in the world.

This appendix examines several of the characteristics that enabled plants to invade the land and describes a few examples of modern plants. Today, plants adapted to life on land are found in most habitats, from prairies to tundra and in the forests and deserts of the world. They come in the widest possible range of sizes. The largest organisms in the world are the giant redwoods in California, such as those shown in figure F2.1. Some of these trees are more than 100 m tall and 7 m in diameter. From a simple multicellular green alga in the ancient sea to a giant California redwood tree with billions of cells, the differences in structure and function are tremendous. The evolution of such a complex land organism from a relatively unspecialized multicellular ancestor did not occur overnight. It took millions of years.

The Evolution of Land Plants

Two Major Groups of Land Plants Evolved

Most plants that we observe around us today are multicellular, photosynthetic organisms adapted primarily for life on land. The evolution of multicellular green algae that had some division of labor among their cells produced many adaptations that allowed plants to live in this environment. The oldest known plant fossils are simple, branched structures that had several important adaptations.

Because the climate on land may vary from hot to freezing and from wet to dry, water is the major limiting factor for any organism. On land, the only reliable source of water is underground, where it is too dark for photosynthesis. What possible advantages of living on land could overcome such a major disadvantage?

Without the presence of other organisms, plants that were able to tolerate the dry conditions that exist on land would have had little competition for food, minerals, living space, and the other necessities of life. Another advantage would have been the abundance of carbon dioxide and oxygen present in the air. A third advantage for these plants would have been increased light levels, because the air does not absorb as much light as water does. Finally, there would have been plenty of space, and minerals would have been readily available in the soil.

A multicellular alga tossed onto the shore would have had a better chance of surviving than would a one-celled alga. The outer cells of the organism could provide some protection against drying out for the inner cells, and the inner cells might have been efficient at photosynthesis. Other cells of the same organism might have specialized in collecting water or nutrients from the environment. The resulting division of labor would have enabled the alga to exploit the resources in its new environment. A specialized multicellular alga, the *Chara* in figure F2.2, is a modern day example of plants that seem to have made this adjustment to land, living and reproducing on land where ocean spray and tides keep it moist.

There is ample evidence that two plant groups share common ancestry with multicellular green algae. The less complex group, the **bryophytes**, are not aquatic, yet they possess few adaptations for life on land. This group includes the true mosses, hornworts, and liverworts. Bryophytes are called nonvascular plants because they do not have vascular tissue. The more complex group, which includes fossils of the oldest land plants, has many adaptations to life on land, including **vascular tissue**—cells joined into tubes that transport water and minerals throughout the body of the plant. These plants are called vascular plants. Vascular plants likely evolved from a common ancestor shared with bryophytes.

Figure F2.2 The multicellular green alga *Chara*. The fossil record of *Chara*-like algae extends back about 400 million years. Why is *Chara* considered to be a model of the ancestors of land plants?

Figure F2.3 **The evolutionary history of plants.** What evidence was used to construct this history?

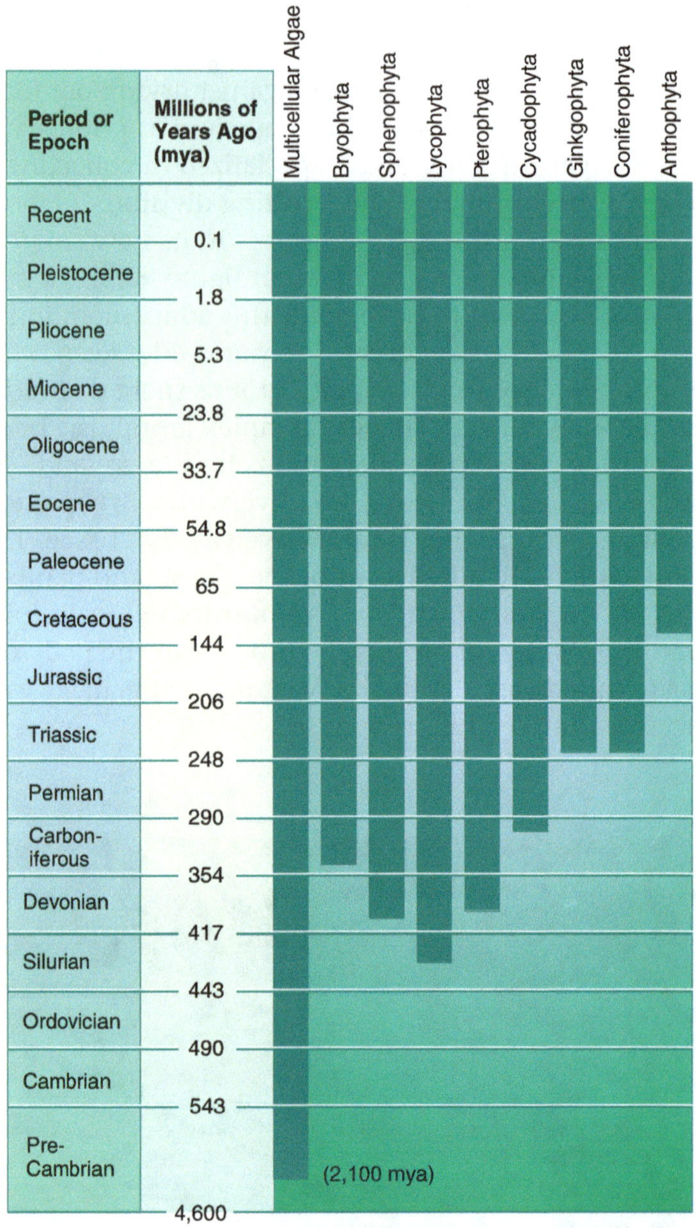

Vascular plants can be divided into two groups: those that produce seeds, and seedless plants such as ferns that reproduce with spores. Vascular plants that produce seeds can be divided further into two groups. One group is those, such as pine trees, that produce naked seeds in cones (the Cycadophyta, Ginkgophyta, and Coniferophyta). The second group is flowering plants (Anthophyta), which produce seeds enclosed in a fruit. Figure F2.3 represents an evolutionary history of plants.

Vascular Plants Have Adaptations That Conserve Water and Permit Gas Exchange

The success of plants on land depends largely on their ability to absorb and hold water. Structures that enable them to do so include roots, vascular tissue, and an outer covering that retards water loss.

Vascular plants such as ferns, conifers (cone-bearing plants), and flowering plants have well-developed root systems that penetrate into many parts of the soil. An extensive root system provides an efficient way to collect water and minerals from the soil and bring them to the main body of the plant. Rooted plants offer a good example of division of labor: The cells in the roots collect water and minerals, and the aboveground cells absorb sunlight and produce food through photosynthesis.

Figure F2.4 Plants such as this stonecrop (*Sedum adolphic*) have very thick cuticles. How are thick cuticles adaptive?

Tall plants require cells that can transport water from the roots to the leaves and cells that can support an upright body. Because vascular tissue serves both these functions, vascular plants can grow taller than bryophytes and thus capture more sunlight. Bryophytes never grow very tall because they lack an efficient water transport system.

In bryophytes, water can evaporate from the entire surface, so a moss plant, for example, dries out quickly. Vascular plants produce a waxy covering, or **cuticle**, that covers the plant body above the ground, reducing the amount of water that can evaporate from its surface. The cuticle is often thick on the leaves of plants living in dry places, such as the stonecrop in figure F2.4. The covering does not prevent gas exchange with the environment, because vascular plants have slitlike openings, or **stomates**, in the surface of their leaves. Stomates (figure F2.5), found in the oldest fossil plants, permit carbon dioxide and oxygen to enter or leave the plant. Roots that absorb water, vascular tissue that supports the plant and conducts water, a cuticle that prevents evaporation, and stomates that permit gas exchange are characteristics that enable plants to live on land.

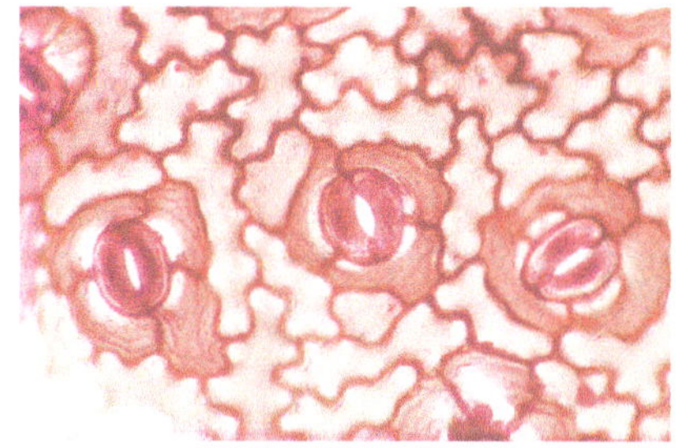

Figure F2.5 Stomates are found on the surface of a leaf.

Bryophytes Require Water for Reproduction

Reproductive adaptations also enabled plants to survive on land. Bryophytes are restricted to moist areas because their sperm are flagellated (have whiplike appendages), as are those of animals and algae. For sexual reproduction to occur, the plants must be bathed in water so the sperm can swim to the egg. Thus, bryophytes reproduce sexually only where water sprays them or after they are wet with dew or rain. Like all sexually reproducing plants, bryophytes have a life cycle in which a haploid (n) phase alternates with a diploid ($2n$) phase. This type of life cycle is called **alternation of generations**.

The carpet of moss shown in figure F2.6a actually is many individual plants. These small plants are haploid, and each is called a **gametophyte**. As the name implies, gametophytes produce gametes, usually in special structures near the tips of the plants. In wet conditions, sperm are produced in male gametophytes. The sperm swim in a film of water to the egg cell in the female gametophyte, where fertilization occurs and a diploid zygote is

Figure F2.6
(*a*) Mosses thrive in moist conditions such as streambanks. (*b*) The whitish structures are sporophytes of the moss *Polytrichium*, growing from the female gametophyte plants.

produced. In some species, both eggs and sperm are produced on the same gametophyte.

The zygote divides by mitosis and develops into a diploid embryo. Eventually, the embryo grows out of the female gametophyte into a stalk-like structure, the **sporophyte**. The diploid sporophyte is visible above the small haploid individuals shown in figure F2.6b. Meiosis occurs within the capsule at the end of the sporophyte, and haploid spores are formed. The capsule helps protect the spores. Once the spores are released, they are carried wherever wind or water transports them. Most spores fall onto unfavorable habitats and die. If a spore reaches a favorable environment, usually a moist soil surface, its wall can burst open. The cell within begins to divide by mitosis, producing long green threads that resemble the filaments of many aquatic algae. The gametophyte moss plant develops from these threads. When a spore germinates and grows into another gametophyte plant, the life cycle diagrammed in figure F2.7 is complete.

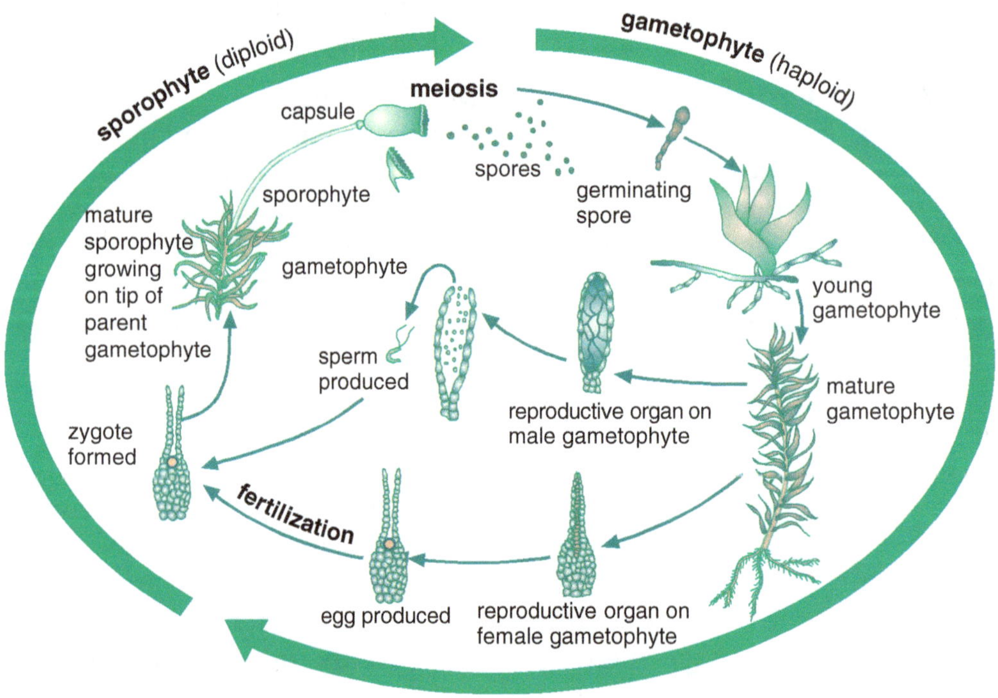

Figure F2.7
Which generation is predominant in the life cycle of a moss?

Flowering Plants Have Special Reproductive Adaptations

Unlike bryophytes and seedless vascular plants, which require water to be present for fertilization, seed plants produce special structures called **pollen grains** (figure F2.8), in which nonflagellated sperm develop. The pollen grains may be blown by wind or may be carried by animals from one plant to another. This efficient means of transferring sperm to the egg under dry conditions is most highly developed in the flowering plants. Brightly colored or scented flowers attract a variety of animals, such as hummingbirds, bats, and insects. Many brightly colored flowers contain nectar that these animals drink and use as an energy source. As the animal drinks, it picks up pollen from one flower and carries it to another flower while searching for more nectar. Because seed plants are not restricted to moist conditions, sexual reproduction can occur whenever the sperm and egg are fully developed. Like bryophytes, flowering plants have a life cycle with alternation of generations, but the gametophyte generation of flowering plants is smaller and protected, at least while it is attached to the sporophyte generation.

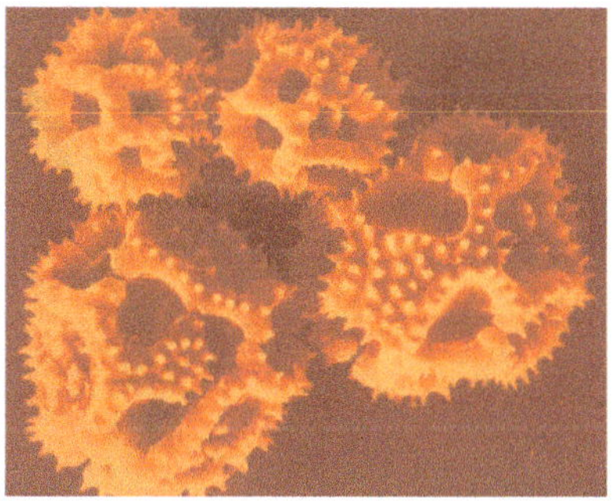

Figure F2.8 Scanning electron micrograph of pollen grains from dandelion (2,500×).

Flowering plants are considered to be the most complex of the vascular plants. Their reproductive structures are found in flowers. A flower is actually a short branch bearing groups of specialized leaves. Some of these leaves may resemble ordinary leaves in many ways, but others are so different in structure that it is hard to think of them as leaves at all. If you closely examine a flower such as the buttercup illustrated in figure F2.9, you will see a number of green, leaflike structures called **sepals** on the underside of the flower. Before the bud opens, the sepals cover and protect the other parts of the flower. The most conspicuous parts of a flower are the colorful **petals.** Although petals are often leaflike in shape, they are not usually green.

Just inside the circle of petals of a typical flower is a ring of male reproductive structures, the **stamens**. In the center of the flower is the **carpel**, the female reproductive organ. Although most plants have both male and female organs within the same flower, a few plants produce flowers with only female parts or only male parts. Stamens usually have an enlarged tip, the anther, whereas the carpel tip (the **stigma**) is more pointed. Despite their shape, however, both stamens and carpels are thought to be modified leaves that have been adapted for reproductive roles.

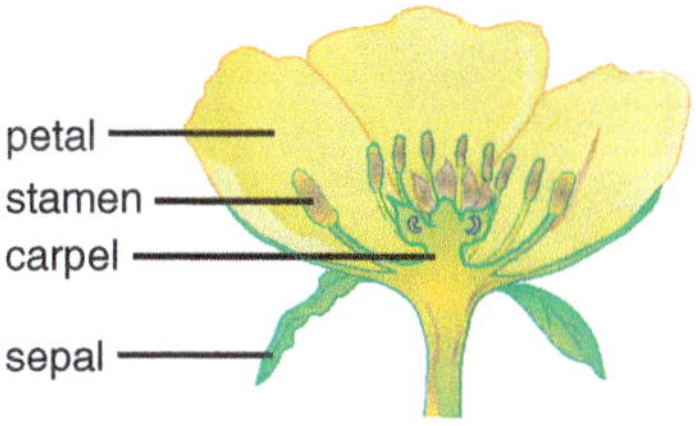

Figure F2.9 The flower structure of a buttercup is shown in a cutaway diagram.

Figure F2.10 shows the relationship between reproductive structures and the life cycle of one type of flowering plant. At the base of the carpel is an enlarged portion, the ovary, that contains one or more small structures called **ovules**. Meiosis occurs in a special cell in each ovule, resulting in the formation of four haploid cells, the female spores. These spores do not separate from the sporophyte as they do in mosses. Instead, three of the spores disintegrate. The fourth spore divides three times by mitosis, forming eight nuclei. The nuclei, with their surrounding cytoplasm, form seven cells, one of which is the egg cell. (One of the cells contains two nuclei, the polar nuclei.) Figure F2.11a shows these seven cells, which constitute the female gametophyte.

Figure F2.10 Which generation predominates in the life cycle of a flowering plant? (The parts are drawn to different scales.)

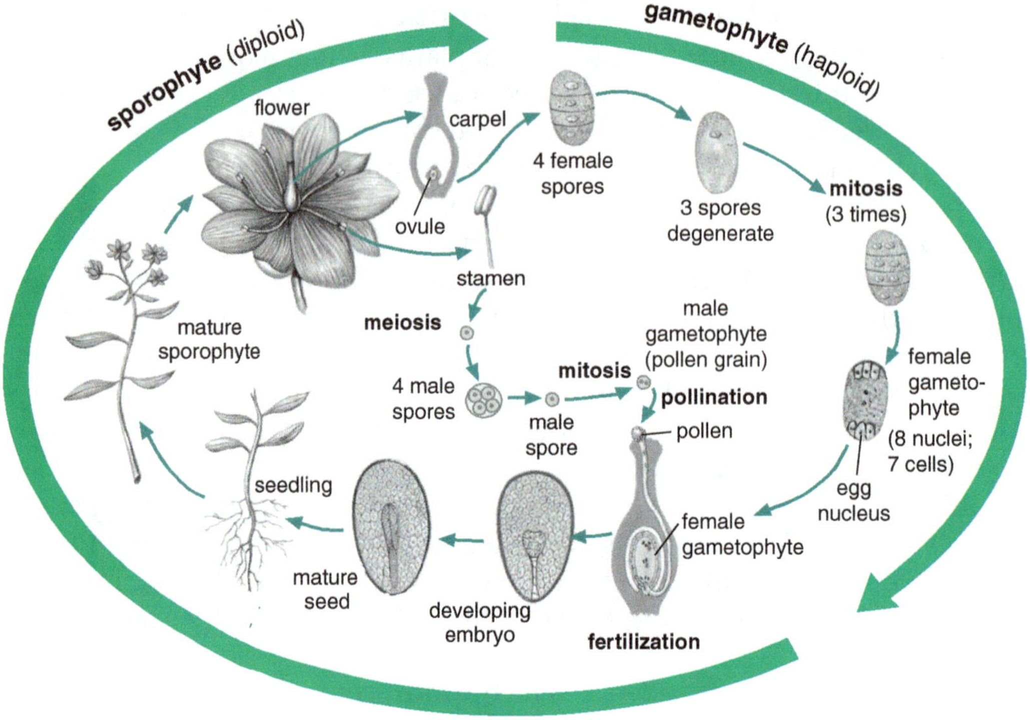

In the stamens, cells in the anthers undergo meiosis, each giving rise to four haploid cells, the male spores. Each spore contains one haploid nucleus that divides by mitosis, forming two nuclei (see figure F2.11b). A spore wall thickens around each nucleus, forming a pollen grain. A single stamen may contain thousands of pollen grains. Each pollen grain is a single-celled male gametophyte containing two nuclei.

Pollination is the transfer of pollen from the stamens to the carpel, either within a flower, between flowers of the same plant, or between plants of the same species. The sticky stigma at the top of the carpel can trap pollen grains carried to it by wind, water, or a visiting animal. Only pollen grains from flowers of the same species are useful in fertilization.

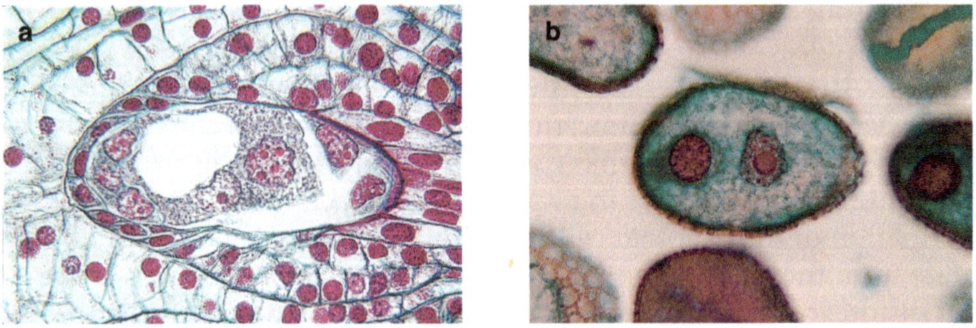

Figure F2.11 These two micrographs show the gametophytes of a lily. (*a*) We see the female gametophyte of a lily. This is a mature ovule ready to be fertilized; the egg nucleus is one of those on the left. (*b*) A pollen grain, the male gametophyte of a seed plant, is shown at the two-nucleate stage.

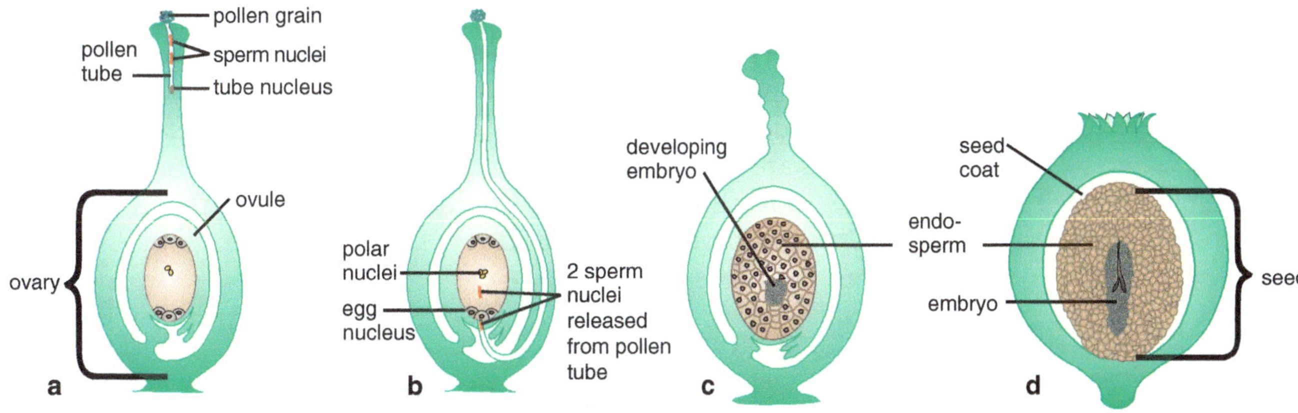

Figure F2.12 The events of pollination (*a*), fertilization (*b*), and development of the embryo (*c*) and seed (*d*) in a flowering plant are drawn here.

The hard wall of the pollen grain protects the haploid cell until the grain lands on a stigma. Once there, a thin finger of tissue, the pollen tube, grows from the grain into the carpel. Within the pollen tube, one nucleus, the tube nucleus, leads the way. The other nucleus divides to form two sperm nuclei. The **pollen tube** grows down the carpel, transporting the sperm nuclei to the ovule. Fertilization occurs when one sperm nucleus unites with the egg, forming a zygote. This diploid cell gives rise to the embryo. The other sperm nucleus unites with the polar nuclei, leading to the formation of **endosperm**, a mass of food-storing cells that will later nourish the developing embryo. The endosperm is triploid; it has three sets of chromosomes: one set from the sperm and two from the polar nuclei. Figure F2.12 summarizes the stages of fertilization and embryo development in a flowering plant.

The life cycle of a flowering plant is similar to that of a moss in two ways. First, meiosis occurs just before spore formation. Second, there is alternation of generations between the sporophyte and gametophyte portions of the life cycle. There are, however, several differences. First, abundant surface water is not necessary for fertilization in flowering plants. Second, the gametophytes are smaller than the sporophytes. Third, the gametophyte that produces an egg and the spore that produces the female gametophyte do not separate from the sporophyte plant. Thus, these structures are better protected from the environment than are their counterparts in an embryonic moss. Finally, the embryo (young porophyte) grows for a short time, then becomes dormant. The embryo and its endosperm are surrounded by a protective coat formed from ovule tissues. This package is a **seed**, diagrammed in figure F2.13. The seed protects the young sporophyte, which remains dormant until environmental conditions are suitable for germination. Moss embryos, on the other hand, cannot tolerate dry conditions. Also, moss spores contain no embryos and only a small amount of food.

Figure F2.13 A bean seed, cut in half, shows the embryo and endosperm. The endosperm of the bean seed, as shown here, is found in structures known as the seed leaves.

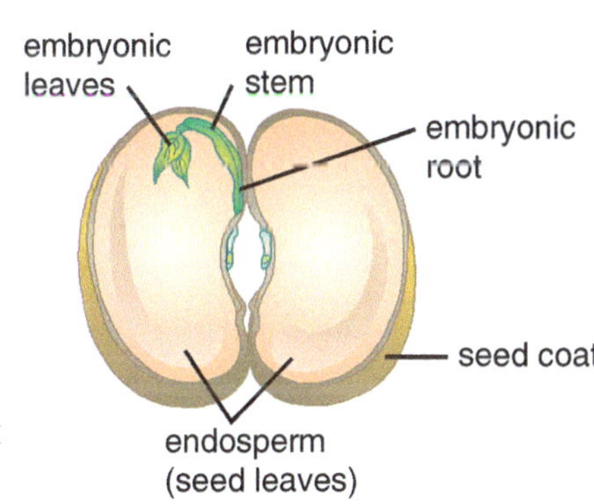

Land plants show a trend toward increasing specialization of the sporophyte and decreasing specialization of the gametophyte, as shown in figure F2.14. Seed plants are more abundant and more diverse than mosses because of their adaptations to the changing environment on land. Flowering plants are well suited to life on land because of the cuticle, vascular tissue, root systems, sperm carried in pollen, spores protected on sporophytes, and embryos protected in seeds.

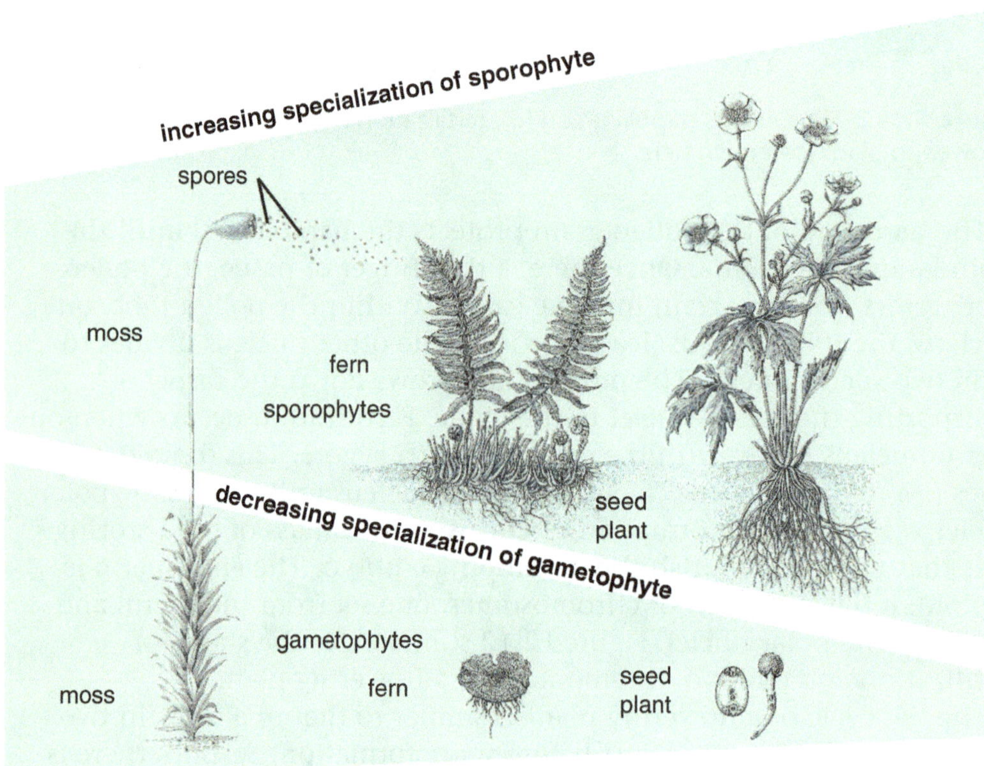

Figure F2.14 The differences between mosses and land plants can be shown by diagramming the evolution and specialization in sporophytes and gametophytes.

Analysis

1. Describe the limiting factors and advantages of life on land for the first land plants.
2. What types of adaptations were necessary for plants to live on land, and why were they important?
3. Briefly distinguish between the different types of land plants.
4. Compare and contrast gametophytes and sporophytes.
5. Compare and contrast the life cycles of bryophytes and flowering plants.
6. Explain the roles of the reproductive structures of flowering plants.
7. Describe the origin of the embryo, the endosperm, and the seed.

Bryophytes and Seedless Vascular Plants

Bryophytes Have No Roots, Stems, or Leaves

Most bryophytes (division *Bryophyta*) are relatively small. Few of them exceed 20 cm in height. Although they have structures resembling stems and leaves, these terms are not used in describing bryophytes because they lack the vascular tissue of other land plants. There are three classes of bryophytes: true mosses, the largest class; liverworts; and hornworts. Figure F2.15 shows several bryophytes.

Figure F2.15 These are examples of varieties of bryophytes: (*a*) moss, *Polytrichium;* (*b*) liverwort, *Marchantia polymorpha;* (*c*) hornwort, *Anthoceros punctatus.* How do bryophytes and seedless vascular plants differ in structure and habitat?

True mosses often grow in clumps or small clusters in rock crevices and on the shady side of trees. An individual moss plant from such a clump is simply an upright, green, stem-like stalk with threadlike structures called **rhizoids**, which perform the function of roots by aiding in absorption and helping to hold the plant in place. Many flat, green, leaf-like structures are attached spirally along the stalk. Water and nutrients are absorbed throughout the body of the bryophyte, so most grow in fairly damp places, and a few grow in water. Under dry conditions, many mosses become dormant. When dormant, the life processes slow down, and the plant appears dead. Normal activities resume when the plant comes in contact with water. Because many bryophytes can photosynthesize in limited light, they often are found on the ground in forest ecosystems, where other plants cannot grow.

One important group of mosses is found in boggy places in the cold and temperate parts of the world. These plants, from the genus *Sphagnum,* form peat bogs—small lakes and ponds completely filled with living and dead mosses. Sphagnum produces a very acidic condition in the water that keeps decomposers such as bacteria and fungi from growing, thus allowing these plants to build up through time, layer on layer. People in Ireland and other countries cut blocks of peat, dry it, and use it for fuel or to build small enclosures. Dry peat absorbs water quickly and holds the water well, characteristics that make peat attractive to gardeners, who add it to their soil to lower the pH and to increase the water-holding capacity.

Liverworts and hornworts grow in very moist areas, such as on the banks of streams where water spray keeps the soil wet. Bryophytes are less complex than vascular plants and are likely the ancestors of vascular plants. The fossil record for bryophytes is small. The earliest bryophyte fossils are only about 350 million years old, but genetic reconstructions of the evolution of plants would predict that older fossils should exist. The first fossils of vascular plants appear 50–100 million years earlier than bryophytes in the fossil record.

Club Mosses and Horsetails Are Seedless Vascular Plants

The roots, stems, and leaves of vascular plants contain the vascular system through which water, sugar, and dissolved nutrients move from one place to another in the plant.

Rhynia major, shown in figure F2.16, best represents the oldest land plants. *Rhynia* had an underground stem that probably anchored the plant and absorbed water. From this underground stem grew upright branched stems that had stomates. At the tips of the stems were sporangia, which split open to release thick-walled spores. A living relative of *Rhynia* is the whisk fern, *Psilotum*, shown in figure F2.17.

The club mosses (division *Lycophyta*) are evergreen plants that seldom grow more than 40 cm tall. Although the word "moss" is a part of their name, they are not true mosses. Their branching, horizontal stems grow on the surface of the soil or just below it. The most noticeable part of a club moss plant is an upright branch growing from one of these stems. Club mosses reproduce by spores, which are produced on modified, specialized leaves. In many species, these leaves form club-shaped cones at the tips of short, upright stems. The name "club moss" is derived from this feature. Club mosses are rather common in the eastern and northwestern United States and often are used to make Christmas wreaths. They rarely grow in the dry states of the Southwest.

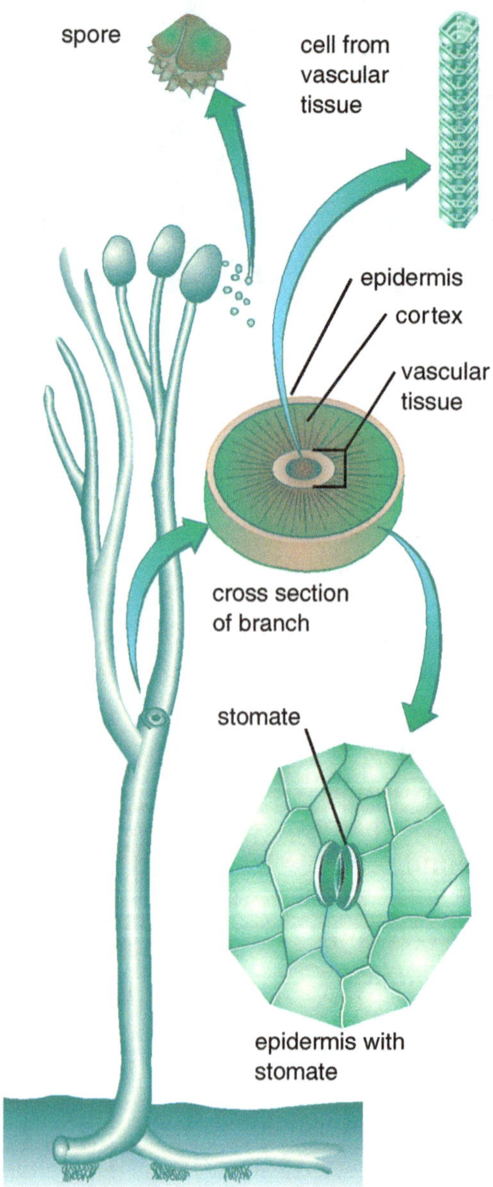

Figure F2.16 The extinct plant *Rhynia* is reconstructed here. It had no leaves or roots and stood about 30 cm high. Sporangia were produced at the tips of the upper branches. The presence of stomates indicates that photosynthesis occurred in the branches. The cells of vascular tissue (upper right) were oblong and sometimes hollow. Thick-walled spores were produced in fours in the sporangia, an indication that the spores were produced by meiosis.

Figure F2.17 Note the sporangia along the stems of the whisk fern, *Psilotum*.

Horsetails (division *Sphenophyta*) have hollow, jointed, upright branches that grow from horizontal underground stems. Their small, scalelike leaves grow in a circle around each stem joint. Spores are produced in conelike structures at the tips of some of the upright branches, such as those shown in figure F2.18. In middle latitudes, horsetails rarely reach a height of 2 m, but in the American tropics one species may grow several meters tall. They are found in moist places, such as along streams. Horsetails are harsh to the touch; their tissues contain silica, a compound present in sand. Because Native Americans and the pioneers scrubbed pots and pans with them, they are commonly called scouring rushes.

Relatives of the club mosses and horsetails can be traced back about 430 million years. During the Coal Age, about 300 million years ago, great parts

Figure F2.18 (*a*) Spores of *Equisetum* (horsetail) are produced in these conelike structures. The vegetative stage of *Equisetum* (*b*) is characterized by hollow, jointed, upright branches.

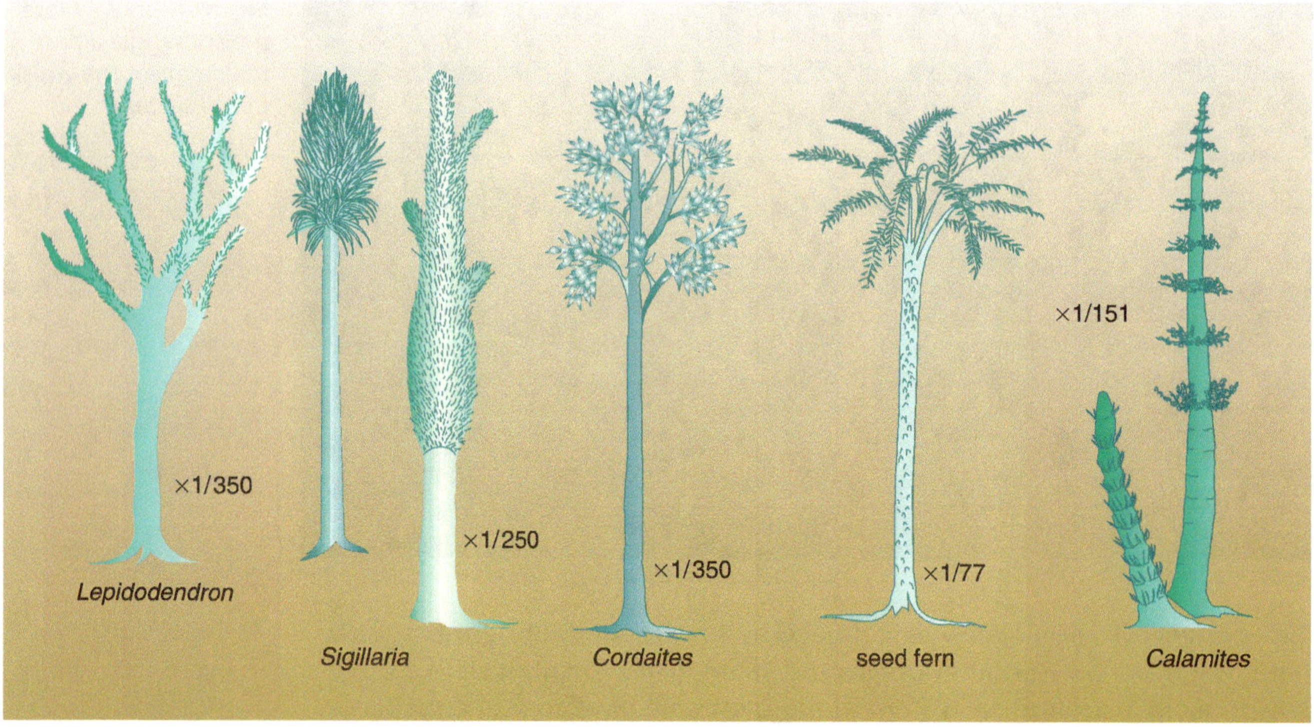

Figure F2.19 These are some examples of the trees of the Coal Age forests: *Lepidodendron* and *Sigillaria* were club mosses. *Cordaites* were primitive cone-bearing plants. The seed ferns have no living species. *Calamites* were horsetails.

of North America were covered by shallow swamps and seas. The warm and wet environment allowed plants to grow year-round. Under these conditions, giant relatives of today's club mosses, horsetails, and ferns, as well as seed-producing plants, covered the land. Some of these plants were more than 20 m tall (figure F2.19). As they died, their large stems were covered with mud and soil before they completely decayed. A tremendous number of plants from the Coal Age were compressed over long periods of time and under high temperature and great pressure. Eventually, they became fossil fuels, mainly coal and some natural gas.

Fern Leaves Grow from Underground Stems

Ferns (division *Pterophyta*), like the club mosses and horsetails, reproduce by spores. At certain times of the year, small brown spots develop on the undersides of fern leaves, as shown in figure F2.20. Each spot consists of a cluster of sporangia. Each sporangium produces a large number of spores, which are almost microscopic in size.

Figure F2.20 The dusty-looking spots on the underside of a Christmas fern leaf, *Polystichum acrostichoides*, are sporangia.

When a sporangium is mature, it opens and the spores are thrown out into the air. Spores are very light and can be carried for

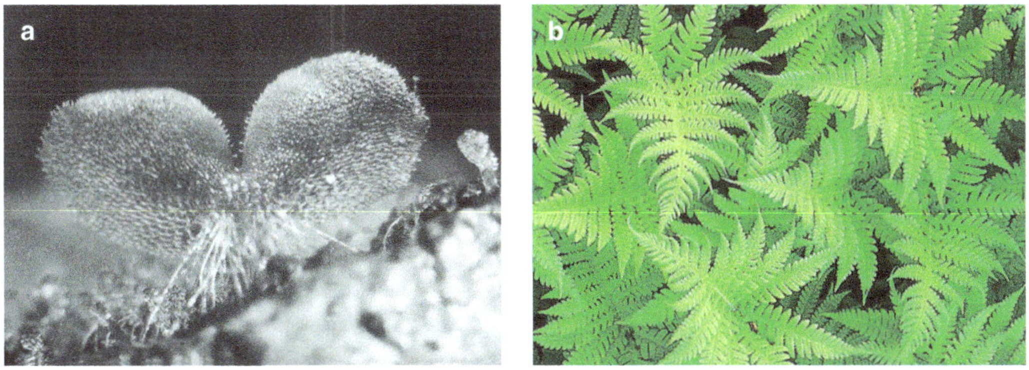

Figure F2.21 You may never have seen a fern gametophyte (*a*), but you are probably familiar with the fern sporophyte (*b*).

incredible distances by wind. If a spore falls in a suitably moist place, it germinates and develops rapidly into a thin, green, heart-shaped plant that is rarely more than 1 cm in diameter (see figure F2.21a). This small gametophyte plant, which is completely different from the familiar fern with its large leaves, is seldom noticed in the woods. The gametophyte produces the sperm or eggs, or in some cases, both. Like the moss sperm, the flagellated fern sperm must swim in a film of water to fertilize the egg. The zygote produced by that fertilization eventually grows into the conspicuous spore-bearing fern plant that most people recognize. This large plant is the sporophyte generation of the fern, seen in figure F2.21b.

The ferns native to most of the United States are shade-dwelling plants with underground roots and stems. From these stems, roots grow downward and new sets of upright leaves appear above the ground each spring. In Hawaii and elsewhere in the tropics, many species of ferns have stems that grow upright. These tree ferns may reach a height of 20 m with leaves 5 m long, as figure F2.22 illustrates. Most fern species are found in the tropics, but many can be found in forest ecosystems around the world.

Figure F2.22 Tree ferns grow on the island of Sumatra.

Analysis

1. How do bryophytes differ from vascular plants?
2. How does reproduction in seedless vascular plants differ from that in bryophytes?
3. What events must occur for a fern sporophyte to be produced from a gametophyte?
4. In what type of environment must ferns and horsetails grow in order to reproduce?

Pioneers

ENGLISH SPANISH

Evolutionary Geologist

Dr. Geerat Vermeij (pronounced *ver may*) is a paleontologist and a professor at the University of California, Davis. As a paleontologist, he studies life-forms from the ancient past by examining fossils. Dr. Vermeij specializes in the natural history of molluscs and their relationships with ancient predators. According to his research, molluscs have adapted to the presence of ever-more-efficient predators by developing progressively heavier and more protective shells. Dr. Vermeij feels that this reaction to natural enemies provides a better explanation of how molluscs evolved than the usual explanations based on climatic changes and other large-scale physical factors.

Dr. Vermeij was born in the Netherlands and raised in New Jersey. His proximity to the Atlantic seaboard allowed him to pursue his early interest in shell collecting. He has developed this interest through worldwide travel and extensive field study. He has traveled widely in the South Pacific, Central and North America, and Africa. His research on predatory patterns on molluscs has involved investigating shell geometry, breakage patterns, scarring, hole patterns, and evidence of repair following encounters with their marine enemies.

A graduate of Princeton and Yale Universities, Dr. Vermeij holds a particular distinction among paleontologists. He has been completely blind since age three and has never actually seen a single organism he has studied. Instead, he uses his fingers to feel the subtle features of the ancient marine creatures he investigates. His sense of touch is so refined that he often is able to identify molluscs to the subspecies based on the most minor changes in the shapes, textures, and thickness of shells. Dr. Vermeij works closely with his wife and professional colleague, Dr. Edith Zipser, also of the University of California, Davis. Together, they have collected shells from their extensive travels and formulated their ideas on the evolutionary patterns of molluscs. Dr. Vermeij has risen above his physical restrictions and opposition from skeptical peers to be a premier-level paleontologist.

Figure F2.23 These are examples of conifers and related plants: (*a*) ponderosa pine, *Pinus ponderosa*; (*b*) cycad, *Dioon edule*; and (*c*) maidenhair tree, *Ginkgo biloba.* Seeds of these plants are not enclosed in tissues as are those of flowering plants. How are seed plants adapted to life on land?

Seed Plants

Many Conifers Are Evergreens

Humans have used plants for food, clothing, shelter, and medicines. The conifers (division *Coniferophyta*) are woody vascular plants with seeds borne in cones. Conifers provide most of the paper pulp and much of the lumber used in home construction and furniture. They include pines, firs, spruces, and junipers, among many other species. Figure F2.23 shows several conifers and related plants.

Almost all conifers are trees or shrubs, and all are at least somewhat woody. Many have leaves that are like needles or scales, such as those in figure F2.24, and most of these plants are evergreen. An evergreen tree or shrub appears green throughout the year because it always maintains most of its leaves. A few leaves die at different times of the year and drop to the ground. Although the number of conifer species is small, the number of individual conifers is enormous.

Figure F2.24 Many conifer leaves are needlelike or scalelike. (*a*) The needles of Douglas fir, *Pseudotsuga menziesii*, are single and needlelike, while the leaves of juniper, *Juniperus chinensis pfitzerii*, (*b*) are scalelike.

Many common conifers are well adapted to life in dry habitats. For example, although pine trees may grow where there is much snow, the snow is really frozen water and is not available for growth. In the spring, much of the snow evaporates, and the melted snow may run off into streams before it soaks into the soil. The leaves of pines are well adapted for growth in dry places. The long, narrow needles reduce the amount of water lost by evaporation. In addition, a pine needle often is covered by a thick, waxy cuticle that further reduces water loss.

Conifers reproduce by seeds that are attached to the upper surface of cone scales. A seed developing in a cone may be protected by the scales. If two scales are separated slightly, you can see the seed between them, as shown in figure F2.25. Thus, the seeds in cones are not completely covered as they are in the fruits of flowering plants.

Conifer spores are of two types and are produced in different cones, illustrated in figure F2.26. Typically, pollen develops from spores in the

Figure F2.25 The winged seed of this pine tree is tucked inside the scales of the pinecone.

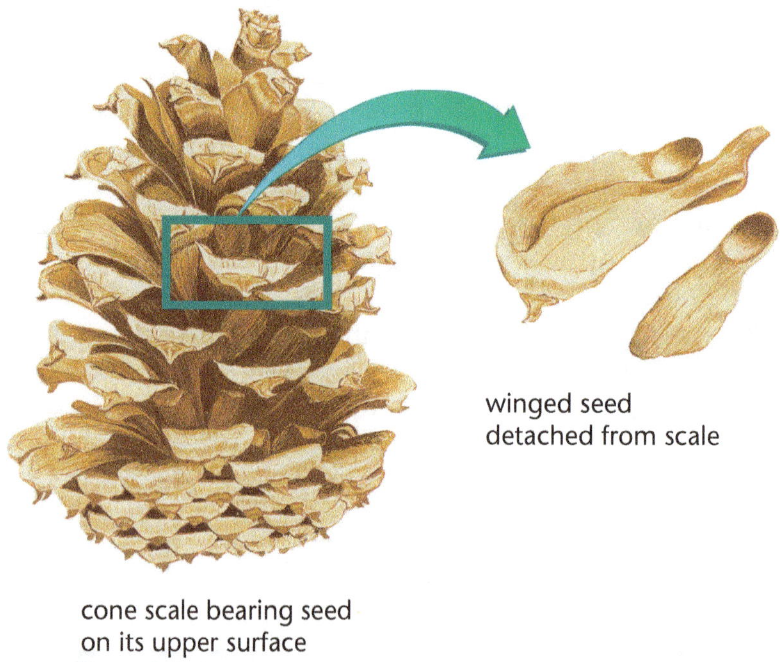

Figure F2.26 These photographs show the difference between the male cone (*a*) and the female cone (*b*) of the piñon pine, *Pinus edulis.*

Table F2.1
Seed Development in a Pine.

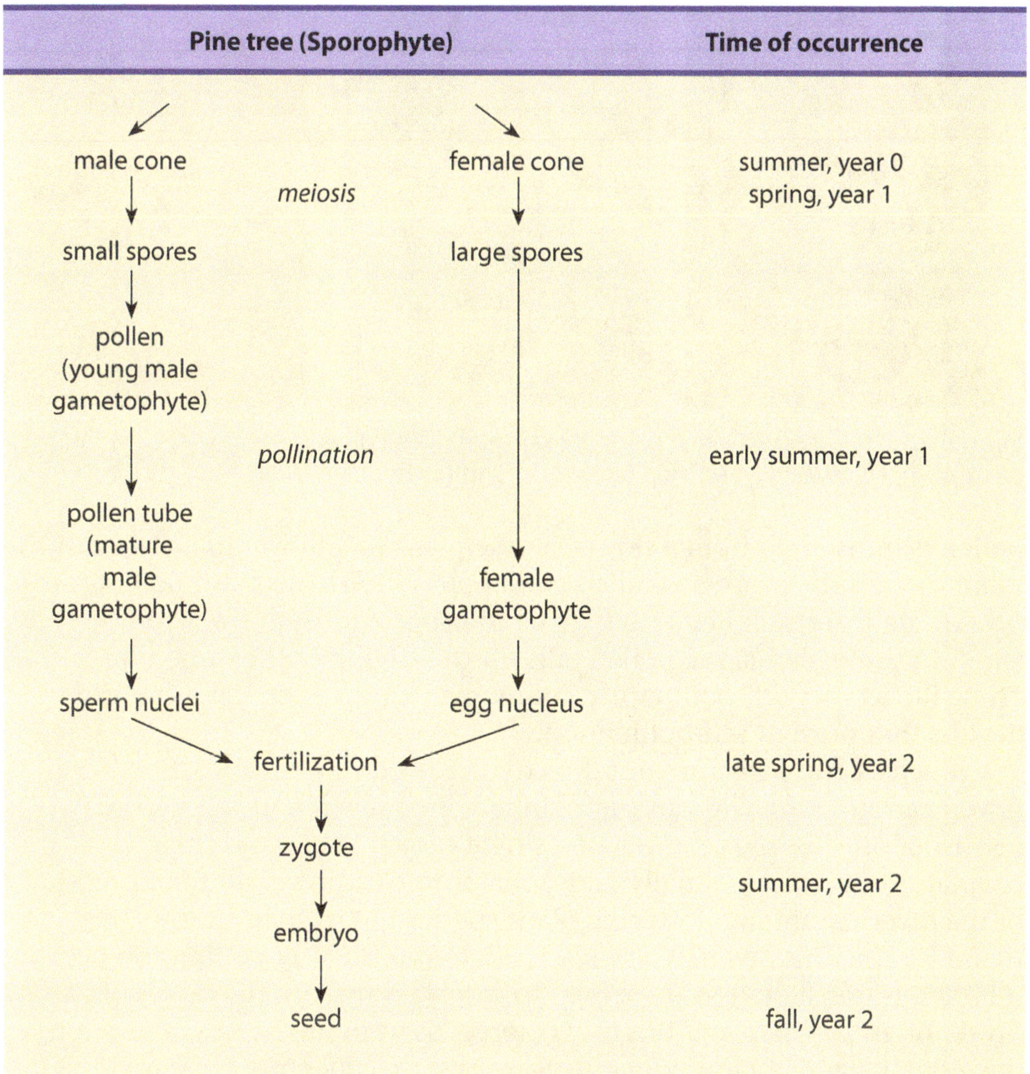

Source: Weier, Stocking, Barbour, Rost, eds., *Botany: An Introduction to Plant Biology*, 6th ed. (New York, NY: John Wiley & Sons), copyright © 1982. Reprinted by permission of John Wiley & Sons, Inc.

small male cone, and pollination occurs in the spring when the pollen is blown onto a female cone. The larger, more familiar female cones contain the ovules. Pollen grains, the male gametophytes, lodge in a sticky substance secreted by the ovule. Once lodged, the grains develop pollen tubes within which the sperm are formed. Within the ovule, the female gametophytes develop and produce eggs. Fertilization occurs approximately a year after pollination, and the seed requires an additional year to mature. Table F2.1 summarizes these events.

Many Flowering Plants Have Special Pollinators

Flowers are the distinguishing feature of the most successful division in the plant kingdom, the *Anthophyta*. Although we commonly appreciate flowers for their beauty, their major role is in reproduction. When a flower opens, it reveals the reproductive structures, the stamens and carpels. Insects, birds, bats, or other animals that visit the flower may pick up pollen from the anther and carry it to the next flower they visit. When

Figure F2.27 Here are a few colorful examples of the diversity of flowers. (*a*) Indian paintbrush; (*b*) water lily; (*c*) elephantheads; and (*d*) bougainvillea.

pollen is transferred from a flower of one plant to a flower of another in the same species, the process is called cross-pollination. The main pollinators of flowers are insects. Pollen also may be transported from flower to flower by wind. In some plants, the pollen falls on the stigma of the same flower. These flowers are self-pollinating. Many plants, however, have evolved devices that prevent self-pollination.

The sepals and petals are not directly involved in seed formation, so a flower can function without them. In fact, in a few plants a flower may consist of only a single stamen or a single carpel. Petals and their adaptations, however, usually play a major role in flower pollination. Much of the diversity among flowering plant species lies in their flowers (see figure F2.27). This diversity usually is related to the way pollination occurs. If pollen is transferred from stamen to carpel by insects, the petals of the flower often are large and brightly colored, as in figure F2.28a, and often have small glands that produce a sugar solution called **nectar**. These adaptations attract pollinating insects. On the other hand, flowers in which

Figure F2.28 Insect-pollinated flowers such as the aster in (*a*) usually are brightly colored. Windpollinated flowers often lack petals and sepals and produce an abundance of pollen. Shown in (*b*) is the male catkin of a willow, *Salix*; note the many stamens.

pollen is transferred by wind usually have small sepals and petals, or none at all. These flowers often are located high on the plant, where they are accessible to wind currents and produce an abundance of pollen (figure F2.28b). Their carpels commonly have large, long, or feathery structures at the tips, which are covered with a sticky fluid. These adaptations increase the likelihood that some pollen will stick to the carpels.

The great variety of flowers has come about, in part, by the coevolution of flowers and pollination agents. For example, hummingbirds have evolved with plants that produce large quantities of nectar. Research has shown that while hummingbirds do not have a good sense of smell, they can see the color red very well. Flowers pollinated by hummingbirds usually are well adapted to their pollinators, as you see in figure F2.29. The columbine flowers shown in the illustration are red, have little or no scent, and produce copious amounts of nectar. The nectar is found at the bottom of a long tube formed by the red petals, a shape that makes it difficult for other organisms to rob the flower of nectar.

Figure F2.29 The columbine and hummingbird are uniquely adapted to each other.

Hummingbirds, however, have long beaks that can probe the flower and reach the nectar. The stamens stick out in such a position that the hummingbird's head is dusted with pollen when it visits these flowers. When the hummingbird flies to another flower of the same species, the tip of the carpel is in a perfect position to have pollen from the hummingbird's head scraped onto it. Thus, the flower is pollinated while the hummingbird drinks nectar. Interactions of this type, in which both organisms become uniquely adapted to each other, have helped shape the way flowering plants have evolved.

Flowering Plants Produce Fruits with Seeds

After pollination and fertilization, seeds begin to develop. The carpel (or carpels in a compound ovary), often with other parts of the flower, develops into a protective fruit around the seed, as shown in figure F2.30. There may be many seeds in a fruit. Each seed began its development when an egg cell in one ovule was fertilized by a sperm from one pollen grain. In flowering plants, then, an embryo is protected within a seed, and seeds are protected within a fruit.

Part of the embryo in the seed consists of one or two modified leaves called **cotyledons**. Another part is a beginning of a root. Each seed also contains a supply of food that is used when the embryo starts to grow. The food may be stored in the endosperm, or it may be stored in the embryo itself, usually in the cotyledon (look back at figure F2.13). Beans and peas are

Figure F2.30 These stages occur as tomato fruit develops from flowers.

examples of seeds. They are enclosed in protective pods, which are just one of many types of fruits. Each bean or pea contains a small embryo and a supply of stored food for its early development. The entire bean or pea can act as food for humans. Beans and peas can be germinated easily, and each embryo gives rise to a plant that, in turn, gives rise to new flowers and fruits. Apples and oranges are examples of fruit that contain a number of seeds. Under natural conditions, these fruits eventually decay, or are eaten by animals leaving their seeds behind to germinate and give rise to the next generation.

Fruits show as much diversity as flowers, as indicated in figure F2.31. This diversity is related to the method of dispersal of the fruits and their seeds, that is, how they are scattered from the parent plant. In Investigation F2.3, *Fruits and Seeds*, you will observe some of the structures that aid in seed dispersal. In many cases, part of the carpel becomes thick and fleshy, as in the fruits of the peach, plum, and tomato. Fleshy fruits, often red in color, may be eaten by birds or mammals. The seeds in many such fruits have thick coats that permit the seeds to pass through an animal's digestive system unharmed. The seeds are dropped later at some distance from the parent plant. Many fruits are not fleshy but have other adaptations that aid in scattering their seeds. These fruits may have spines that catch on the fur of an animal that brushes up against the plant. The fruit is carried from the plant and later falls off or is brushed off by the animal. Many fruits and seeds are lightweight and have special winglike projections that help them to be carried away from the plant by wind. An entire plant, such as the tumbleweed, can be broken off near the ground and blown about by the wind. As the tumbleweed bounces about, it drops its fruits (and the seeds within) all along its path.

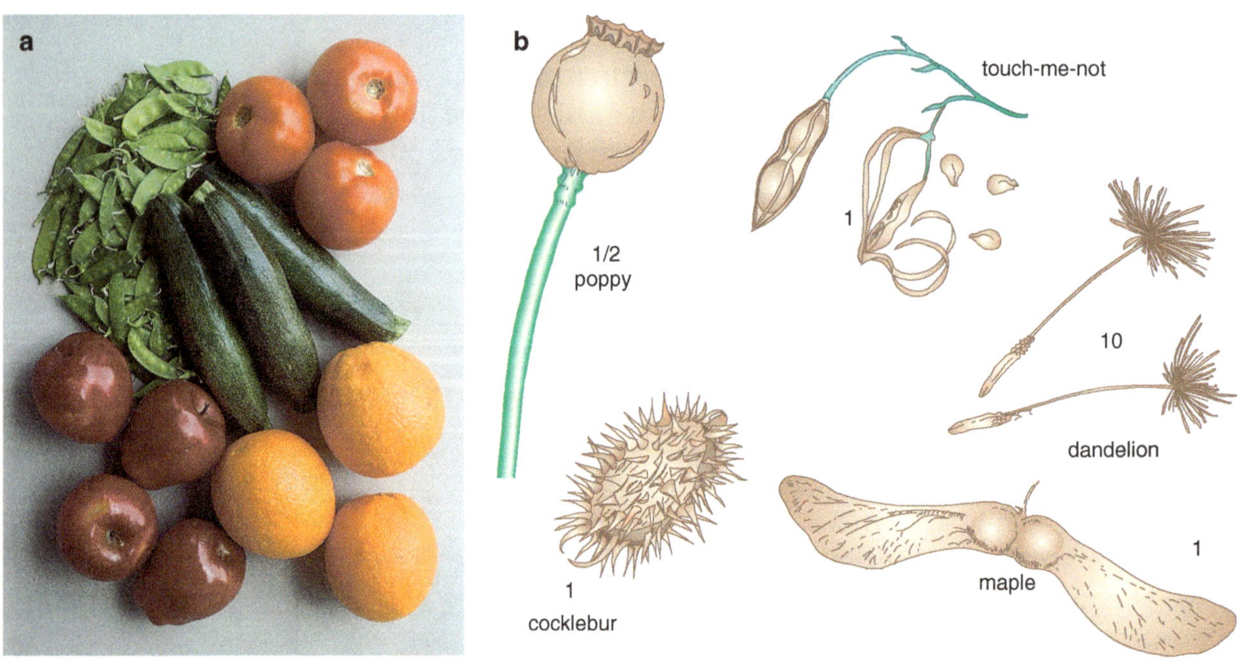

Figure F2.31 (*a*) Each edible fruit is a mature ovary containing the seeds of the plant. (*b*) Many fruits have structural adaptations for seed dispersal. Can you describe them?

Flowering Plants Are the Most Diverse Group of Land Plants

There is great diversity in the size of flowering plants and in the life span of their shoots—the parts that appear above the ground. Many flowering plants are trees. A tree bears leaves well above the ground where they are likely to receive more light than do the leaves of shorter plants. Because of their size, trees can store large reserves of food in trunks and roots and can survive through bad years. Trees have relatively long life spans. A tree species may survive even if a year's seed crop is destroyed. Most species of flowering plants are not trees, however. Some, such as roses and raspberries, are woody shrubs. Others, such as ivy, grapes, and hundreds of tropical species, are woody vines that grow on rocks, walls, or other plants. Most, however, are neither trees, shrubs, nor vines. Instead, they are nonwoody, or **herbaceous**, plants such as those in figure F2.32.

Figure F2.32 Herbaceous plants include marigolds, common garden flowers.

Flowering plants are divided into two large classes—the **monocots** and the **dicots**. In monocots (monocotyledons), the embryo contains a single cotyledon. The monocots include grasses and grain-producing plants such as wheat, rice, and corn—the chief food plants of the world. The pasture grasses that feed cattle, another source of human food, are also monocots. Evidence suggests that the human population could not have evolved to its present state without monocots.

The seeds of the dicots (dicotyledons) have two seed leaves. This class is larger than the monocot class. Most fruits and vegetables, such as carrots, lettuce, apples, and grapes, are dicots. In addition, the so-called hardwoods used in furniture, flooring, hockey sticks, and baseball bats come from dicot trees. Almost all shade trees are dicots, also. Figure F2.33 illustrates the major differences between monocots and dicots.

Analysis

1. How are conifers adapted to dry conditions?
2. How do conifers differ from flowering plants?
3. What role does flower color play in the reproduction of flowering plants?
4. How are fruits related to seed dispersal?
5. In what ways do monocots and dicots differ?

Figure F2.33 Monocots and dicots have distinct characteristics.

Biology Today

Biological Illustrator

Marjorie Leggit is a freelance biological illustrator. Biological illustrators illustrate reports, journal articles, and books prepared by researchers. Many illustrators work in natural history museums, painting and constructing the background scenery for different types of displays. These artists may recreate Alaskan ice floes to show off a group of polar bears or a South American hillside for a display of llamas. Marjorie's family is composed of writers, sculptors, painters, and woodcarvers, so an interest in art comes naturally to her. Her interest in science, Marj thinks, may have developed because she could do so much drawing in science classes.

Although a biological illustration course in college sparked her interest, Marjorie continued to study fine arts. After studying for a year in Europe, however, she realized she missed science and the technical art that goes with it. She returned to college and designed her own program of independent studies that led to a degree in scientific art. Her program included an internship at the Denver Museum of Natural History in Denver, Colorado. While searching for a full-time job after graduation, Marj illustrated a professor's botanical field guide. Her persistence was rewarded eventually with a position at the Field Museum of Natural History in Chicago, Illinois. Marj later returned to Denver.

Reestablishing her career there proved to be difficult because opportunities for biological illustration were far fewer than in very large cities where more research is conducted. For seven years, she worked at art-related jobs, including geological drafting and computer-aided graphics. Although she learned a whole new field of art, Marj was not satisfied with the direction of her career.

To keep in touch with the field of biological illustration, she contacted the Guild of Natural Science Illustrators (GNSI) in Washington, DC. Knowing Colorado had nothing similar, she founded the Colorado Chapter of GNSI and made friends with a group of local illustrators.

In 1986, Marj went into business for herself as a biological illustrator and graphic artist. She contacted old and new local clients as well as publishers throughout the country. She spent many hours researching the possibilities for assignments and marketing her skills. Marj did many of the illustrations used in this textbook, including figure F2.33. To ensure accuracy in her work, Marj does a great deal of research and uses many references and photos. She consults extensively with authors when planning an illustration. It is essential that she understand the author's intent, what the illustration is meant to portray, and how it relates to the text and to other illustrations. Proper interpretation depends on good communication between author and artist. Marj strongly believes that the hard work involved in creating illustrations has its rewards. At times, the hours are long, the deadlines close to impossible, and the challenges seemingly insurmountable. However, she is doing exactly what she wants to be doing: full-time biological illustration.

Summary

Plants share common ancestry with multicellular algae. A multicellular organism has many advantages over a single cell, including size, division of labor, and ability to conserve water. Other adaptations that permit plants to absorb and hold water are roots, vascular tissue, and the cuticle. All plants have a life cycle that alternates between two different generations, a gametophyte and a sporophyte. In flowering plants, however, the egg cell (the gametophyte that produces the egg), and the spore that produces the gametophyte all remain protected in the flower. Mosses are nonvascular plants that lack roots, stems, and leaves. They possess some adaptations to life on land, but they still require a moist environment. Club mosses, horsetails, and ferns are seedless vascular plants, and many of their ancestors produced today's fossil fuels. Conifers and flowering plants are seed-producing vascular plants, but flowering plants have a fruit that encloses their seeds. The great diversity of flowers and fruits is the result of the coevolution of plants and their agents of pollination or dispersal.

Applications

1. Which organisms in this appendix would you consider to be less complex? Which would you consider to be more complex? Explain your answer.
2. Why are nonvascular plants short?
3. The fossil record reveals that the diversity of flowering plants and the diversity of certain insect groups increased at the same time. How might you explain this?
4. Explain the following statement: The plant parts that furnish the greatest amount of human food are either seeds, roots, or underground stems.

Problems

1. Desert plants are land plants adapted to extremely dry conditions. What adaptations allow these plants to live with little water? Make a collection of desert plants or find out about them from a plant ecology text.
2. Fruits come in many sizes and shapes and with diverse appendages. Collect different fruits and try to identify their plants. Identify the methods of dispersal for the different fruits and their seeds.
3. Bring to the classroom different pieces of furniture, tools, sculptures, or other objects made from wood. What qualities of the wood make it useful for each object? Can you identify which tree provided the wood? Are some woods better than others for certain tasks?
4. Throughout history, plants have played religious and cultural roles in different societies. Many of these roles have been captured in paintings, sculptures, and tapestries. Find artwork that contains a plant or plants as an important part of the composition. What does the plant signify?

Investigation F2.1

Increasingly Complex Characteristics

Biologists sometimes use pairs of terms such as "less complex" and "more complex" when discussing diversity among organisms. A species that has changed little from its ancestors is said to be less complex than organisms that differ greatly from their ancestors. Conversely, a species that has few of the characteristics of its ancestors is said to be more complex. After studying many types of evidence in the fossil record and in living organisms, scientists have reached fairly general agreement about which characteristics have been in existence for a long time and which are more recent. Table F2.3 in step 2 of the investigation is based on such studies. Because there may be many degrees of complexity, the terms "less complex" and "more complex" are not absolute, but they are useful for making the types of comparisons you will make in this investigation, *Increasingly Complex Characteristics*.

Materials (per team of 6)

stereomicroscope or 10× hand lens
10 labeled specimens of organisms of various kingdoms and divisions
microscope slides
compound microscope
coverslips
pens

PROCESS AND PROCEDURES

1. In your science notebook, draw a table similar to table F2.2, with enough lines for all 10 specimens.

Name of organism	Numerical values of choices made	Total complexity score	Rank
1.			
2.			
etc.			

Table F2.2

2. Determine the complexity score for each of the labeled specimens. More-complex organisms such as plants are represented at each station, as well as some less-complex organisms from other kingdoms. Start at the left of table F2.3. Arrows from the starting point lead to 2 descriptions. Choose the one that fits the organism you are scoring.

Table F2.3 **Key for determining the complexity score for an organism**

BEGIN HERE

- Plant kingdom. Has structures that look like roots, stems, or leaves. (4)
  - No true roots, stems, or leaves. May bear hairlike rhizoids instead of roots. Not more than 10 cm tall. (4)
    - Leaflike structures with midribs; spaced equally around stalk. (5)
    - Leaflike structures without midribs, in two rows. (4)
  - True roots, stems, or leaves with vascular tissue. Usually more than 10 cm tall. (4)
    - No seeds. Reproduction by spores. (6)
    - Produce seeds. (5)
      - Flowers present. (5)
        - Flower parts in threes. Leaves parallel-veined. (4)
          - trees or shrubs (1)
            - trees (1)
            - shrubs (2)
          - woody vines or herbs (2)
            - woody vines (1)
            - herbs (2)
        - Flower parts in fours or fives. Leaves net-veined. (4)
          - trees or shrubs (1)
            - trees (1)
            - shrubs (2)
          - woody vines or herbs (2)
            - woody vines (1)
            - herbs (2)
      - No flowers. Seeds borne in cones. Leaves usually needlelike. (5)
        - trees (1)
        - shrubs (2)
- Not in plant kingdom. Does not have structures that look like roots, stems, or leaves. (1)
  - Not green. Usually white, gray, brown, or yellow. (1)
    - All individual parts microscopic in size. Occurring singly or in chains. (1)
    - Some or all individual parts not microscopic in size. (2)
      - Umbrella-like, shelflike, or spherical. Usually from 1 to 15 cm high. (3)
      - Hairlike, slender, usually fuzzy or powdery. Usually less than 1 cm high. (2)
  - Green, blue-green, or gray-green. (2)
    - Body crustlike and flat or upright and branched. (3)
      - Bright green. Generally on damp soil. (5)
      - Gray-green. Generally on rocks or tree trunks. (4)
    - Body not flat or leaflike. Usually aquatic. (2)
      - Blue-green. Prokaryotic. Chorophyll diffused throughout cells, not in distinct structures. (3)
      - Grass-green. Eukaryotic. Chlorophyll in definite structures only. (4)

3. Proceed across table F2.3 by following the arrows and choosing in each column the description that best fits each organism. Continue as far as the arrows go.
4. With each description there is a number. As you proceed, record the numbers of your chosen descriptions in the second column of your data table. The complexity score for the organism is the sum of all the numbers appearing after the descriptions you used in working through table F2.3. The more alike 2 organisms are, the more alike their scores will be. The greater the difference between 2 organisms, the greater will be the difference in their scores. More-complex organisms, such as plants, will have high scores (maximum 26), and less-complex organisms, such as prokaryotes, will have low scores (minimum 3).
5. When you have the complexity score for each of the organisms, give the organism with the lowest score a rank of 1 and the organism with the highest score a rank of 10. Then rank the rest of the organisms according to their scores. Record the rankings in the column at the right side of your data table.

Analysis

1. Basing your conclusions on the way the complexity score key was designed, list some of the most important differences among the organisms you observed.
2. What are some of the less important differences?
3. Using the information in table F2.3, list the characteristics you would expect to find in one of the less-complex organisms.
4. Do the same for one of the more-complex plants.
5. Evidence exists that today's land plants evolved from water-dwelling ancestors. Plants that live on land are in constant danger of drying out. Suggest how each of the following characteristics found in a more-complex plant would help the plant live on land: (1) roots; (2) stems that contain vascular tissues; (3) seeds; and (4) flowers.

Investigation F2.2

Reproductive Structures and Life Cycles

Although their reproductive organs differ, as do the environments in which they live and reproduce, the basic principles of sexual reproduction are the same in a moss, a flower, a bee, and a human. In this investigation, *Reproductive Structures and Life Cycles,* you will learn how the structures of a moss and a flower serve reproductive functions in their respective environments.

Mosses form mats on logs and on the forest floor, growing best in damp, shaded environments. Sporophytes grow out of the tops of the gametophytes and often look like hairs growing out of the mat of moss. Mosses cannot reproduce sexually unless they are wet. Flowering plants, on the other hand, are found in many different environments and climates. They need water to live but not to reproduce.

Materials (per team of 2)

compound microscope
stereomicroscope or 10× hand lens
3 microscope slides
3 coverslips
dissecting needle
scalpel
forceps
15% sucrose solution
moss plant with sporophyte
prepared slide of filamentous stage of moss
fresh moss
prepared slide of moss male and female reproductive organs
gladiolus flower
fresh bean or pea pod
other simple flowers for comparison
modeling clay
water in a dropping bottle
petri dish with moistened cotton
pens

PROCESS AND PROCEDURES

Part A Moss

1. Examine a moss plant with sporophyte attached. The sporophyte consists of a smooth stalk terminated by a little capsule. Separate the 2 generations by pulling the sporophyte stalk out of the leafy shoot of the gametophyte.

2. Using a dissecting needle, break open the capsule of the sporophyte into a drop of water on a slide.

CAUTION: Needles are sharp; handle with care.

3. Add a coverslip and examine under the low power of a compound microscope. What structures do you observe? How are these structures dispersed? How are they adapted for life on land?
4. Most moss spores germinate on damp soil and produce a filamentous stage that looks like a branching green alga. Examine a prepared slide of this stage.
5. The filamentous stage gives rise to the leafy shoot of the gametophyte. Using forceps, carefully remove a leafy shoot from the fresh moss. How does this shoot obtain water and nutrients for growth?
6. The reproductive organs of the gametophyte are at the upper end of the leafy shoot. Examine a prepared slide of these organs under the low power of a compound microscope. The male sex organs are saclike structures that produce large numbers of sperm cells. The female sex organs are flask-shaped and have long, twisted necks. An egg is formed within the base of the female organ. How does a sperm reach the egg? Would you expect to find moss plants growing where there was little or no water? Explain. The union of the egg and sperm results in a cell called the zygote. Where is the zygote formed? What grows from the zygote?

Part B Flowers

1. Examine the outside parts of a gladiolus flower. The outermost whorl of floral parts may be green and leaflike. These green sepals protected the flower bud when it was young. In some flowers, such as lilies, the sepals look like an outer whorl of petals. Petals are usually large and colored and lie just inside the sepals. Both sepals and petals are attached to the enlarged end of a branch. These parts of the flower are not directly involved in sexual reproduction. What functions might petals have?
2. Strip away the sepals and petals to examine the reproductive structures. Around a central stalklike body are 5–10 delicate stalks, each ending in a small sac, or anther. Together, the stalks and the anthers form the male reproductive organs, or stamens. Thousands of pollen grains are produced in the anther. The number of stamens varies according to the type of flower. How many stamens are present in the flower you are using? How is pollen carried from the anthers to the female part of the flower?
3. If the anthers are mature, shake some of the pollen into a drop of 15% sucrose solution on a clean slide. Add a coverslip and examine with the low power of a compound microscope. What is the appearance of the pollen? How is the pollen adapted for dispersal?
4. Make another pollen preparation on a clean coverslip. Use modeling clay to make a 5-mm-high chamber, slightly smaller than the coverslip, on a clean slide. Add a small drop of water to the chamber

and invert the pollen preparation over it. Examine after 15 minutes and again at the end of the lab period. What, if any, changes have occurred? (If no changes have occurred, store the slide in a covered petri dish containing a piece of cotton moistened with water, and examine it the next day.)

5. The central stalk surrounded by the stamens is the female reproductive organ, or carpel. It is composed of an enlarged basal part, the ovary, above which is an elongated part, the style, ending in a stigma. How is the stigma adapted to trap the pollen grains and to provide a place for them to grow?
6. Use a scalpel to cut the ovary lengthwise.

CAUTION: Scalpels are sharp; handle with care.

7. Using a hand lens or stereomicroscope, look at the cut surface. How many ovules can you see? Each ovule contains 1 egg. To what stage of the moss life cycle is the ovule comparable? How close to the egg can the pollen grain get? If the pollen grain cannot get to the egg, how do the sperm produced by the pollen reach it? To what stage of the moss life cycle is a pollen grain comparable?
8. The union of egg and sperm causes extensive changes in the female reproductive parts. Fertilization of the egg stimulates the growth of the ovary and the enclosed ovules. Carefully examine a fresh bean or pea pod. Open the pod to find the seeds. What part of the female reproductive apparatus is the pod of a bean or pea? What is the origin of a seed? If you plant ripe bean or pea seeds and water them, to what will they give rise? What can you conclude develops within a seed as a result of fertilization?
9. If time permits, examine other types of flowers. Compare the numbers of various parts and the ways the parts are arranged with respect to one another.
10. Wash your hands thoroughly before leaving the laboratory.

Analysis

1. In alternation of generations in a moss, which is the predominant, independent generation? Which is the less conspicuous generation?
2. Compare the life cycle of a moss (with alternation of generations) with your life cycle (with no alternation of generations).
3. Would you expect the most variation in flowering plants or in those plants reproducing by asexual means? Explain.
4. Compare and contrast the life cycle of a moss with that of a flowering plant.
5. Do flowering plants represent more or less adaptation to a land environment than mosses? Explain.

Investigation F2.3

Fruits and Seeds

The survival of plants depends on their ability to reproduce. In seed producers, the most-complex plants, reproductive ability is enhanced by mechanisms that protect and disperse the seeds (and fruits) so the seeds do not compete with the parent plants for nutrients, light, and water. Dispersal increases the likelihood that some seeds will not be eaten. In flowering plants, seeds are protected by the tissues of the mature ovary, or fruit. (Conifers, which lack an ovary, produce naked seeds on cone scales.)

In this investigation, *Fruits and Seeds,* you will try to determine how seeds of various fruits are dispersed. You also will use the fruits to make a dichotomous key, which can help you distinguish between objects by focusing on their similarities and differences.

Materials (per team of 3)

set of fruits
pens

PROCESS AND PROCEDURES

Part A Seed Dispersal

1. Blow gently at the fruits. What happens? Why? What parts of the fruits are important in allowing them to be wind-dispersed?
2. Gently rest the sleeve of your blouse or shirt on the fruits and then lift up your arm. Which fruits are lifted? What parts of these fruits are important in allowing them to be dispersed by animals?
3. Some fruits attract birds and other animals that eat the fruits but do not damage the seeds inside. What characteristics of the fruits might serve as attractants?
4. Some fruits contain a chemical that acts as a laxative. How might this function in seed dispersal? How might the contents of bird droppings assist in the survival of the new seedlings?

Part B Dichotomous Key

1. Design a dichotomous key using the fruits from part A. Your dichotomous key should separate all the available fruits into individual categories. Assemble the fruits and review the characteristics you observed in part A.
2. To form a dichotomous key, use characteristics that some fruits have and some do not, rather than characteristics that all the fruits share. For example, *fruits with spines* versus *fruits without spines* might be a good characteristic to use in building your key. On the other hand, *fruits that contain seeds* would not be a good characteristic because almost all fruits contain seeds.

3. To divide the fruits into 2 groups, choose 1 major characteristic not shared by all the fruits. Separate your fruits into 2 groups, one group that possesses the characteristic and another group that does not. In your science notebook, prepare a table similar to table F2.3 in Investigation F2.1, *Increasingly Complex Characteristics,* but with the lines only. Write the characteristic on the table as shown below:

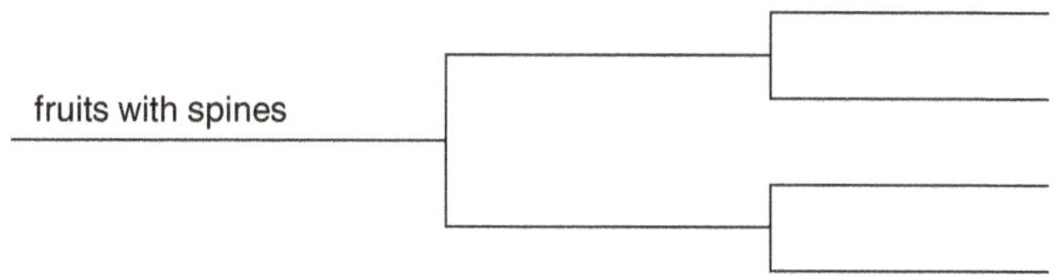

4. Focus on all the fruits in just 1 of your 2 groups. Choose another characteristic that separates those fruits into 2 groups. Write this characteristic in the second column of the key as shown below:

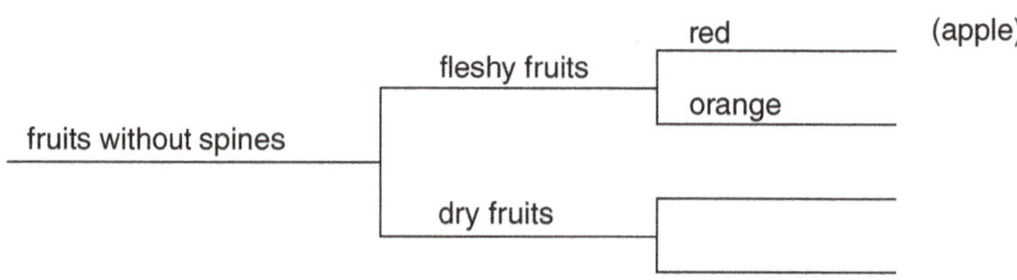

5. Continue to select characteristics and write them on the key until you have produced a key that separates all the fruits into individual categories. In other words, your key should split the fruits into smaller and smaller groups until each fruit is in a group by itself. Once this is done, place the name of the fruit (maple, ash, etc.) next to the appropriate line on the right side of the key.
6. Exchange keys with another team. See if you can follow their key and they can follow yours.
7. Wash your hands thoroughly before leaving the laboratory.

Analysis

1. What are some of the ways seeds/fruits are dispersed?
2. Describe how a dispersal mechanism that relies on the presence of other organisms might develop in a plant species.
3. What would happen to the distribution of plants that produce cockleburs if the plants lived on an island having no animals?
4. Explain how poplar trees might come to inhabit an island in the middle of a large lake.
5. What is the purpose of a dichotomous key?
6. What characteristics did you use to develop your dichotomous key? What characteristics did your class as a whole use?
7. Could you develop a dichotomous key for organisms that looked identical to one another? Explain.

APPENDIX F3

The Body's Organization

We can think of the entire body as a large compartment that is physically separate from the outside environment. But your body also contains smaller compartments, with still smaller compartments inside of them. Let's examine the body's organization by looking at the structure of some of these compartments. We begin by considering the structure and significance of one of the body's smallest compartments—the cell.

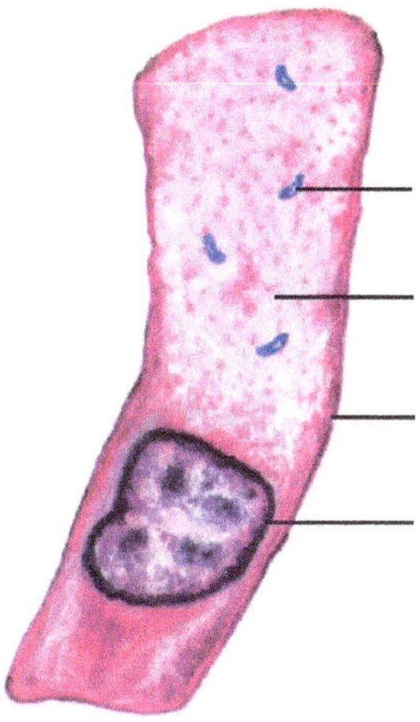

Figure F3.1
Epithelial cell. This epithelial cell contains all of the parts of a typical eukaryotic cell.

The **cell** is the basic unit of structure and function in living things (figure F3.1). All organisms are made of one or more cells. The exterior border of the cell is the **cell membrane**, a structure that separates the inside of the cell from the outside. The basic material within the cell is called **cytoplasm**, which consists of a complex mixture of water, salts, proteins, and other substances.

Suspended in the cytoplasm are numerous smaller compartments called **organelles**. Different types of organelles perform different functions in the cell. Chief among these is the **nucleus**, which contains the information required to operate the cell. Other important organelles include **mitochondria**, which provide most of the energy that powers activities within the cell.

The human body contains several hundred different types of cells, such as muscle cells, nerve cells, liver cells, and blood cells. Different types of cells have different structures and functions.

A **tissue** is a group of similar cells that are organized together and perform a specific function. The human body contains four types of tissues (figure F3.2). The cell in figure F3.1 is from epithelial tissue; the other three types are connective tissue, nerve tissue, and muscle tissue.

An **organ** is a group of tissues that are organized together to form a structural and functional unit. Every organ contains all four types of body tissue (figure F3.3). Examples of organs include the heart, the kidney, and the stomach.

The tissues of an organ work together, bringing about the organ's function. For example, the stretching that occurs when food enters the stomach activates nerve tissue. The nerve tissue, in turn, stimulates epithelial tissue cells, which then secrete digestive substances that help break down the food. Contractions of muscle tissue cells mix food with digestive substances and move it into the next organ in the system. The connective tissue provides an elastic and supportive framework that holds all of the tissues together.

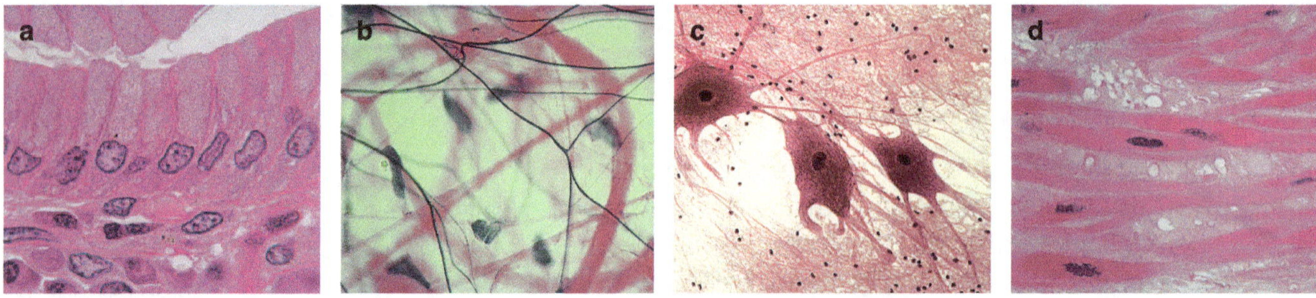

Epithelial tissue covers external and internal body surfaces. Examples of epithelial tissue include the skin and the linings of the digestive tract. Epithelial tissues protect the body's surfaces and produce secretions such as sweat and mucus.

Connective tissue binds, supports, and protects body structure. Connective tissue cells secrete nonliving material that forms a matrix. The matrix may be liquid (as in blood) or solid (as in bone and cartilage).

Nerve tissue is made up of nerve cells, or neurons, which are specialized to transmit nerve impulses (essentially information in the form of electrical signals) from one part of the body to another. The brain and spinal cord are made of nerve tissue.

Muscle tissue consists of cells that are able to contract in response to stimulation. Muscle tissue gives shape and support to the body and produces heat by shivering. It also helps move the whole body, as well as its individual parts.

Figure F3.2 **Tissue types.** The human body contains four types of tissues: (*a*) epithelial tissue, (*b*) connective tissue, (*c*) nerve tissue, and (*d*) muscle tissue. The structures of different cells support the function of the tissues they are a part of.

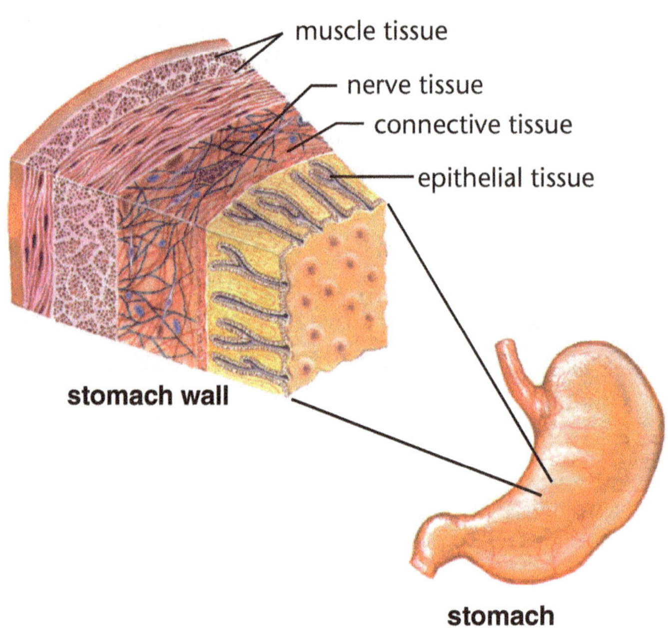

Figure F3.3 **Organ.** An organ such as the stomach functions as a result of all four types of body tissue working together and organized in a specific way.

Groups of organs form **organ systems** that carry out major body activities such as circulation, gas exchange, and digestion. These activities are accomplished by the coordinated action of the organs that make up the system. The digestive system, for example, breaks down food into small molecules (figure F3.4). This breakdown begins with mechanical and chemical processes in the mouth and continues with further chemical processes in the stomach and small intestine. Several organs in the digestive system contribute the chemicals needed for digestion. The nutrients that result are absorbed from the small intestine into the blood and transported by the circulatory system to all cells of the body.

Thinking about the structure of the body in this way should help you see that the function of a whole organism is the result of the coordinated action and interaction of all of its parts. For example, the function of a cell results from the coordinated action of its organelles, the function of a tissue results from the coordinated action of the cells that make it up, and so on. Thus, the function of the body results from the action and interaction of all of its parts.

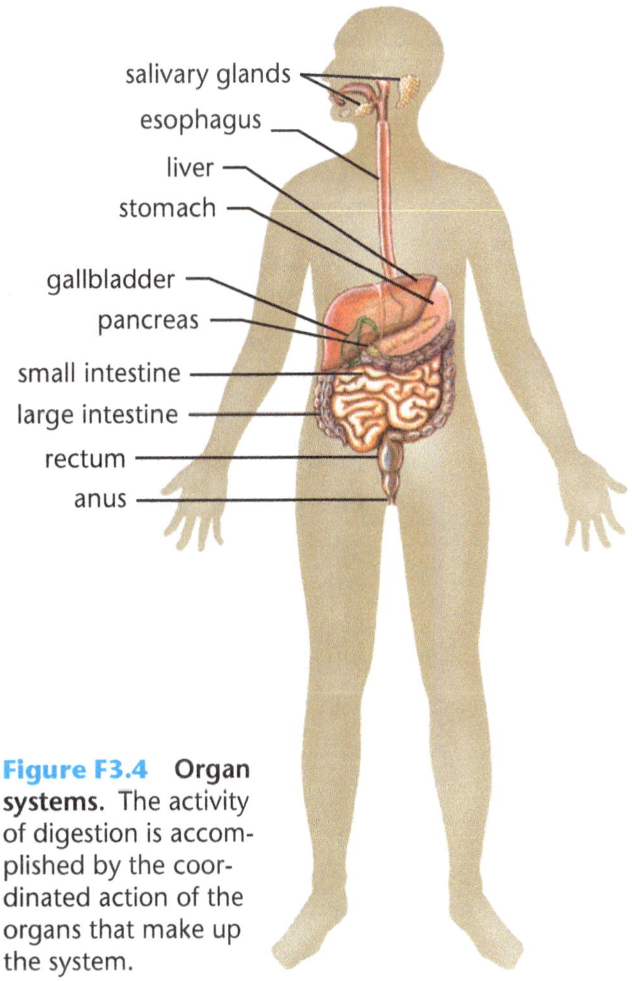

Figure F3.4 Organ systems. The activity of digestion is accomplished by the coordinated action of the organs that make up the system.

GLOSSARY

A

abiotic [abiótico]: Nonliving. In ecosystems, this is used to refer to nonliving components such as sunlight, wind, nutrients, and minerals.

acrosome [acrosoma]: The structure at the tip of a sperm cell, containing enzymes used to dissolve away the jellylike substance surrounding the egg cell.

activation energy [energía de activación]: The smallest amount of energy necessary to allow a chemical reaction to occur.

active transport [transporte activo]: The movement of a substance across a biological membrane against its concentration gradient with the help of energy input and specific transport proteins.

adaptation [adaptación]: In natural selection, a hereditary characteristic of some organisms in a population. The characteristic improves the organisms' chances for survival and reproduction in their environment compared with the chances of other organisms in the population.

adult stem cells [células madre adultas]: Cells derived from adult tissues that, under specific conditions, can differentiate into more than one cell type.

aerobic (eh ROH bik) [adaptación]: Taking place or living in the presence of free or dissolved oxygen.

allele (al LEEL) [alelo]: One of two or more possible forms of a gene. Each affects the hereditary trait somewhat differently.

allergies [alergia]: An overreactive immune response to foreign matter also known as allergens, which can be touched, inhaled, eaten, or injected.

amino acids [aminoácidos]: The molecular building blocks of proteins, which are each composed of an amino group, a carboxyl group, and a side chain. There are 20 different amino acids.

amylase (AM el ase) [amilasa]: A digestive enzyme found in saliva.

anaerobic (an eh ROH bik) [anaeróbico]: Taking place or living in conditions without free oxygen.

analogous structures [estructuras análogas]: Structures that have similar function, but do not derive their similarities from common ancestry. Frequently, analogous structures arise due to convergent evolution.

anaphase [anafase]: A stage in mitosis or meiosis, during which chromosomes move to opposite sides of the dividing cell.

antibody [anticuerpo]: A blood protein produced in response to an antigen, with which it combines specifically. Antibodies block the ability of pathogens or foreign material to injure the body.

antibody-mediated response [respuesta inmune mediada por anticuerpos]: Immune response in which B cells produce antibodies to bind with and mark antigens for destruction by nonspecific defenses; initiated by helper T cells.

antigen (AN tih jen) [antígeno]: Any material, usually a protein, that is recognized as foreign and raises an immune response.

antihistamine (an tih HIST uh meen) [antihistamínico]: A drug found in medicines used to counteract the histamine response.

artery [arteria]: A blood vessel through which blood flows away from the heart.

artifact [artefacto]: Material remains, such as pieces of pottery, tools, and textiles, that are cultural evidence.

artificial selection [selección artificial]: A process whereby humans allow only those variants of a population or species carrying desirable traits to reproduce.

asexual reproduction [reproducción asexual]: Any method of reproduction that requires only one parent or one parent cell. This type of reproduction leads to offspring that are genetically identical to one another.

atom [átomo]: The smallest particle of an element. In turn, an atom is made of smaller particles that do not separately have the properties of the element.

ATP (adenosine triphosphate [uh DEN oh seen try FOS fayt]) [ATP (trifosfato de adenosina)]: A compound that has three phosphate groups and is used by cells tostore energy.

autoimmune disease [enfermedad autoinmune]: A response in which antibodies are produced to attack some of the body's own cells.

autosome (AW tow sowm) [autosoma]: A chromosome that is not directly involved in determining sex.

autotroph (AW toh trohf) [autótrofo]: An organism able to make and store food, using sunlight or another nonliving energy source.

axon [axón]: A structure that extends out from a neuron and conducts impulses away from the cell body.

B

B cell [linfocito B]: A type of lymphocyte that develops in the bone marrow and later produces antibodies.

behavioral responses [respuestas conductuales]: Actions taken by an organism in response to some stimulus.

binary fission [fisión binaria]: A type of asexual reproduction where a cell divides into two identical cells. It is the chief means of asexual reproduction in prokaryotic cells.

binomial nomenclature [nomenclatura binomial]: The scientific system of naming organisms, where each species is given a Latin genus and species name.

biodiversity [biodiversidad]: The number and variability of organisms found within a community, ecosystem, or Earth.

biological classification [clasificación biológica]: The scientific system of classifying organisms into nested groups, generally along the lines of their evolutionary relationships.

biomass [biomasa]: The dry weight of organic matter that makes up a group of organisms in a particular habitat.

biosphere (BY oh sfeer) [biosfera]: The outer portion of Earth—air, water, and soil—where life is found.

biosynthesis (by oh SIN thuh sis) [biosíntesis]: The process of putting together or building up the large molecules characteristic of a particular type of cell or tissue.

biotic [biótico]: Living. In ecosystems, biotic components include all living things.

bipedal [bípedo]: Capable of walking erect on the hind limbs, freeing the hands for other uses.

bladder [vejiga]: An organ in the urinary system that collects urine produced by the kidneys. Urine leaves the bladder through the urethra.

blood pressure [presión sanguínea]: The fluid pressure created by heart contractions; allows blood to circulate.

blood type [grupo sanguíneo]: A class of inherited traits that describe a set of proteins or antigens on the outside of blood cells. Different blood types are frequently not immunologically compatible. Also known as blood group.

body temperature [temperatura corporal]: The internal temperature of an animal's body; for humans, this is typically 37°C (98.6°F).

brain stem [tronco encefálico]: A portion of the base of the brain that controls many involuntary body functions such as breathing and heart rate and some reflexes.

breathing rate [frecuencia respiratoria]: The number of times a terrestrial animal inhales and exhales per minute.

budding [gemación]: A type of asexual reproduction where new individuals begin as outgrowths on the body of a single parent. The offspring eventually separates from the parent and becomes independent. Offspring are genetically identical to the parent.

buffer [tampón]: A solution of weak acids and bases that resists changes in its pH level when an acid or base is added to it.

C

calorie [caloría]: The amount of heat required to raise the temperature of one gram of water 1°C.

calorimeter [calorímetro]: A scientific device used to measure heat.

cancer [cáncer]: Malignancy arising from cells that are characterized by profound abnormalities in the plasma membrane and in the cytosol, and by abnormal growth and division.

capillary [capilar]: A small blood vessel found between the end of an artery and the beginning of a vein. Capillaries are the sites of much gas and nutrient exchange in the human body.

carbohydrates [carbohidratos]: A group of organic compounds composed of carbon, oxygen, and hydrogen that form major portions of plants and are important food sources for all living things.

carbon [carbono]: A element found in all organic molecules that frequently forms bonds with hydrogen and oxygen, having an atomic weight of 12.011 and the atomic number 6.

carbon cycle [ciclo del carbono]: The chemical cycle in which carbon compounds made by some organisms (such as plants) are digested and decomposed by others. The carbon is released in small inorganic molecules that can be used again by more organisms to synthesize carbon compounds.

carbon fixation [fijación del carbono]: Incorporating atmospheric carbon into carbohydrates through photosynthesis.

carrier [portador]: A person who has one typical and one affected allele for a recessive disorder. These people do not show symptoms of the disorder, but can pass it to their offspring.

carrying capacity [capacidad de carga]: The maximum population size that can be supported by the available resources of a given area.

cell culture [cultivo celular]: The process of maintaining and growing cells outside an organism's body, usually supported by a substrate, dissolved gases, and nutrients in liquid form.

cell cycle [ciclo celular]: An ordered sequence of events in the life of a dividing cell, composed of the M, G_1, S, and G_2 phases.

cell differentiation [diferenciación celular]: The process by which embryonic cells develop specialized structures and functions.

cell division [división celular]: The process by which the genetic material and cytoplasm of a cell are divided between daughter cells. Meiosis and mitosis are two types of cell division.

cell-mediated response [respuesta inmune mediada por células]: Immune response in which highly specialized lymphocytes circulate in the blood and lymphoid organs and attack and destroy cells that carry specific surface antigens.

cellular respiration [respiración celular]: The series of chemical reactions by which a living cell breaks down food molecules and obtains energy from them.

cell wall [pared celular]: A nonliving covering around the plasma membrane of certain cells, as in plants, many algae, and some prokaryotes. In plants, the cell wall is constructed of cellulose and other materials.

cerebellum [cerebelo]: A structure of the brain that is important for coordination and motor control, as well as attention and the processing of language and music. It is located at the base of the brain and is characterized by a pattern of small folds.

cerebrum [cerebro]: The largest portion of the brain, which controls voluntary thought and action. It is made of a left and right hemisphere and is characterized by large folds.

cervix [cuello uterino]: The narrow, lower portion of the uterus, which connects the uterus to the vagina.

chemical bond [enlace químico]: The attraction between two atoms that results from the sharing or transfer of outer electrons from one atom to another.

chemical digestion [digestión química]: The breakdown of food molecules in the body with the help of chemical and enzymatic secretions, such as stomach acid.

chlorophyll (KLOR uh fil) [clorofila]: The green pigments of plants and many microorganisms; converts light energy (via changes involving electrons) to chemical energy that is used in biological reactions.

chloroplast (KLOR oh plast) [cloroplasto]: An organelle found only in plants and photosynthetic protists; contains chlorophyll, which absorbs the light energy used to drive photosynthesis.

chromatography (KROM uh toh gra fee) [cromatografía]: Various techniques that scientists can use to separate mixtures of molecules based on their mass, charge, or ability to bind to other molecules.

chromosome (KROH moh sohm) [cromosoma]: A long, threadlike group of genes found in the nucleus of all eukaryotic cells and most visible during mitosis and meiosis; chromosomes consist of DNA and protein.

circulatory system [sistema circulatorio]: An organ system consisting of a muscular pump (heart), blood vessels, and blood; the means by which materials are transported to and from cells. In many animals, it also helps stabilize body temperature and pH.

clade [clado]: A group or lineage of organisms that shares a single common ancestor.

class [clase]: The fourth most inclusive grouping, after domain, kingdom, and phylum or division, in the biological classification system.

classification [clasificación]: An organizing process that focuses on the characteristics that different living systems share by virtue of their common ancestry. This places the millions of different types of life into large categories of similar organisms.

climate [clima]: The prevailing weather conditions of a region, including temperature and precipitation, averaged over many years.

clitoris (KLI tor iss) [clítoris]: An external reproductive organ of the female situated at the front of the vulva. It is homologous to the male penis.

clone [clon]: A cell or organism that is genetically identical to the parent organism from which it was made.

cloning [clonación]: A type of asexual reproduction that produces genetically identical individuals from a single cell or set of cells. This type of reproduction is used in prokaryotes and plants.

closed system [sistema cerrado]: A region or area that is isolated; no matter or energy moves into or out of it.

codon (KO don) [codón]: The basic unit of the genetic code made up of a sequence of three adjacent nucleotides on one strand of DNA or mRNA.

cognition [cognición]: Acts or processes related to perceiving, knowing, or learning.

cognitive growth [crecimiento cognitivo]: An individual's development of knowledge and the ability to learn.

community [comunidad]: All the organisms that inhabit a particular area.

competition [competencia]: Interaction between members of the same population or of two or more populations to obtain a mutually required resource in limited supply.

complementary base pairing [apareamiento de bases complementarias]: Predictable interactions between nitrogen bases on opposite strands of DNA and between DNA and RNA; consists of adenine-thymine and guanine-cytosine (adenine-uracil in RNA) base pairing.

components [componentes]: The collection of things that makes up an ecological system, including biotic and abiotic factors.

concentration gradient [gradiente de concentración]: A difference in the concentration of certain molecules over a distance.

conditioning [condicionamiento]: Training that modifies a response so that it becomes associated with a stimulus different from the stimulus that originally caused it.

conservation of energy [conservación de energía]: The principle that energy within a closed system is constant, though it undergoes changes in form.

consumable biomass [biomasa consumible]: The portion of biological matter at a trophic level that is available for consumption by organisms at higher trophic levels.

consumer [consumidor]: A heterotroph; an organism that feeds on other organisms or on their organic wastes.

continental drift [deriva continental]: A hypothesis proposed in 1912 suggesting that Earth's landmasses had at one time been joined in a supercontinent that has broken up to form the present continents. Continental drift is now considered to be part of a broader theory of plate tectonics.

continuous variation [variación continua]: Occurs when the possible phenotypes for a trait occur as a series of intermediates rather than falling into two distinct categories. One example is human height.

contraceptive [anticonceptivo]: A substance or device that can prevent pregnancy.

control [control]: A treatment within a scientific experiment, also known as the experimental control, that features all characteristics of the experiment except the experimentally manipulated variable.

controlled experiment [experimento controlado]: A scientific experiment that is designed so that the influence of the experimental variable can be detected by preventing other conditions from varying across treatments. This design helps to eliminate alternative explanations of the results generated.

corpus callosum (KOR pus KAL o sum) [cuerpo calloso]: A bundle of nerve fibers that connects the right and left hemispheres of the brain.

corpus luteum (KOR pus LOOT ee um) [cuerpo lúteo]: The structure that forms from the tissues of a ruptured ovarian follicle and secretes female hormones.

covalent bond [enlace covalente]: A chemical bond formed by two atoms sharing a pair of electrons.

Cowper's (KAU perz) gland [glándula de Cowper]: Gland in the male reproductive system that produces an alkaline mucus. This mucus is secreted before ejaculation to protect sperm from the acidic vagina.

crossing over [encruzamiento cromosómico]: During prophase I of meiosis, it is the breakage and exchange of corresponding segments of chromosome pairs at one or more sites along their length. This results in genetic recombination.

cultural adaptation [adaptación cultural]: An adaptation to new pressures or situations resulting from cultural innovation.

cultural evolution [evolución cultural]: The change of cultural features over time, which are transmitted not through inherited traits but through social structures and learning. Cultural evolution operates through different processes than biological evolution.

culture [cultura]: A system of learned behaviors, symbols, customs, beliefs, institutions, artifacts, and technology characteristic of a group and transmitted by its members to their offspring.

D

decomposer [descomponedor]: An organism that lives on decaying organic material, from which it obtains energy and its own raw materials for life.

deletion mutation [deleción (o delección)]: Mutation that results when replication enzymes mistakenly skip a base. The new DNA strand forms missing a base.

dendrite [dendrita]: A structure that extends out from a neuron and transmits impulses toward the cell body.

deoxyribonucleic (dee OK sih ry boh noo KLEE ik) acid (DNA) [ácido desoxirribonucleico (ADN)]: The hereditary material of most organisms. DNA makes up the genes and chromsomes. The subunits that make up DNA are nucleotides, which contain the sugar deoxyribose, a triphosphate group, and one of four bases.

development [desarrollo]: Cell division, growth, and differentiation of cells from embryonic layers into all the tissues and organs of the body; later changes with age, including reproductive maturity, with its effects on appearance and body function.

developmental biologist [biólogo del desarrollo]: A scientist who studies the development of organisms.

dialysis tubing [tubo de diálisis]: A tube made of material that allows some substances, such as water, to pass across while preventing substances of a certain molecular size from crossing. This material is used in dialysis treatments for patients with kidney problems because it can be used to filter out toxic materials from the blood that their kidneys are no longer able to process.

diastole (dy AS tuh lee) [diástole]: The stage of the cardiac cycle in which the heart muscle is relaxed, allowing the chamber to fill with blood. (Compare with *systole*.)

diastolic pressure [presión diastólica]: The blood pressure that occurs during diastole; this is the lowest pressure that occurs during a cardiac cycle.

differentiation [diferenciación]: The process by which cells develop specialized structures and functions, also known as cell differentiation.

diffusion [difusión]: The movement of a substance down its concentration gradient from a more concentrated area to a less concentrated area.

digestion [digestión]: Processes that break down food molecules into smaller units.

dihybrid (DI HI brid) cross [cruza dihíbrida]: A genetic cross between individuals differing in two alleles.

diploid (DIH ployd) [diploide]: A cell containing two copies of every chromosome. The copies may not be identical.

discrete variation [variación discreta]: Occurs when the phenotypes for a trait fall into two distinct categories. One example is human sex, where the phenotype may be male or female.

diuretic (di yer RET ik) [diurético]: Any substance or agent that promotes the increased formation and excretion of urine.

division [división]: The third most inclusive grouping, after domain and kingdom, in the biological classification system for plants.

DNA [ADN]: See *deoxyribonucleic acid*.

DNA ligase [ADN ligasa]: Enzyme that functions during the replication and repair of DNA molecules to bond nucleotides together.

domain [dominio]: The most inclusive grouping in the biological classification system.

dominant trait [rasgo dominante]: A trait that is visible in a heterozygous organism.

E

ecology [ecología]: The study of living and nonliving components of the environment and of the interactions that affect biological species.

ecosystem [ecosistema]: A biological community in its abiotic environment.

ectotherms [ectotermos]: Organisms whose body temperature tends to be similar to their surroundings; also known as poikilotherms. These organisms can regulate their temperature behaviorally, but not physiologically.

ejaculation [eyaculación]: Expulsion of semen (sperm-bearing fluid) from the urethra.

electron transport system [cadena de transporte de electrones]: The process in which electrons are transferred from one carrier molecule to another in photosynthesis and in cellular respiration; results in storage of some of the energy in ATP molecules.

electrophoresis (eh LEC tro for EE sis) [electroforesis]: Any of several techniques for separating large molecules, such as proteins or DNA fragments, based on their ability to move through a gel medium subjected to an electric field.

embryo (em BREE o) [embrión]: An organism in its earliest stages of development after ferilization.

embryonic stem cells [células madre embrionarias]: Cells derived from embryonic tissues that have the potential to differentiate into any cell type of an organism's body.

emigration [emigración]: Departure of individuals from a population; decreases the size of the population.

endocrine (EN doh krin) system [sistema endocrino]: The system of glands that secretes their products from their cells directly into the blood.

endoskeleton (EN doh SKEL eh tun) [endoesqueleto]: A hard skeleton buried in the soft tissues of an animal, such as the spicules of sponges and bony skeletons of vertebrates.

endothermic reaction [reacción endotérmica]: A chemical reaction that involves the absorption of heat energy.

endotherms [endotermos]: Organisms capable of generating their own heat to maintain a stable body temperature, primarily through physiological but also through behavioral mechanisms. Also known as homeotherms.

enzyme (EN zime) [enzima]: A protein or part-protein molecule made by an organism and used as a catalyst in a specific biochemical reaction.

epididymis (eh pih DID ih mus) [epidídimo]: A coiled structure along the surface of the testis that provides for the storage, transmission, and maturation of sperm.

essential elements [elementos esenciales]: Elements that are necessary for the growth or maintenance of organisms.

estrogen (ES troh jen) [estrógeno]: A hormone that stimulates the development of female secondary sexual characteristics.

ethical analysis [análisis ético]: A process by which situations related to right and wrong can be examined to facilitate decision making and the development of understanding.

ethical dilemma [dilema ético]: A situation in which a decision related to right and wrong is needed, but which is complicated by other nonethical concerns, such as survival or well-being.

ethologist [etólogo]: Scientist who studies animal behavior.

eukaryote (yoo KAIR ee oht) [eucariota]: An organism whose cells have a membrane-enclosed nucleus and organelles; a protist, fungus, plant, or animal. (Compare with *prokaryote*.)

evidence [evidencia]: Reliable information that can be used to support a conclusion; in science, this information has been generated in accordance with scientific practices and methods.

evolution [evolución]: Changes in the genetic information and the traits encoded by genetic information across generations of a population or species of organisms due to the processes of mutation, selection, or genetic drift.

evolutionary biologist [biólogo evolutivo]: A scientist who studies how organisms change over time and are related to one another through common ancestry.

evolutionary trade-off [compensación evolutiva]: A change in a trait that offers some advantages for an organism's survival or reproductive success, but which is linked to changes in a second trait that results in some disadvantages for survival or reproduction.

evolutionary tree [árbol filogenético]: A branching diagram that illustrates patterns of relationships between a group of living and/or extinct organisms, where common ancestors are indicated at the nodes between branches. These diagrams are generated from data from physical characteristics of living or fossil organisms, such as molecular sequences or body structures.

exoskeleton (EK soh SKEL eh tun) [exoesqueleto]: A hard encasement deposited on the surface of an animal, such as the shell of a mollusk, that provides protection and points of attachment for muscles.

exothermic reaction [reacción exotérmica]: A chemical reaction that is characterized by the release of heat energy.

exponential growth [crecimiento exponencial]: A type of population increase observed when every individual is able to reproduce, characterized by a constant rate of growth and a rapidly increasing population size.

extinction [extinción]: The process by which a population or species dies out, eliminating the genetic material unique to that group.

F

facilitated diffusion [difusión facilitada]: The process by which molecules are transported across membranes with the help of carrier molecules.

family [familia]: The sixth most inclusive grouping after domain, kingdom, phylum or division, class, and order in the biological classification system; a group of related genera.

feedback system [sistema de realimentación]: A relationship in which one activity of an organism affects another, which in turn affects the first, yielding a regulatory balance.

fermentation (fer men TAY shun) [fermentación]: The incomplete breakdown of food molecules, especially sugars, in the absence of oxygen.

fertilization (fer til iz AY shun) [fertilización]: The union of an egg nucleus and a sperm nucleus. The result is one nucleus in a zygote.

fetus [feto]: A vertebrate embryo in later stages of development when it has attained the recognizable structural plan and features of its type.

filtration [filtración]: In vertebrate kidneys, filtration takes place when blood pressure forces the blood into the glomerulus of the nephron where blood cells and plasma proteins are separated from the blood's water, nitrogenous wastes, and ions. Most of the liquid filtrate is reabsorbed, but some wastes are secreted from the body in the urine.

follicle [folículo]: A cluster of cells found in the ovary that contain a single egg. Generally, one follicle grows and develops each month during the menstrual cycle.

follicle-stimulating hormone (FSH) [Hormona foliculoestimulante (FSH)]: A substance secreted by the anterior lobe of the pituitary that stimulates the development of an ovarian follicle in a female or the production of sperm cells in a male.

food [alimento]: A substance containing energy-rich organic compounds made by organisms and used as a source of energy and matter for life.

food web [red alimentaria]: Food chains in an ecosystem taken collectively, showing partial overlapping and competition for many food organisms.

fragmentation [fragmentación]: A type of asexual reproduction. Here, a piece broken from a parent organism grows into a new individual.

G

gamete (GAM eet) [gameto]: A sex cell, either an egg or a sperm, formed by meiosis, having half the number of chromosomes as body cells.

gas exchange system [sistema de intercambio de gases]: A body system that primarily functions to draw in oxygen for use in the energy-releasing processes of cells, and to remove the carbon dioxide by-product of these reactions from the body. In humans and other mammals, the primary organ of this system is the lungs.

gene [gen]: The fundamental physical unit of heredity, which transmits a set of specifications from one generation to the next. A segment of DNA that codes for a specific product.

gene expression [expresión genética]: The process by which genetic information is used to synthesize a functional product, such as an RNA or protein.

gene flow [flujo genético]: Changes in the frequency of alleles or entire genes in a population due to crossbreeding with other populations; the immigration of genetic material into a population.

gene therapy [terapia génica]: The introduction of a gene into a cell to correct a hereditary disorder.

genetic code [código genético]: The "language" of the genes in which the nucleotide sequence of DNA (in codons) specifies the amino acid sequences of proteins.

genetic drift [deriva genética]: Changes in the frequency of alleles in a population due to the chance survival or reproduction of individuals with particular traits. This mechanism of evolution acts independently of natural selection and is particularly important for small populations.

genetic engineering [ingeniería genética]: The experimental technology developed to alter the genome of a living cell for medical or industrial use.

genome [genoma]: The full set of genetic information in an organism.

genotype (JEE noh tipe) [genotipo]: The genetic makeup of an organism.

genus (JEE nus) [género]: A group of related species; the seventh most inclusive grouping after domain, kingdom, phylum or division, class, order, and family in the biological classification system.

geologic time [tiempo geológico]: Time periods related to the formation of Earth and its approximately 4.5-billion-year history.

gestation (je STAY shen) [gestación]: Development period between conception and birth that takes place in the uterus.

glans [glande]: Rounded gland structure at the tip of the penis, covered by the foreskin.

glial (GLEE ul) cells [células gliales]: Protection, support, and insulation for neurons.

glomerulus (glah MER yoo lus) [glomérulo]: A ball of capillaries surrounded by a capsule in the nephron and serving as the site of filtration in the kidneys.

glucose [glucosa]: A common 6-carbon sugar.

glucose test strips [tiras reactivas de glucosa]: Diagnostic materials used to quickly determine the pH of a liquid or semiliquid substance.

glycogen (GLY ko jen) [glucógeno]: The chief carbohydrate used by animals for energy storage.

glycogenesis [glucogénesis]: The process by which glycogen, an energy storage molecule, is formed.

glycolysis (gly KA luh sis) [glucólisis]: The initial breakdown of a carbohydrate, usually glucose, into smaller molecules at the beginning of cellular respiration.

gonads [gónadas]: The gamete-producing organs. These are the testes in males and ovaries in females.

growth [crecimiento]: Increase in size. In development, the successive rounds of cell division that produce a multicellular organism.

growth factors [factores de crecimiento]: Molecules that trigger cell division or the organization or maintenance of tissues. These factors are frequently expressed during the development of an embryo but are also found in specific adult tissues.

H

habitat (HA bih tat) [hábitat]: Place where an organism lives. Even in the same ecosystem, different organisms differ in their habitats.

haploid (HAP loyd) [célula haploide]: A cell containing only one member (n) of each chromosome pair characteristic of a species.

helper T cell [linfocitos T colaboradores]: A type of T cell required by some B cells to make antibodies, by other T cells to respond to antigens, or by other T cells that secrete lymphokines.

hemoglobin (HEE moh gloh bin) [hemoglobina]: The pigment in red blood cells responsible for transporting oxygen.

heritable trait [rasgo heredado]: A feature of an organism that is at least partly determined by genetic material inherited from parents; also known as an inherited trait.

heterotroph (HET er oh trohf) [heterótrofo]: An organism that obtains carbon and all metabolic energy from organic molecules previously assembled by autotrophs; a consumer.

heterozygous (HET er oh ZY gus) [heterocigoto]: Having differences between the two alleles for a given trait.

histamine (HISS tuh meen) [histamina]: A substance released by injured cells that causes blood vessels to dilate during an inflammatory response.

homeostasis (hoh mee oh STAY sis) [homeostasis]: A fundamental characteristic of living systems; maintaining a stable number of individuals within a population (social); the tendency of an organism to maintain a stable, constant internal environment (physiological).

hominid (HOM ih nid) [homínido]: A primate of the family Hominidae, which includes modern humans, earlier subspecies, and australopithecines.

homogenate (ho MAH jeh nate) [homogenato]: A material with uniform consistency.

homologous structures [estructuras homólogas]: Organismal traits, structures, or molecular sequences that share characteristics due to common ancestry, but may not necessarily have the same function.

homozygous (HOH moh ZY gus) [homocigoto]: Having two identical alleles for a given trait.

hormone [hormona]: A substance, secreted by cells or glands, that has a regulatory effect on cells and organs elsewhere in the body; a chemical messenger.

hyperthermia (HI pur THUR mee AH) [hipertermia]: An abnormally elevated body temperature; fever.

hypertonic (HI pur TAHN ik) [hipertónico]: A solution where the concentration of solutes outside a cell is greater than the concentration inside.

hypothalamus (HI po THAL uh muss) [hipotálamo]: A part of the brain. In humans, it links the nervous system to the endocrine system. (The endocrine system controls the hormones that regulate many body functions.)

hypothermia (HI po THIR mee ah) [hipotermia]: An abnormally lowered body temperature.

hypothesis (hy PA thih sis) [hipótesis]: A statement that suggests an explanation for an observation or an answer to a scientific problem.

hypotonic (HI po TAHN ik) [hipotónico]: A solution where the concentration of solutes outside the cell is less than the concentration inside.

I

immigration (im ih GRAY shun) [inmigración]: Arrival of new individuals into a population; increases the size of a population.

immunity [inmunidad]: Disease-resistance, usually specific for one disease or pathogen.

impermeable (im PUR mee uh bul) [impermeable]: Not permeable; not possible to pass through.

incomplete dominance [dominancia incompleta]: Pattern of gene expression in which the phenotype of a heterozygous individual is intermediate between those of the parents.

independent assortment [surtido independiente]: Independent inheritance of the alleles for a trait.

indicator [indicador]: A chemical that has a variable physical property, such as color, depending on the materials with which it interacts.

induction [inducción]: A developmental process by which one part of an embryo (or a molecule generated in one embryonic region) influences the differentiation of another embryonic region.

inductive reasoning [razonamiento inductivo]: The process of developing a generalization or general principles from detailed facts.

inference [inferencia]: Conclusions that follow logically from some form of direct evidence.

inherit [heredar]: To receive from genetic material.

innate behavior [comportamiento innato]: Actions that organisms take that are not learned, such as instincts. These tend to be governed by inherited traits of the nervous system.

inquiry [inquirir]: A way of finding out information, such as through an investigation.

insertion mutation [mutación por inserción]: A change in a DNA sequence involving adding a new nucleotide or sequence of nucleotides.

interphase [interfase]: The phase of the cell cycle in which division is not occurring.

in vitro (in VEE tro) [in vitro]: Literally means "in glass"; refers to laboratory procedures done in test tubes or in petri dishes.

ion [ión]: An atom or a molecule that has either gained or lost one or more electrons, giving it a positive or negative charge.

ionic bond [enlace iónico]: A chemical bond formed by the attraction between oppositely charged ions.

isotonic (I suh TAH nik) [isotónico]: A solution where the concentration of solutes outside a cell equals the concentration inside.

K

kidney [riñón]: An organ of the urinary system that regulates the balance of water and ions, and produces urine.

killer T cell [linfocito T citotóxico]: A lymphocyte that is produced in the bone marrow but matures in the thymus. These cells recognize and destroy infected cells, limiting the spread of infection.

kilocalorie (kil o KAL er ee; kcal) [kilocaloría]: A measure of food energy equal to 1,000 calories.

kinetic (KIN et ik) energy [energía cinética]: Active energy.

kingdom [reino]: The second most inclusive grouping in the biological classification system, following domain.

Krebs cycle [ciclo de Krebs]: The energy-releasing cycle in cellular respiration that completes the breakdown of intermediate products of glycolysis; also a source of carbon skeletons for use in biosynthesis reactions.

L

labia (LAE be uh) [labios (genitales)]: Folds of skin that protect the female genitals and cover the clitoris and the vaginal and urinary openings.

larva (LAR vuh) [larva]: An immature stage of development in offspring of many types of animals.

learned behavior [comportamiento aprendido]: The responses an organism develops as a result of specific interactions with the environment.

LH surge [pico de LH]: An increase in luteinizing hormone that occurs near the middle of the menstrual cycle and signals the follicle to release the ovum.

limiting factor [factor limitante]: An environmental condition such as food, temperature, water, or sunlight that restricts the types of organisms and population numbers that an environment can support.

lineage [linaje]: A group or clade of organisms that is related by common ancestry.

linkage [ligamiento]: In inheritance, the association of different genes due to their physical proximity on chromosomes.

lipid bilayer [bicapa lipídica]: The arrangement of molecules in cellular membranes.

lipids [lípidos]: A group of organic compounds that includes fats, oils, waxes, triglycerides, and steroids, which do not dissolve in water and make up important parts of cells, such as portions of the cell membrane.

logistic growth [crecimiento logístico]: A type of population increase frequently observed in natural conditions.

luteinizing (LEW tee ih NY zing) hormone (LH) [hormona luteinizante (LH)]: A hormone secreted by the anterior lobe of the pituitary gland that controls the formation of the corpus luteum in females and the secretion of testosterone in males.

lymphocyte (LIM foh site) [linfocito]: A type of white blood cell important in the immune response.

lysosome (LY soh zohm) [lisosoma]: A cell vesicle that contains digestive enzymes.

M

macrophage (MAK roh fayj) [macrófago]: A large white blood cell that ingests pathogens and dead cells.

meiosis (my OH sis) [meiosis]: Special process of cell division in eukaryotes that produces the gametes. Two successive nuclear divisions (with corresponding cell divisions) produce gametes that have one-half of the genetic material of the original cell.

melanin [melanina]: A dark-colored pigment that can cause darker coloration or spots in some organisms.

memory cell [linfocito de memoria]: B or T lymphocyte produced in response to a primary immune response. The lymphocyte remains in the circulation and can respond rapidly if the same antigen is encountered in the future.

menopause (MEN oh pawz) [menopausia]: In human females, the cessation of menstruation, usually taking place between the ages of 45 and 60.

menstrual (MEN strew al) cycle [ciclo menstrual]: The female reproductive cycle that is characterized by regularly recurring changes in the uterine lining. One cycle lasts approximately 28 days.

menstruation (men strew AY shun) [menstruación]: Periodic sloughing of the blood-enriched lining of the uterus when pregnancy does not occur.

meristem [meristemo]: A portion of a plant that contains embryonic tissue and actively dividing cells, from which new roots, stems, leaves, or reproductive structures can grow.

messenger RNA (mRNA) [ARN mensajero (ARNm)]: The RNA complementary to one strand of DNA. It is transcribed from genes and translated by ribosomes into protein.

metabolism (meh TAB oh liz um) [metabolismo]: The sum of all the chemical changes taking place in an organism.

metamorphosis [metamorfosis]: A major change in the body of an organism as it passes from one life history stage to another.

metaphase [metafase]: A stage in mitosis or meiosis, during which chromosomes are lined up along the middle of the dividing cell.

metastasis [metástasis]: The spread of cancerous cells from the tumor where they developed to other parts of the body.

metastasize [metastatizar]: To spread, as in the spread of cancer cells.

miscarriage [aborto espontáneo]: The natural and spontaneous expulsion of a fetus before it is capable of surviving independently, usually between the third and seventh month of pregnancy.

mitochondria (my toh KON dree uh) [mitocondrias]: The organelles in eukaryotic cells that carry on cellular respiration, releasing energy from food molecules and storing it in ATP.

mitosis (my TOH sis) [mitosis]: The replication of the chromosomes and the production of two nuclei in one cell that is usually followed by cell division. Each new cell that results from a mitotic division has the same genetic makeup as the original cell. This process is used to make new cells to replace damaged or dying cells.

model [modelo]: In science, these are representations of specific aspects of a natural phenomenon, built from available evidence and used to gain understanding of the phenomenon or make predictions about it.

molecules [molécula]: The smallest unit of a substance made of two or more atoms that retains the chemical and physical properties of the substance.

monogamy (moh NOG ahm ee) [monogamia]: A bond between two individuals.

monohybrid (mon o HI brid) cross [cruza monohíbrida]: A genetic cross between individuals differing in one allele.

mons pubis (MONS PYOO bis) [monte de Venus]: A pad of tissue covering the female pubic bone that becomes covered with hair.

mortality (mor TAL ih tee) [mortalidad]: Death rate, measured as the proportion of deaths to total population over a given period; often expressed as number of deaths per 1,000 or 10,000 individuals.

motor cortex [corteza motriz]: The region of the cerebral cortex of the brain which transmits information to voluntary muscles to generate movement.

motor neuron [motoneurona]: A nerve cell that conducts information to muscles or glands.

mRNA [ARNm]: See *messenger RNA*.

mutation (myoo TAY shun) [mutación]: A hereditary change to the nucleotide sequence of a gene. A mutation may result in a new allele or a change in a regulatory sequence for a gene. Changes can lead to an increase or decrease in the amount of RNA or protein product made by the affected gene; can alter the function of the gene's product; can eliminate the production of the gene's product; or can have no effect on the gene's product. Whether a mutation is beneficial, detrimental, or has no effect on an organism depends on the change to the gene product and the organism's current environmental conditions.

mutualism (MYOO tyoo ul is um) [mutualismo]: A positive relationship between two species that benefits both species. The relationship may enhance growth, survival, and/or fitness for each species.

N

negative controls [controles negativos]: A type of experimental treatment in which all of the conditions of the experiment are replicated except the experimental variable. A negative control treatment is expected to not show the condition predicted in the experimental treatment.

negative feedback [realimentación negativa]: A kind of feedback that creates equilibrium between input and output in a system or process.

nephron (NEF rahn) [nefrona]: The functional unit of a kidney. It consists of a long, coiled tubule. One end forms a cup that encloses a mass of capillaries. The other end opens into a duct that collects urine. The entire nephron is surrounded by a network of capillaries.

nervous system [sistema nervioso]: A coordinating mechanism in all multicellular animals, except sponges, that regulates internal body functions and responses to external stimuli. In vertebrates, it consists of the brain, spinal cord, nerves, ganglia, and parts of receptor and effector organs.

neurobiologist [neurobiólogo]: A scientist who studies the nervous system of animals.

neuron (NOOR ahn) [neurona]: A nerve cell; a name usually reserved for nerve cells in animals that have a complex brain and specialized associative, motor, and sensory nerves.

neurotransmitter (NOOR oh TRANS mit er) [neurotransmisor]: A chemical messenger that diffuses across the synapse and transmits a nerve impulse from one neuron to another.

niche (NITCH) [nicho]: The sum total of all the adaptations an organism uses to survive in its environment. This includes its role in the community, what it eats, and what interactions it has with other organisms and with its environment.

nondisjunction (non dis JUNG shun) [no disyunción]: The failure of a pair of homologous chromosomes to separate during meiosis or mitosis.

nonspecific barriers [barreras no específicas]: In immunity, these include general barriers to infection such as the skin, mucus, earwax, and colonies of harmless bacteria.

nucleic (noo KLAY ik) acid [ácido nucleico]: DNA or RNA. An organic compound composed of nucleotides, it is important in coding instructions for cell processes.

nucleotide (NOO klee oh tide) [nucleótido]: A subunit or building block of DNA or RNA. It is chemically constructed of a 5-carbon sugar, a nitrogen base, and a triphosphate group.

nucleus (NOO klee us) [núcleo]: In eukaryotic cells, it is the membranous organelle that houses the chromosomal DNA. In atoms, the central core containing positively charged protons and (in all but hydrogen) electrically neutral neutrons.

O

oncogene [oncogén]: A gene found in viruses or as part of the normal genome that is crucial for triggering cancerous characteristics.

order [orden]: A group of related families; the fifth largest grouping, after domain, kingdom, phylum or division, and class, in the biological classification system.

organelle (or guh NEL) [orgánulo]: An organized structure within a cell, with a specific function. Chloroplasts and mitochondria are examples.

organic molecules [moléculas orgánicas]: Matter naturally found within organisms, typically composed of a carbon backbone or ring connected to atoms of other elements such as oxygen, nitrogen, and hydrogen.

organ system [sistema orgánico]: A group of organs that carries out an interrelated set of functions, such as gas exchange, the elimination of wastes, or the circulation of blood.

orgasm [orgasmo]: Sensory and motor events at the peak of sexual stimulation that result in ejaculation for the male and involuntary contraction of the muscles surrounding the vagina in the female.

osmosis (os MOH sis) [ósmosis]: The movement of water across a selectively permeable membrane.

ova (singular: ovum) [óvulos]: Mature female gametes. Also known as eggs, ova are larger than sperm.

ovarian cycle [ciclo ovárico]: The events of the menstrual cycle that occur in the ovary, including the maturation and release of ova in response to hormonal changes.

ovary (plural: ovaries) [ovarios]: The primary reproductive organs of a female; egg-cell-producing organs.

oviduct (OH vi dukt) [trompa de Falopio]: A tube leading from an ovary to the uterus.

ovulation (ahv yoo LAY shun) [ovulación]: In vertebrates, the release of one or more eggs from an ovary.

P

paleobiologist [paleobiólogo]: A scientist who studies relationships between fossil and living organisms.

paleontologist (PAY lee on TOHL oh jist) [paleontólogo]: A scientist who studies fossils to gain insight into the history of life on Earth.

parasitism (PAIR uh sih tiz um) [parasitismo]: An ecological niche in which one organism is the habitat and the food for another. A parasite lives and feeds on the host organism, usually without killing it.

passive transport [transporte pasivo]: The diffusion of a substance across a biological membrane through a transport protein in the membrane.

pathogen (PATH oh jen) [patógeno]: A disease-causing organism.

penicillin (pen ih SIL lin) [penicilina]: Any of several antibiotic compounds obtained from penicillium mold and used to prevent or treat a wide variety of diseases and infections.

penis [pene]: In vertebrates, the male organ through which sperm are passed to the female and through which wastes from the kidneys—in the form of urine—are discharged outside the body.

peptide bond [enlace peptídico]: The type of bond that links amino acids together to form a protein.

permeable [permeable]: Open to passage or penetration.

pH [pH]: A measure of how acidic or basic a solution is, based on the concentration of hydrogen ions in that solution.

phenotype (FEE noh tipe) [fenotipo]: The expression of a genotype in the appearance or function of an organism; the observable traits of an organism.

pheromone (FARE oh mohn) [feromona]: A small chemical signal that animals produce that can stimulate and influence the physiology and behavior of other individuals of the same species; acts much like hormones.

photosynthesis (fo toe SIN thus sis) [fotosíntesis]: The process by which living cells that contain chlorophyll use light energy to make organic compounds from inorganic materials.

phylum (FY lum) [filo]: The third most inclusive grouping, after domain and kingdom, in the biological classification system. A phylum is assigned to all organisms except plants, which are classified in divisions.

physical anthropologist [antropólogo físico]: A scientist who studies the biological evolution of humans.

physical digestion [digestión física]: The breakdown of food molecules through the movement of the mouth, stomach, and other parts of the digestive system.

physiological (fiz ee OH laj ih kul) processes [procesos fisiológicos]: Related to functions of an organism or its parts.

physiological response [respuesta fisiológica]: An automatic change generated by the body of an organism that is triggered by an external or internal stimulus.

pituitary (pih TOO ih ter ee) [pituitaria]: A part of the brain that produces and secretes hormones that regulate a variety of body functions. The pituitary also stores and then releases two hormones produced by cells in the hypothalamus.

placenta (pluh SEN tuh) [placenta]: A structure in the pregnant uterus for nourishing a fetus with the mother's blood supply; formed from the uterine lining and embryonic membranes.

plasma (PLAZ muh) [plasma]: The liquid portion of the blood in which cells are suspended.

plasma cell [célula plasmática]: An antibody-producing cell that forms as a result of the proliferation of sensitized B lymphocytes.

plate tectonics (tek TON iks) [tectónica de placas]: The theory and study of the movement of the plates that make up Earth's crust.

polygamy (puh LIG uh me) [poligamia]: Mating strategy whereby one individual mates with more than one individual of the opposite sex.

polymerase (pah LIH mur ayse) chain reaction (PCR) [reacción en cadena de la polimerasa (PCR)]: A technique for copying a DNA molecule. After 20 cycles of PCR amplification, more than a million copies of a DNA molecule will have been made. Extremely important in biotechnology and research.

polysaccharide [polisacárido]: A carbohydrate molecule made up of three or more sugar or monosaccharide subunits.

population [población]: A group of organisms of one species that lives in the same place at the same time.

population density [densidad de población]: Number of organisms per unit of habitat area.

positive controls [controles positivos]: A type of experimental treatment in which all of the conditions of the experiment are replicated plus a variable known to reliably elicit the expected experimental results. Positive controls are used to test that the experimental setup worked as designed.

positive feedback [realimentación positiva]: A kind of feedback that disturbs or prevents equilibrium between input and output in a system or process.

potential energy [energía potencial]: The energy of an object deriving from the position of that object in space.

predation (pred EH shun) [depredación]: The killing and consumption of prey.

predator-prey relationship [relación depredador-presa]: The relationship between organisms in which one, the predator, feeds on the other, the prey.

principle of independent assortment [principio de distribución independiente]: The inheritance of alleles for one trait does not affect the inheritance of alleles for another trait.

principle of segregation [principio de segregación]: During meiosis, chromosome pairs separate so that each of the two alleles for any given trait appears in a different gamete.

probability [probabilidad]: The relative possibility that an event will occur or is likely to occur.

process of science [proceso de ciencia]: The set of practices that help a person think about and learn about the natural world.

producer [productor]: An autotroph. Any organism that produces its own food using matter and energy from the nonliving world.

progesterone (proh JES tuh rohn) [progesterona]: A female hormone secreted by the placenta and the corpus luteum that acts to prepare and maintain the uterus for pregnancy and to prepare the breasts for lactation.

prokaryote (pro KAIR ee oht) [procariota]: An organism whose cells do not have membrane-enclosed organelles, such as nuclei, mitochondria, and chloroplasts; a bacterium. (Compare with *eukaryote*.)

prophase [profase]: A phase of mitosis or meiosis in which the nuclear membrane breaks down and chromosomes condense.

prostate (PRAH stayt) gland [próstata]: Male reproductive gland located near the bladder and urethra that produces and adds more fluid to the semen.

protein [proteína]: An organic compound composed of one or more polypeptide chains of amino acids. Most structural materials and enzymes in a cell are proteins.

protocol [protocolo]: A plan or procedure for a basic investigation that can answer a testable question; serves as a standard for controlling variables.

puberty (PYOO bur tee) [pubertad]: The stage of development in which the reproductive organs become functional.

pulse [pulso]: The regular pressure change in the arteries generated by the contractions of the heart.

R

radioactive isotope [radioisótopo]: A form of a chemical element that emits energetic particles (such as electrons) by the decay of its atomic nucleus.

reabsorption (REE ab ZORP shun) [reabsorción]: In the kidney, this takes place in capillaries outside the glomerulus (in the renal tubule). Water and some dissolved substances that had been filtered from the blood are returned (reabsorbed) to the filtered blood.

reactants [reactivos]: The substances that undergo a chemical change in a chemical reaction.

receptor (ree SEP ter) [receptor]: A specialized sensory cell, as in the eye or the skin, that is sensitive to a particular type of stimulus.

recessive (ree SESS iv) [recesivo]: A term used to describe an allele or trait that is masked by a dominant allele or trait.

recessive trait [rasgo recesivo]: A trait that is only observable when a person inherits two recessive alleles from his or her parents.

recombinant (re KOM bin ent) DNA [ADN recombinante]: DNA that incorporates parts of different parent DNA molecules, as formed by natural recombination mechanisms or by recombinant DNA technology.

recombination (re KOM bin EH shun) [recombinación]: The regrouping of genes in an offspring caused by the crossing over of chromosomes during meiosis.

reflex [reflejo]: An involuntary reaction or response to a stimulus.

reflex arc [arco reflejo]: A nerve pathway that forms the structural and functional basis for a reflex.

replication (rep li KAY shun) [replicación]: The process of making a copy of the genetic information in a cell nucleus.

reproduction [reproducción]: The process by which organisms generate new individuals.

reproductive isolation [aislamiento reproductivo]: A state in which two populations of organisms are unable or strongly limited in their ability to reproduce with each other. Reproductive isolation can, over time, lead to speciation.

reproductive strategy [estrategia reproductiva]: The combination of characteristics used by an organism to ensure success in reproduction.

restriction enzyme [enzima de restricción]: An enzyme that recognizes specific nucleotide sequences in DNA and breaks the DNA chain at those points.

ribonucleic (ry boh noo KLEE ik) acid (RNA) [ácido ribonucleico (ARN)]: The hereditary material coded by the DNA of cells to carry out specific genetic functions; for example, messenger RNA and transfer RNA. Also the hereditary material of some viruses.

ribosomal (RY boh SOHM ul) RNA (rRNA) [ARN ribosómico (ARNr)]: A class of RNA molecules found, together with characteristic proteins, in ribosomes.

ribosome (RY boh SOHM) [ribosoma]: The cellular organelle that serves as the site for protein synthesis.

rRNA [ARNr]: See *ribosomal RNA*.

S

scientific theory [teoría científica]: A general principle or system of principles that provides the best logical explanation for a group of related natural phenomena. Scientific theories generally develop from the work of many scientists through the synthesis of multiple lines of experimental and observational evidence, and can be used to predict natural phenomena and provide direction for new areas of investigation.

scrotum [escroto]: A pouch of skin that encloses the testes.

secretion [secreción]: In the kidney, this takes place near the end of the renal tubule. Unfiltered wastes are passed from the blood (secreted) into the filtrate, a process that adjusts the blood pH.

selectively permeable [selectivamente permeable]: A property of biological membranes that allows some substances to cross and prevents others from crossing.

semen (SEE men) [semen]: In mammalian males, the thick fluid that transports sperm.

seminal vesicles (SEM in ul VES ih kuls) [glándulas seminales]: In the male reproductive system, two small sacs that contribute to semen production.

seminiferous tubules [túbulos seminíferos]: Long, coiled tubes found in the testes that serve as the site of sperm production.

senescence [senescencia]: Aging, or growing old.

sensory cortex [corteza sensorial]: The region of the cerebral cortex of the brain that receives and integrates sensory information from the body.

sensory neurons [neuronas sensoriales]: Nerve cells that transmit information from a sense organ to the brain.

sequence [secuencia]: The order of the subunits that make up certain molecules; for example, the order of nucleotides that make up DNA or the order of amino acids that make up a protein.

sexually transmitted disease (STD) [enfermedad de transmisión sexual (ETS)]: A disease that may be contracted through sexual contact; for example, herpes, chlamydia, gonorrhea, HIV, and others.

sexual reproduction [reproducción sexual]: Reproduction involving the contribution of genetic material from two parents. Because each parent contributes information, the offspring of sexual reproduction are not genetically identical to either parent.

sexual selection [selección sexual]: Selection based on variation in secondary sexual characteristics, leading to the enhancement of individual reproductive fitness.

shock [shock]: A condition of profound disturbance of the body's vital processes, characterized by failure of the circulatory system to deliver adequate amounts of blood to vital organs.

solute (SAWL yoot) [soluto]: The dissolved substance in a solution.

solution [solución]: Two or more substances combined to make a homogeneous mixture.

solvent [solvente]: The liquid in which the solute is dissolved.

somatic cell [célula somática]: Any nonreproductive cell in an organism.

speciation [especiación]: The process by which a new species evolves from an ancestral species due to factors that generate the new population reproductively isolated from the original population.

species (SPEE sheez) [especie]: All individuals and populations of a particular type of organism, maintained by biological mechanisms that result in their breeding mostly with their type.

specific immune response [respuesta inmune específica]: A portion of the immune system of vertebrates that responds to specific foreign matter, known as antigens, in such a way that usually eliminates that matter from the body. Components of the specific immune response include B cells and the antibodies they generate; memory B cells, which are stored following the generation of a new antibody type; and T cells, which attack and remove identified foreign matter.

sperm [espermatozoides]: The gametes produced by males. They are often motile and consist of a compact nucleus surrounded by a membrane, a flagellum for propulsion, mitochondria that provide energy, and a small sac at the tip containing enzymes that help penetrate the ovum.

spores [esporas]: One-celled reproductive bodies that are usually resistant to harsh environmental conditions and may remain dormant, in a dry covering, for long periods. In some organisms, spores are asexual and may initiate the growth of a new organism under favorable conditions. In other organisms, spores are sexual and must unite with spores of the other sex before producing a new organism.

sporulation [esporulación]: The process of reproducing asexually by producing spores that grow directly into new individuals.

starch [almidón]: A polysaccharide carbohydrate derived from plants and composed of glucose.

stem cell [célula madre]: A cell that can divide continuously and generate cells capable of differentiating into any cell type of an organism's body.

stimulus (STIM yoo lus; plural: stimuli) [estímulo]: A change or signal in the internal or external environment that causes an adjustment or reaction by an organism.

stomate (stow MATE; plural: stomates or stomata) [estoma]: The opening between two guard cells in the epidermis of a plant leaf through which gases are exchanged with the air.

stop codon (KO don) [codón de terminación]: A special codon in messenger RNA that does not specify an amino acid. When present, the synthesis (translation) of the growing amino acid chain stops.

strata (STRA tuh; singular: stratum) [estratos]: Layers, usually of deposited earth sediments carried by erosion. Many strata become mineralized into rock layers and they may contain fossils.

stratigraphy (stra TIG ra fee) [estratigrafía]: A branch of geology that is concerned with the systemized study, description, and classification of stratified rocks.

stressor [situación estresante]: A factor capable of stimulating a stress response.

stroma (STROH muh) [estroma]: The colorless substance in a chloroplast surrounding the thylakoids. The enzymes of the Calvin cycle also are in the stroma.

substitution mutation [mutación por sustitución]: Mutation caused by substituting one nucleotide for another nucleotide during replication.

substrate (SUB strayt) [sustrato]: A molecule on which enzymes act.

symbiosis (sim by OH sis) [simbiosis]: An ecological relationship between organisms of two different species that benefit from living together in direct contact.

synapse (SIN aps) [sinapsis]: An open junction between neurons, across which an impulse is transmitted by a chemical messenger, a neurotransmitter.

systems analysis [análisis de sistemas]: A process by which limited portions of a complex phenomenon are studied, perhaps multiple times with variation, to gain insight about the phenomenon.

systole (SIS toh lee) [sístole]: The stage of the cardiac cycle in which the heart muscle contracts and the chambers pump blood. (Compare with *diastole*.)

systolic pressure [presión sistólica]: The blood pressure that occurs during systole; this is the highest pressure that occurs during a cardiac cycle.

T

T cell [linfocito T]: A lymphocyte that matures in the thymus stimulated by the presence of a particular antigen. It differentiates and divides, producing offspring cells (killer cells) that attack and kill the cells bearing the antigen.

technology [tecnología]: The application of scientific knowledge for practical purposes.

telomere [telómero]: A specific segment of DNA found at the end of a chromosome.

telophase [telofase]: The last phase of a mitotic or meiotic cell cycle in which daughter cells are divided and the nuclear membrane forms around the genetic material.

testis (plural: testes) [testículo]: The primary reproductive organ of a male; sperm-cell-producing organs.

testosterone (tes TOS ter ohn) [testosterona]: A male sex hormone secreted by the testes.

theory [teoría]: See *scientific theory*.

thylakoid (THY luh koyd) [tilacoide]: A flattened sac in a chloroplast. Many of the thylakoids are arranged in stacks known as grana. The pigments and enzymes for the light reactions of photosynthesis are embedded in the sac membrane.

toxin [toxina]: A substance produced by one organism that is poisonous to another.

trachea (TRAY kee uh) [tráquea]: The windpipe of an air-breathing vertebrate, connecting the air passage in the throat with the lungs.

transcription (tran SKRIP shun) [transcripción]: The assembly of an RNA molecule complementary to a strand of DNA. The product may be messenger RNA, transfer RNA, or ribosomal RNA.

transfer RNA (tRNA) [ARN transferente (ARNt)]: The RNA responsible for transporting amino acids to the ribosome during protein synthesis.

transgenic (trans JEN ik) [transgénico]: Plants or animals that contain genes from unrelated species.

translation [traducción]: The assembly of proteins on ribosomes, using messenger RNA to direct the order of amino acids.

triage (TREE azh) [triage]: A sorting out and classification of patients according to the seriousness of injuries, urgency of treatment, and place for treatment. Carried out as an emergency procedure during an influx of large numbers of victims of a disaster.

tRNA [ARNt]: See *transfer RNA*.

trophic level [nivel trófico]: A group of organisms in a community that occupies the same position in the food web, such as producers or primary consumers.

U

umbilical (um BIL ih kul) cord [cordón umbilical]: In placental mammals, a tube connecting the embryo with the placenta.

ureter (yoo REE ter) [uréter]: A muscular tube that carries urine from the kidney to the urinary bladder.

urethra (yoo REE thruh) [uretra]: In vertebrates, the tube through which urine is carried from the bladder to the outside of the body.

urinary system [sistema urinario]: An organ system that functions to generate and eliminate urine and includes the kidneys, ureters, bladder, and urethra.

uterine cycle [ciclo uterino]: The events of the menstrual cycle that take place in the uterus, including the shedding of the uterine lining in response to hormonal changes.

uterus (YOO ter us) [útero]: A hollow muscular organ, located in the female pelvis, in which a fetus develops.

V

vaccine [vacuna]: A medical preparation used to stimulate the specific immune response of an organism's body into producing antibodies against a pathogen, without generating an infection.

vacuole (VAK yoo ohl) [vacuola]: A membrane-enclosed structure in the cytoplasm of a cell or a unicellular organism. Different types of vacuoles serve different functions.

vagina (vuh JI nuh) [vagina]: A tubular organ that leads from the uterus to the opening of the female reproductive tract.

variable [variable]: A condition that varies from one organism to another (size, shape, color) or is subject to change for an individual organism (humidity, temperature, light intensity, fatigue).

vas deferens (vas DEF er enz) [conducto deferente]: A tube that leads from the epididymis to the urethra of the male reproductive tract.

vasopressin (vas O press in) [vasopresina]: Hormone released by the pituitary in response to the signal of dehydration sent from the hypothalamus.

vegetative reproduction [reproducción vegetativa]: Asexual reproduction involving parts of an organism that detach and form new organisms. Offspring are genetically identical to the parent. Examples include budding as the "eyes" on potatoes do, sprouting from roots as aspen trees do, or sprouting from runners as strawberries do.

vein [vena]: A blood vessel through which blood flows toward the heart.

vestigial (ve STIJ ee uhl) [rudimentario]: Describes a structure of an organism's body that has little or no apparent function, though homologous structures in related organisms or a common ancestor had some function.

vitamins [vitaminas]: A group of organic substances essential for the metabolism of living organisms. They are found in very small quantities in foods and some are produced synthetically.

vulva (VUL va) [vulva]: External female reproductive organs including the labia and clitoris.

W

white blood cells [glóbulos blancos]: A group including many types of immune system cells that circulate in the blood and provide specific functions related to the response to and elimination of foreign matter.

X

X-linked recessive [recesivo ligado al cromosoma X]: A pattern of inheritance for a trait that is carried on the X chromosome. In females, two atypical alleles are necessary for the trait to be expressed. In males, only one atypical allele is necessary for the trait to be expressed because males have only one X chromosome.

Z

zygote (ZY goht) [cigoto]: The cell that results from the fusion of gamete nuclei when egg and sperm unite. Zygotes contain a full set of genetic information.

Credits

Front Matter

Page vi oneinchpunch/Shutterstock.com; **Page vii** TanyaBird/Shutterstock.com; **Page viii** Pist/Shutterstock.com; **Page ix** dammihai/Shutterstock.com; **Page xi** Goodluz/Shutterstock.com; **Page xii** Nicola Bertonlini/Shutterstock.com; **Page xiv** Catalin Petolea/Shutterstock.com

Unit 0

Opener: (**dogs**) Erik Lam/Shutterstock.com; (**green pond**) Cheng Wei/Shutterstock.com; Page 3 (**flower**) kkaplin/ Shutterstock.com; (**bacteria behind flower**) Dariush M/Shutterstock.com; (**lower left diversity**) Rawpixel.com/ Shutterstock.com.

Chapter 1: Rawpixel.com/Shutterstock.com; **Page 6** Shutterstock/ Christopher Futcher; **Figure 1.1** Hugo Van Lawick/ Goodall Institute; **Page 10** Volt Collection/Shutterstock.com; **Page 11** The Jane Goodall Institute; **Page 12** Ysbrand Cosijn/Shutterstock.com; **Page 13** Image courtesy of the University of California Museum of Paleontology-Understanding Science www.understandingscience.org.; **Page 16** Jane Goodall/Goodall Institute; **Page 17** Toshe Ognjanov/Shutterstock.com; **Page 19** and **20 Figure 1.7a-g** by BSCS by Carlye Calvin; **Figure 1.8** BSCS by Carlye Calvin; **Figure 1.9** Filipe Frazao/Shutterstock.com; **Page 24** AP Photo/ANDREAS FISCHER; **Figure 1.10** gorillaimages/ Shutterstock.com; **Page 26** (flint spearhead) Strannik_fox/Shutterstock.com; (stone disk) Lightspring/Shutterstock.com; (spelt wheat) Polina Petrenko/Shutterstock.com; **Page 27** istanbul_image_video/Shutterstock.com; **Page 28** (boys sports) Monkey Business Images/Shutterstock.com; (walking baby) wong sze yuen/Shutterstock.com; **Page 29** Eric Gevaert/ Shutterstock.com; **Page 30** (kids) Gelpi/Shutterstock.com; (multi-generational) imagedb.com/Shutterstock.com; **Page 31** Rawpixel.com/Shutterstock.com; **Figure 1.14** Carlye Calvin; **Figure 1.17** Chaikom/shutterstock.com; **Page 35** Dudarev Mikhail/Shutterstock.com; **Page 36** (mushrooms) kirych/Shutterstock.com; (wildflowers and fish) Corel; (algae) BSCS by Tom Bishop; (bacteria) 3Dalia/Shutterstock.com; **Figure 1.19** Lebendkulturen.de/Shutterstock.com; **Figure 1.20** BSCS by Carlye Calvin; **Figure 1.21** BSCS by Carlye Calvin; **Figure 1.22** Bruce J. Russell/BioMedia Associates; **Figure 1.23** tamsindove/Shutterstock.com.

Unit 1

Opener (**top**) Svetlana Gajic/Shutterstock.com; (**bottom**) fototrips/Shutterstock.com; **Page 45** Svetlana Gajic/ Shutterstock.com; travelview/Shutterstock.com

Chapter 2: **Page 46** travelview/Shutterstock.com; **Page 48** Arthimedes/Shutterstock.com; **Page 50** Liew Weng Keong/ Shutterstock.com; **Figure 2.1** pixelrain/Shutterstock.com; **Figure 2.3** (**mutualism**) OMMB/Shutterstock.com; (**predator-prey**) kochanowski/Shutterstock.com; (**parasitism**) Evgeny Parushin/Shutterstock.com; (**competition**) Protasov AN/ Shutterstock.com; (**commensalism**) Milosz Maslanka/Shutterstock.com; **Figure 2.4** Kalcutta/Shutterstock.com; **Page 56** Ronaldo Almeida/Shutterstock.com; **Page 57** Imfoto/Shutterstock.com; **Figure 2.5** Courtesy of the University of Texas Libraries, The University of Texas, Austin; **Figure 2.6** Courtesy of Daryl O'Connor; **Figure 2.7** Corel; **Figure 2.10** Lisa Strachan/Shutterstock.com; **Figure 2.11** jo Crebbin/Shutterstock.com; **Figure 2.12** BSCS by Carlye Calvin; **Figure 2.16** Choksawatdikorn/Shutterstock.com; **Page 67** Bryan Busovicki/Shutterstock.com; **Figure 2.17** Brian Lasenby/ Shutterstock.com; **Figure 2.18** USDA Forest Service; **Figure 2.22** NOAA/National Climatic Data Center; **Figure 2.23** daulon/Shutterstock.com; **Figure 2.25** jorisvo/Shutterstock.com; **Figure 2.26** keldridge/Shutterstock.com; **Figure 2.27** Bridget Calip/Shutterstock.com.

Unit 2

Opener Page 82 and 83 (baby elephant) worradirek/Shutterstock.com

Chapter 3: **Page 84** Jool-yan/Shutterstock.com; **Page 86** Pixavril/Shutterstock.com; **Figure 3.1** jiawangkun/Shutterstock. com; **Figure 3.2** (plastic bottles) Moreno Soppelsa/Shutterstock.com, (recycled paper) alterfalter/Shutterstock.com, (steaming compost) Joan Ramon Mendo Escoda/Shutterstock.com; **Page 89** (mushrooms) Phoric/Shutterstock.com, (village in Amazon) Rafal Cichawa/Shutterstock.com, (humans in SF) EQRoy/Shutterstock.com; **Figure 3.3** Huguette Roe/Shutterstock.com; **Figure 3.4** Serg64/Shutterstock.com; **Figure 3.5** Ginger Livingston Sanders/Shutterstock.com; **Figure 3.6** sumikophoto/Shutterstock.com; **Figure 3.8** Alfred Sonsalla/Shutterstock.com; **Figure 3.9** (earthworm) alexsvirid/Shutterstock.com, (earthworm in compost) schankz/Shutterstock.com; **Figure 3.11** Matteo Galimberti/ Shutterstock.com; **Figure 3.13** Jason Kasumovic/Shutterstock.com; **Page 98** (river near Sacramento) Andrew Zarivny/

Shutterstock.com; **Figure 3.16** BSCS by Bill Beaudin; **Page 105** Simun Ascic/Shutterstock.com; **Figure 3.17**(wheat field) gor Strukov/Shutterstock.com, (combine harvesting wheat) oticki/Shutterstock.com, (flour) Antonio Gravante/ Shutterstock.com, (flour tortillas) Dream79/Shutterstock.com; **Figure 3.18** (baby chick) yevgeniy11/Shutterstock.com, (chicken) TippaPatt/Shutterstock.com, (raw chicken) Diyana Dimitrova/Shutterstock.com, (chicken dinner) papkin/ Shutterstock.com; **Figure 3.19** BSCS by Bill Beaudin; **Figure 3.20** (Herbivore) Pyty/Shutterstock.com, (Carnivore) Alta Oosthuizen/Shutterstock.com, (Omnivore) mangojuicy/Shutterstock.com, (Decomposer) puttography/Shutterstock.com; **Figure 3.25** Cathy Keifer/Shutterstock.com; **Figure 3.28** Tom Grundy/Shutterstock.com; **Figure 3.30** Patricia Dulasi/ Shutterstock.com; **Figure 3.31** Jason Vandehey/Shutterstock.com; **Figure 3.32** Evan Lorne/Shutterstock.com; **Page 119** Aleksandra Suzi/Shutterstock.com; **Page 121** Gabor Havasi/Shutterstock.com; **Figure 3.35** Wead/Shutterstock. com; **Figure 3.36** Bernadette Heath/Shutterstock.com; **Page 124** Daan Kloeg/Shutterstock.com; **Figure 3.37** LorraineHudgins/Shutterstock.com

Chapter 4: **Page 126** KPG_Payless/Shutterstock.com; **Page 128** Marioner/Shutterstock.com; **Page 130** Shawn Pecor/ Shutterstock.com; **Figure 4.1** mihalec/Shutterstock.com; **Figure 4.2** Mikadun/Shutterstock.com; **Figure 4.4** BSCS by Carlye Calvin; **Figure 4.8** Dan Hanscom/Shutterstock.com; **Figure 4.10a** mj007/Shutterstock.com; **Figure 4.10b** Colin Stitt/Shutterstock.com; **Figure 4.10c** anyaivanova/Shutterstock.com; **Figure 4.11** Kevin M. McCarthy/Shutterstock. com; **Page 144** (top left margin) Blend Images/Shutterstock.com; **Page 144** (bottom left margin) TORWAISTUDIO/ Shutterstock.com; **Figure 4.16** (top) Joe Gough/Shutterstock.com; **Figure 4.21** Pecold/Shutterstock.com; **Figure 4.22** baranq/Shutterstock.com; **Figure 4.24** irin-k/Shutterstock.com; **Figure 4.25** The Magical Lab/Shutterstock.com; **Figure 4.27 leaf:** Comstock; **chloroplast:** Courtesy of Dr. L. Andrew Staehelin, University of Colorado; **Figure 4.29** olias32/Shutterstock.com; **Figure 4.34** jopelka/Shutterstock.com; **Figure 4.35** Sheila Fitzgerald/Shutterstock.com; **Figure 4.36** worldswildlifewonders/Shutterstock.com; **Figure 4.37** Knorre/Shutterstock.com; **Figure 4.38** Dr. Dennis Kunkel/Visuals Unlimited, Inc.

Unit 3

Opener: Page 178 (blue frog) Aleksey Stemmer/Shutterstock.com; **Page 179** (background strep) Stastny_Pavel/ Shutterstock.com; (corpse flower) Andrei Medvedev/Shutterstock.com

Chapter 5: **Page 180** thomas koch/Shutterstock.com; **Page 182** javarman/Shutterstock.com; **Page 185** © Bettmann/ CORBIS; **Page 186** Odua Images/Shutterstock.com; **Page 191** (top right margin) Volt Collection/Shutterstock.com; **Page 191** (bottom right margin) Charles Shapiro/Shutterstock.com; **Page 194** (top left margin) vasosh/Shutterstock.com (middle page margin) StubblefieldPhotography/Shutterstock.com; **Page 198** Monkey Business Images/Shutterstock.com; **Figure 5.9a** Jan-Nor Photography/Shutterstock.com **Figure 5.9b** Alain Lauga/Shutterstock.com; **Figure 5.10a** Forest & Kim Starr; **Figure 5.10b** Steven Blandin/Shutterstock.com; **Figure 5.14** © Peter and Rosemary Grant; **Figure 5.15** Adapted from page 142 of *Morphological Differentiation and Adaptation in the Galapagos Finches* by Robert I. Bowman, University of California Publication in Zoology Volume 58 (1961); **Figure 5.16** Copyright © Princeton University. Photographer: Denise Applewhite.; **Figure 5.17** Everett Historical/Shutterstock.com; **Figure 5.19** Guido Vermeulen Perdaen/Shutterstock.com; **Figure 5.20** Art Today; **Figure 5.21** Art Today; **Figure 5.22a** almondd/Shutterstock.com; **Figure 5.22b** Ron Rowan Photography/Shutterstock.com; **Figure 5.22c** Greg Amptman/Shutterstock.com; **Figure 5.23a** Egor Rodynchenko/Shutterstock.com; **Figure 5.23b** NinaM/Shutterstock.com; **Figure 5.23c** bonchan/Shutterstock.com; **Figure 5.23d** HandmadePictures/Shutterstock.com; **Figure 5.23e** paulista/Shutterstock.com; **Figure 5.23f** Lotus Images/ Shutterstock.com; **Page 212** (spider) kajornyot/Shutterstock.com; **Page 215** Photo courtesy of Heidi Schutz; **Page 218** Kanea/Shutterstock.com; **Figure 5.27** John Reader/Science Source; **Figure 5.28a** Kjuuurs/Shutterstock.com; **Figure 5.28b** deb talan/Shutterstock.com; **Figure 5.29a** Photo provided by the Essig Museum of Entomology, University of California, Berkeley.; **Figure 5.29d** www.wpclipart.com; **Figure 5.30** BSCS by Carlye Calvin; **Page 226** Slava Gerj/Shutterstock. com; **Figure 5.34** Faviel_Raven/Shutterstock.com; **Figure 5.35** creativemarc/Shutterstock.com; **Page 231** BSCS by Tom Bishop; **Page 232** 3Dalia/Shutterstock.com; **Figure 5.39** Zina Deretsky, National Science Foundation; **Page 238** Lars Zahner/Shutterstock.com

Chapter 6: **Page 242** Photographer: Greg McFall, Gray's Reef NMS, NOS, NOAA; **Page 244** Ethan Daniels/Shutterstock. com; **Page 248** PhotoDisc; **Figure 6.4** Photo by Scott Bauer; **Figure 6.5a** Courtesy of National Park Service; **Figure 6.5b** Alex Emanuel Koch/Shutterstock.com; **Figure 6.5c** Randimal/Shutterstock.com; **Page 253** (fern and moss) Comstock; **Figure 6.8** R. Guerrero/Ward's Natural Science Establishment, Inc./ Five Kingdom Slide Set; **Figure 6.9** Lynn M. Margulis/Dept. of Biology/University of Massachusetts, Amherst; **Figure 6.10** John D. Cunningham/Visuals Unlimited; **Figure 6.11** Bio Disc, Inc., 808 Newbury St., Ripon, WI 54971; **Figure 6.12a** BSCS by Doug Sokell; **Figure 6.12b** KoBoZaa/Shutterstock.com; **Figure 6.19** Adapted in part from art by Cornelia Blik and Patricia J. Wynne; **Figure 6.22** Copyright 2009 Agile Mind, Inc.® Used with permission.; **Figure 6.23a** Carlye Calvin; **Figure 6.23b** E. Musil/Visuals Unlimited; **Figure 6.23c** BSCS by Rich Tolman; **Figure 6.24** Copyright 2009 Agile Mind, Inc.® Used with permission.; **Page 273** USDA. Photo by Scott Bauer.; **Figure 6.25 a-d** Courtesy, Douglas Schemske and Toby Bradshaw; **Page 275** (top right margin) alice-photo/Shutterstock.com; (bottom right margin) PhotoDisc; **Page 277** Courtesy of Scott Edwards; **Page 283** Photo courtesy of Kevin Pang.; **Figure 6.30a** RobHamm/Shutterstock.com; **Figure 6.30b** Patrick

K. Campbell/Shutterstock.com; **Figure 6.30c** frantisekhojdysz/Shutterstock.com; **Figure 6.32** (dogs) Eric Isselee/ Shutterstock.com; (teens) RyFlip/Shutterstock.com; **Figure 6.34** Anatoliy Lukich/Shutterstock.com; **Figure 6.35a** Steve Allen/Shutterstock.com; **Figure 6.35b** Yakov Oskanov/Shutterstock.com; **Figure 6.36a** Bruce Russell/ BioMedia Associates; **Figure 6.36b** David M. Phillips/Visuals Unlimited; **Figure 6.36c** Dr. Terry Beveridge/Visuals Unlimited, Inc.; **Page 296** (top left margin) BSCS by Carlye Calvin; (lower left margin) Luciano Mortula/Shutterstock.com; **Page 297** (top left margin) ODM/Shutterstock.com; (middle left margin) lazyllama/Shutterstock.com; (bottom of page) optionm/ Shutterstock.com; **Page 298** Tischenko Irina/Shutterstock.com

APPENDICES:

Figure F1.1: Everett Historical/Shutterstock.com; **Figure F1.2**: Grandpa/Shutterstock.com; **Figure F1.3**: RioPatuca/ Shutterstock.com; **Figure F1.4**: David M. Phillips/Visuals Unlimited; **Figure F1.5a**: Science VU/Visuals Unlimited; **Figure F1.5b**: Hans Gerlderblon/Visuals Unlimited; **Figure F1.5c**: Cabisco/Visuals Unlimited; **Figure F1.6**: Science VU/ Visuals Unlimited; **Figure F1.8a**:Lee D. Simon/Science Source; **Figure F1.8b**: Biozentrum, University of Basel/Science Source; **Page 369** (poppies) waniuszka/Shutterstock.com; **Figure F2.1**: Jack Stein Grove/www.JSGrove.com; **Figure F2.2**: John D. Cunningham/Visuals Unlimited; **Figure F2.4**: Carlye Calvin; **Figure F2.5**: E.J. Cable/Tom Stack and Associates; **Figure F2.6a**: Doug Sokell/BSCS; **Figure F2.6b**: William S. Ormerod/Visuals Unlimited; **Figure F2.8**: Fred Hossler/Visuals Unlimited; Figure **F2.11a**: © Ripon Microslides. Used with permission from Carolina Biological Supply Company, Burlington, NC; **Figure 2.11b**: John D. Cunningham/Visuals Unlimited**;** **Figure F2.15a**: Carlye Calvin; **Figure F2.15b**: John Shaw/Tom Stack andAssociates; **Figure F2.15c**: Ken Davis/Tom Stack and Associates; **Figure F2.17**: William S. Ormerod/Visuals Unlimited; **Figure F2.18a**: Greg Vaughn/Tom Stack and Associates; **Figure F2.18b**: Kevin Schafer Photography; **Figure F2.20**: © 2002 Steven J. Baskauf; **Figure F2.21a**: Hugh Spenser/BSCS; **Figure F2.21b**: Doug Sokell/ BSCS; **Figure F2.22**: George Loun/Visuals Unlimited; **Figure F2.23a**: Doug Sokell/BSCS; **Figure F2.23b**: BSCS; **Figure F2.23c**: John Cunningham/Visuals Unlimited; **Figure F2.24a, b**: Doug Sokell/BSCS; **Figure F2.26a, b**: Doug Sokell/BSCS; **Figure F2.27 a–d**: Carlye Calvin; **Figure F2.28a, b**: Doug Sokell/BSCS; **Figure F2.31a**: Doug Sokell/BSCS; **Figure F2.32**: Jackie-Ott Rogers/BSCS; **Figure F3.1**: Carolina Biological Supply/Phototake NYC; **Figure F3.2a**: Carolina Biological Supply/Phototake NYC; **Figure F3.2b**: Comstock; **Figure F3.2c, d**: Ed Reschke